FUNDAMENTALS OF EARLY CLINICAL DRUG DEVELOPMENT

FUNDAMENTALS OF EARLY CLINICAL DRUG DEVELOPMENT

From Synthesis Design to Formulation

Edited by

AHMED F. ABDEL-MAGID
and
STÉPHANE CARON

WILEY-INTERSCIENCE
A John Wiley & Sons, Inc., Publication

Published by John Wiley & Sons, Inc., Hoboken, New Jersey
Published simultaneously in Canada

Library of Congress Cataloging-in-Publication Data:

Fundamentals of early clinical drug development : from synthesis design to formulation / edited by Ahmed F. Abdel-Magid and Stéphane Caron.
p. ; cm.
ISBN-13 978-0-471-69278-2 (cloth)
ISBN-10 0-471-69278-6 (cloth)
1. Clinical pharmacology. 2. Pharmaceutical technology. 3. Drugs–Testing. I. Abdel-Magid, Ahmed F., 1947- II. Caron, Stéphane.
[DNLM: 1. Drug Design. 2. Drug Compounding. 3. Pharmaceutical Preparations–chemical synthesis. QV 744 F981 2006]
RM301.27.F86 2006
615′.1–dc22

2006016265

Printed in the United States of America
10 9 8 7 6 5 4 3 2 1

CONTENTS

CONTRIBUTORS

Aikins, James, Chemical Product Research and Development, Lilly Research Laboratories, A Division of Eli Lilly and Company, Indianapolis IN 46285

am Ende, David J., Chemical Research & Development, Pfizer Global R&D, Groton, CT 06340

Andler-Burzlaff, Simone, CarboGen AG, CH-5502 Hunzenschwil, Switzerland

Bertola, Jason, CarboGen and AMCIS, Whitehouse Station, NJ 08888

Bhagavatula, Lakshmi, GPRD Process Research and Development, Department R450, Abbott Laboratories, North Chicago, IL 60064

Briggs, Barbara, Chemical Research and Development, Lilly Research Laboratories, A Division of Eli Lilly and Company, Indianapolis, IN 46285

Caron, Stéphane, Chemical Research & Development, Pfizer Global R&D, Groton, CT 06340

Childers, Jr., Joseph H., Johnson & Johnson Pharmaceutical Research and Development, L.L.C., Chemical Development. Spring House, PA 19477

Davies, Merrill L., Process R&D Department, Bristol-Myers Squibb Co.

Davis, Roman, GlaxoSmithKline, Synthetic Chemistry, Chemical Development, Research Triangle Park, NC 27709-3398

Delaney, Edward J., Process R&D Department, Bristol-Myers Squibb Co.

Doan, Brian D., GlaxoSmithKline, King of Prussia, PA

Grabowski, Edward J.J., Vice President Chemistry—Retired, Westfield, NJ 07090

Henry, Rodger F., GPRD Process Research and Development, Department R450, Abbott Laboratories, North Chicago, IL 60064

Hoerr, David C., Johnson & Johnson Pharmaceutical Research and Development, L.L.C., Spring House, PA 19477

Jennings, Rex, Chemical Research and Development, Pfizer Global Research and Development, Ann Arbor, MI 48105

Karcher, Brent D., Process R&D Department, Bristol-Myers Squibb Co.

Kissel, William S., Chemical Research and Development, Pfizer Global Research and Development, Ann Arbor, MI 48105

Kleinman, Mark H., Pharmaceutical Development, GlaxoSmithKline, King of Prussia, PA 19406

Kruger, Albert W., GPRD Process Research and Development, Department R450, Abbott Laboratories, North Chicago, IL 60064

Le, Tung V., Clinical Research and Development, Pfizer Global Research and Development, Ann Arbor, MI 48105

Lee, Beeah, Pharmaceutical Development, GlaxoSmithKline, King of Prussia, PA 19406

Lenoir, Edward, Chemical Research and Development, Pfizer Global Research and Development, Ann Arbor, MI 48105

Lovelace, Thomas C., GlaxoSmithKline, Synthetic Chemistry, Chemical Development, Research Triangle Park, NC 27709-3398

Marti, Roger E., Zurich University of Applied Sciences, Winterthur, CH-8401 Winterthur, Switzerland

Mulhern, Thomas, Chemical Research and Development, Pfizer Global Research and Development, Ann Arbor, MI 48105

Patterson, Daniel E., GlaxoSmithKline, Synthetic Chemistry, Chemical Development, Research Triangle Park, NC 27709-3398

Quallich, George J., Global Research and Development, Pharmaceutical Research and Development, Pfizer, Inc., Groton, CT 06340

Ragan, John A., Chemical Research and Development, Pfizer Global Research & Development, Eastern Point Road, Groton, CT 06340

Roberts, John C., GlaxoSmithKline, Synthetic Chemistry, Chemical Development, Research Triangle Park, NC 27709-3398

Rohde, Bridget D., GPRD Process Research and Development, Department R450, Abbott Laboratories, North Chicago, IL 60064

Roschangar, Frank, Boehringer Ingelheim, Richmond, VA

Rose, Peter R., Chemical Research & Development, Pfizer Global R&D, Groton, CT 06340

Rosso, Victor W., Process R&D Department, Bristol-Myers Squibb Co.

Rozema, Michael J., GPRD Process Research and Development, Department R450, Abbott Laboratories, North Chicago, IL 60064

Rubin, A. Erik, Process R&D Department, Bristol-Myers Squibb Co.

Shchuka, Maria I., Johnson & Johnson Law Department, Spring House, PA 19477

Shelat, Bhadra, GPRD Process Research and Development, Department R450, Abbott Laboratories, North Chicago, IL 60064

Sickles, Barry, GlaxoSmithKline, Synthetic Chemistry, Chemical Development, Research Triangle Park, NC 27709-3398

Tien, James J., GPRD Process Research and Development, Department R450, Abbott Laboratories, North Chicago, IL 60064

Toczko, Jennifer F., GlaxoSmithKline, Synthetic Chemistry, Chemical Development, Research Triangle Park, NC 27709-3398

Venit, John J., Process R&D Department, Bristol-Myers Squibb Co.

Wade, Robert, Chemical Research and Development, Pfizer Global Research and Development, Ann Arbor, MI 48105

Weaner, Larry E., Johnson & Johnson Pharmaceutical Research and Development, L.L.C., Spring House, PA 19477

Yang, Ju, GlaxoSmithKline, Synthetic Chemistry, Chemical Development, Research Triangle Park, NC 27709-3398

Zhang, Tony Y., Chemical Product Research and Development, Lilly Research Laboratories, A Division of Eli Lilly and Company, Indianapolis, IN 46285

Zhang, Weijiang, GPRD Process Research and Development, Department R450, Abbott Laboratories, North Chicago, IL 60064

Zmijewski, Milton, Chemical Product Research and Development, Lilly Research Laboratories, A Division of Eli Lilly and Company, Indianapolis IN 46285

PREFACE

The discovery of a new drug is the most challenging task in the pharmaceutical industry. It takes the efforts of a large number of medicinal chemists, biologists, toxicologists, and many others for several years to identify just one candidate for a specific biological target in a specific therapeutic area. It is estimated that medicinal chemists prepare and examine over 10,000 new compounds on the average to identify and develop just one commercial drug. Each pharmaceutical company manages to identify a few lead compounds from screening thousands of compounds. From these leads, massive structure/activity studies are initiated to identify even fewer drug candidates. These precious few drug candidates are promoted to the development status. Therefore, by the time drug candidates reach the development status, they are already very valuable commodities resulting from years of hard work and represent a large portion of the company's wealth of intellectual property. However, for us in development, it is just the beginning; the development of a drug is a long and complex process that requires a high degree of teamwork and collaboration among different disciplines. The candidates that survive the early stages of the development continuum and advance further into the later stages will have better chances to be successful as drugs and form what is known as the company's pipeline. The lead process in drug development is chemical process research, also referred to as process chemistry or chemical development, and is responsible for identifying the best synthesis route to prepare the needed amounts of a selected drug candidate for all other development functions (e.g., toxicology, formulation development, clinical studies, etc). Most chemical development departments are responsible for the large-scale preparation of drug substances and the selection of suitable salts and/or solid forms of the drug candidates.

The pharmaceutical industry has enjoyed a very successful period of discovery and development of new drugs in the 1960s, 1970s, and 1980s. During that time, the main goal of development chemists was the design of very efficient syntheses that may be used, not only for providing early supplies but also for supplying enough substance to advance the drug candidate into late development stages and even in manufacturing of the drug substances. The emphasis was always on the development of the perfect chemical process. This approach required a lot of effort and resources before knowing if the project would survive or be terminated due to lack of preclinical safety or undesirable clinical finding.

In the mid-1990s the industry started to confront new challenges. The pharmaceutical companies, in general, started to face declines of their pipelines as well as severe competition from generics. Many of the patents on commercial drugs were expiring, and fewer drugs were coming to replace them. The industry needed to adjust and modify its approach for drug discovery and development to face the new challenges. While the goals of drug research and development have remained in essence the same, the strategies applied to accomplish these goals have changed significantly. To maintain their competitive edge, various companies have adopted many changes aimed at increasing their chances of identifying lead compounds and shortening the development life cycle. Discovery introduced tools and techniques such as combinatorial chemistry, high throughput screening, bioinformatics, and chemoinformatics (just to name a few) to create and screen massive libraries of new compounds in order to increase their chances of identifying lead compounds and speed the discovery process.

Development has also changed philosophy and started to use new innovative techniques to reach its goals quickly and efficiently to keep up with the new demands of drug developments and to shorten the development life cycle. Early introduction of drug candidates into development and at-risk process development to reach a quick-go or no-go decision mark some of the changes that were implemented into the development process. Chemical development shifted its focus from the long development of the "perfect" chemical processes and the production of large quantities of drug substance and adopted a faster approach to quickly improve the existing synthesis and produce smaller quantities of product, thus saving resources and valuable time. Many companies started to resort to outsourcing as a means of supplementing their efforts and to free the time of their scientists to more value-added tasks. Since all development activities do not start until sufficient quantities of drug substance are produced, chemical process research maintained its central role in the early decision to continue or discontinue the development of potential drug candidates. The key stage in the development of new drugs requires the preparation of 2–10 kg of drug substance under current good manufacturing practice (cGMP) conditions. The timely development of fast practical syntheses is of central importance, and it is the rate-limiting step for all other development activities moving forward toward clinical evaluation. As process chemists we tend to emphasize our role, but we also realize the importance of all other activities in the development process and the role of interdisciplinary interactions, which are the backbone of any successful development function. The changing nature of our industry and the challenges we face can only be handled effectively when all professionals from different disciplines work together to achieve the ultimate goal, the development of a drug. Giving the respect and appreciation to other functions starts with understanding the nature of their job, their challenges, and the importance of their achievements. In this book we hope to introduce many of these other functions in the development process to chemists to add to their understanding and appreciation of the role and responsibilities of several development professionals.

The idea of this book project originated from the symposium "*The Role of Organic Synthesis in Early Clinical Drug Development II*," which was held at the ACS Fall National Meeting in New York City on September 8, 2003. The symposium, co-sponsored by the Organic Chemistry and the Medicinal Chemistry divisions of the ACS, was the second with the same title; the first was held in a Chicago ACS National Meeting, August 2001. The symposium featured 10 speakers chosen from the chemical process research divisions of several pharmaceutical companies to provide recent synthesis examples of development candidates to highlight new synthetic methodologies, the modification of existing methods into practical, more efficient ones, and the application of new techniques (such as automation) in selecting the optimal reaction conditions.

This book is different from symposia proceedings because it does more than merely gather several synthesis topics. We expanded its scope beyond process chemistry to present a more comprehensive look at other aspects of drug development and to cover additional subjects of interest. The book contains 15 chapters. The first six highlight some of the most recent advances in chemical process research written by lead researchers in the field from different pharmaceutical companies. The following nine chapters discuss other aspects of early clinical drug development such as automation, chemical engineering, solid form selection, and formulations, all of which are crucial to the success of the early clinical drug development. We also included chapters on the use of radioisotopes, the current trends of outsourcing, and the importance of intellectual property. The list of chapters shows the diversity of topics presented by accomplished scientists in their fields. None of these chapters was intended for the experts in these fields, but instead were written for organic chemists in general and process chemists in particular to increase their awareness of these topics and encourage the interaction with other development professionals.

The opening chapter is of particular interest to process chemists everywhere. One of the most noted and admired process chemists, Edward J. J. Grabowski, who retired recently from Merck Research Laboratories, writes the second of his reflection chapters looking back at his long and illustrious career at Merck, his thoughts on an ever-changing industry, and an overview of several drug development processes. The next five chapters describe specific and unique processes that demonstrate the abilities of process chemists to accomplish their tasks and achieve their goals. These chapters are written by John Ragan of Pfizer; James Aikins with coauthors Tony Zhang, Milton Zmijewski, and Barbara Briggs of Lilly Research; John Roberts and coauthors Roman Davis, Brian Doan, Thomas, Lovelace, Daniel Patterson, Frank Roschangar, Barry Sickles, Jennifer Toczko, and Ju Yang of GlaxoSmithKline; Albert Kruger and coauthors Michael Rozema, Bridget

Rohde, Bhadra Shelat, Lakshmi Bhagavatula, James Tien, Weijiang Zhang, and Rodger Henry of Abbott Laboratories; and William Kissel and coauthors Rex Jennings, Tung Le, Edward Lenoir, Thomas Mulhern, and Robert Wade of Pfizer. They are all highly accomplished scientists with excellent track records in chemical process research. Chapter 7 is a distinctive one written by Stéphane Caron of Pfizer to bridge the earlier synthesis chapters with the following ones by discussing the role of process chemists that extends beyond synthesis and the importance of regulatory aspects of our jobs as process chemists. Automation in development has been slowly emerging as an excellent technique to meet the demands of faster development cycles. An outstanding discussion of automation and the design of experiments is presented with examples by one of the true leaders in automation, Edward Delaney, and his coauthors Merrill Davies, Brent Karcher, Victor Rosso, Erik Rubin, and John Venit of Bristol-Myers Squibb. Next is a discussion by Simone Andler-Burzlaff, Jason Bertola, and Roger E. Marti on the importance of the growing role of outsourcing as a new aspect in the changing drug development from their own experience at CarboGen. The role of chemical engineering in large-scale synthesis has always been a crucial one; however, it remains a mystery to many chemists. An expert in the field of chemical engineering, Joseph Childers of Johnson & Johnson PRD, writes on the engineering perspective and fundamentals of chemical engineering. The use of radioisotopes is another subject that has a distinctive role in both pharmaceutical discovery and development, but its role and challenges are not fully understood by many chemists. Larry Weaner and David Hoerr, proficient radiochemists at Johnson & Johnson PRD, present the fundamentals of radiochemistry with working examples. We follow with three chapters dealing with the final solid form of the drug substance. The drug form identification and selection is a major part of the development process that adds significantly to understanding the drug properties to aid in formulation, and it represents an important intellectual property that extends the life cycle of a drug. George Quallich of Pfizer, whose broad chemistry background enables him to give a complete perspective of the subject, writes the basic concepts of this topic. The control of particle size of a drug substance is another very important, yet unfamiliar subject to most chemists. The effect of particle size on solubility and bioavailability may be obvious to many, but how you control that is really a fascinating subject. David am Ende and Peter Rose of Pfizer explain the subject clearly with a general overview of the different approaches and techniques. All the previous activities led to the final challenge, converting drug substances into drug products, the end result that determines the delivery method and the form to be given to a patient—that is, formulation. Mark Kleinman and Beeah Lee of GlaxoSmithKline present a superb and clear chapter on the topic written especially for chemists. We end with one of the most important topics across the full spectrum of drug discovery and development, namely, the importance of patents and intellectual property in our business. The subject is written clearly and concisely with emphasis on early development issues by Maria Shchuka of the Johnson & Johnson law department; she is a highly qualified patent manager with extensive patent experience and an advanced degree in chemistry.

We put this book together with the hope that it will be helpful to synthetic organic chemists including process and medicinal chemists in the pharmaceutical industry, as well as those in academia who develop new synthetic methods and/or apply existing methods to the synthesis of natural products and new materials. It will also be of interest to chemists who want to learn and understand the drug development process beyond synthesis. The book provides a reference for current topics in chemical process research including different styles and structures and the application of new and different techniques. It also provides a basic look at many of the other development functions written by experts in their field specifically for chemists.

ACKNOWLEDGMENTS

The idea of organizing a process chemistry symposium came from the collaboration of the Organic Chemistry and the Medicinal Chemistry Divisions of the American Chemical Society. We thank the two divisions for sponsoring the symposium.

We thank each of the authors whose outstanding contributions to this book made it a worthy project and a valuable contribution to the literature. We appreciate their professional courtesy, patience, and kindness, which made this an enjoyable experience. We would also like to extend our thanks and appreciation to Jodi Gaynor for her work in the final editing of the chapters.

We thank and appreciate our respective companies, Johnson & Johnson Pharmaceutical Research and Development, L.L.C. and Pfizer Global Research & Development for their encouragement and support.

Last, but not least, we thank the staff at John Wiley & Sons, Inc., particularly Amy Byers, Associate Editor, for her exceptional help and outstanding efforts in realizing the idea of this book into a reality, as well as Kellsee Chu, Senior Production Editor, for her exceptional effort in producing this book.

AHMED F. ABDEL-MAGID

Johnson & Johnson Pharmaceutical Research and Development,
L.L.C. Chemical Development
Spring House, PA

STÉPHANE CARON

Pfizer Global Research & Development,
Chemical Research & Development
Groton, CT

1

REFLECTIONS ON PROCESS RESEARCH II[1]

EDWARD J. J. GRABOWSKI

Vice President Chemistry—Retired, Westfield, NJ 07090

Among the various groups that are responsible for early drug development in the pharmaceutical industry, the Process Research Groups are among the most important because the drug development process cannot begin until sufficient bulk drug has been prepared. Those who are most removed from organic chemistry and process research typically associate bulk drug needs with the illusive term "scale-up." They generally believe that because the medicinal chemists have devised a synthesis that can produce 2–20 grams of the candidate, straightforward "scale-up" by the process chemists will readily produce hundreds, if not thousands, of grams of product. Indeed, many in this group believe that process research and "scale-up" are synonymous. For some drug candidates, where the medicinal chemistry is relatively simple, this is indeed true, and today these candidates are most likely targets for outsourcing. However, this has not generally been the case in my 39-year career in the industry. Numerous candidates on which my groups have worked have required high levels of scientific creativity, novel synthetic methods, complex organic chemical development, and a bit of serendipity, all against tight timelines, to rapidly bring forward practical syntheses and large quantities of bulk drug for early clinical development. Given the explosive growth and development of novel synthetic organic methods in the past generation, medicinal chemists are now able to devise drug candidates of unprecedented complexity, and the role of organic synthesis in early drug development is more important than ever before.

In thinking about how to illustrate the importance of organic synthesis in early drug development, I thought to once again look back on some of the more

Fundamentals of Early Clinical Drug Development: From Synthesis Design to Formulation, edited by Ahmed F. Abdel-Magid and Stéphane Caron

interesting programs during my Merck career[1] and then choose a number of examples where organic synthesis proved to be the key element in being able to rapidly bring a program forward. Each of the examples selected explores the general thesis from different points of view. Most of the candidates discussed did not make it to product status (what's new?). Often getting a program going requires novel chemistry or complex chemical development or both. Although higher management would prefer that we quickly devise the ultimate process, this is most often not the case, and we must crawl before we walk, and walk before we run.

PROCESS RESEARCH AND PHYSICAL ORGANIC CHEMISTRY

In beginning this tale, I thought that I might highlight the importance of physical organic chemistry in process research, as well as emphasize its impact on early synthetic development. During my graduate training at the University of Rochester, the labs were not arranged by research groups; instead, the new students simply occupied available lab spaces. Thus, while I was trained as a synthetic organic chemist, I quickly came under the influence of the physical organic chemistry of Bill Saunders and the physical photochemistry of Jack Kampmeier via their students with whom I shared a lab. My quiet transition to a closet physical organic chemist with a desire to understand synthetically important reactions from the mechanistic perspective began from these simple roots. I think this has been frequently reflected in my programs. In fact, were I now considering a career in teaching, I would formulate a research program around studying the mechanisms of synthetically important reactions, along with their applications in the development of practical processes. Each time we have turned to physical organic chemistry during my Merck career, the results have always justified the effort we have put in. We have thus brought the chemistry to new levels and some excellent publications have resulted. It was also to my great joy to have worked with Dave Hughes during the early part of his Merck career. I consider him to be one of the Industrial Deans of physical organic chemistry. He continued to be a regular consultant to all of our chemists when reaction mechanisms and kinetics came to the table.

While I could cite many of our programs that have relied on physical organic chemistry and mechanistic considerations for solutions, I have been particularly fond of some of Dave's work from the early 1980s in which he proposed a simple yet elegant solution to a long-standing problem and then subsequently developed the most straightforward method for the selective *O*-alkylation of amino acids without protection. We were developing a prodrug of methyldopa, Merck's first major antihypertensive agent. The medicinal chemists chose to make the pivaloyloxyethyl (POE) ester (note that many said this stood for Poor Old Ed because the problem ended up in my group) via the synthesis shown in Figure 1.1. It was satisfactory for making tens of grams of product to initiate the program, but it was not serviceable beyond that scale. With a double protection (classical Cbz protection on nitrogen and diphenylmethyl protection on oxygen via an unscalable fusion reaction with dichlorodiphenylmethane), followed by the alkylation and double deprotection,

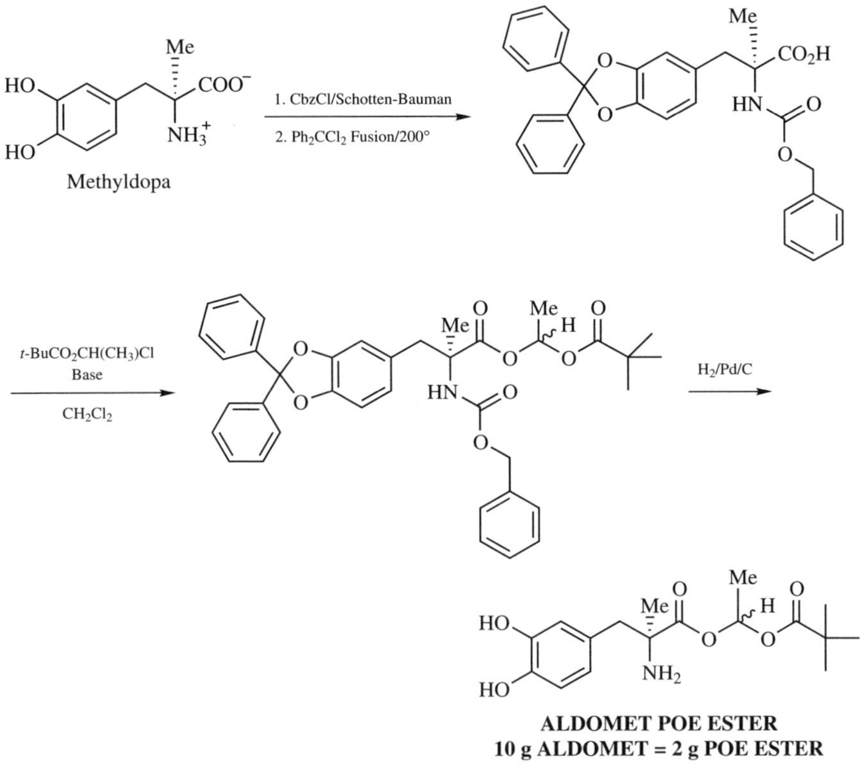

Figure 1.1. The medicinal chemistry synthesis of methyldopa POE ester.

the overall yield of the POE ester was poor. In addition, there were significant isolation problems since the product was a mixture of diastereoisomers.

This route was untenable at a preparative level, and the development program could not begin until we defined a new route. In thinking about how to approach this problem, we were guided by a comment from John Chemerda, who headed Merck Process Research until the late 1970s: "The best protecting group is NO protecting group!" Indeed we found that the direct alkylation of methyldopa in HMPA, TMU (tetramethylurea), or other polar aprotic solvents with 1-chloroethyl pivalate (pivaloyloxyethyl chloride: POECl) afforded a 65:35 ratio of isomeric *O*- versus *N*-alkylation adducts. The *O*-adducts could be separated from the product mixture by acid extraction. We conducted a formidable amount of phenomenological work on trying to improve this ratio, all to no avail. However, even at a modest yield we were able to prepare sufficient quantities of material to begin safety studies via a co-crystallization of the diastereomeric salts.[2] While the direct alkylation of the unprotected amino acid was a good idea, we were not able to improve the selectivity of the *O*- versus *N*-alkylation. This illustrates a point often seen in process research: A great idea is brought to the table, but the team is not able to bring it completely home. At this point Dave Hughes joined the program, and he brought

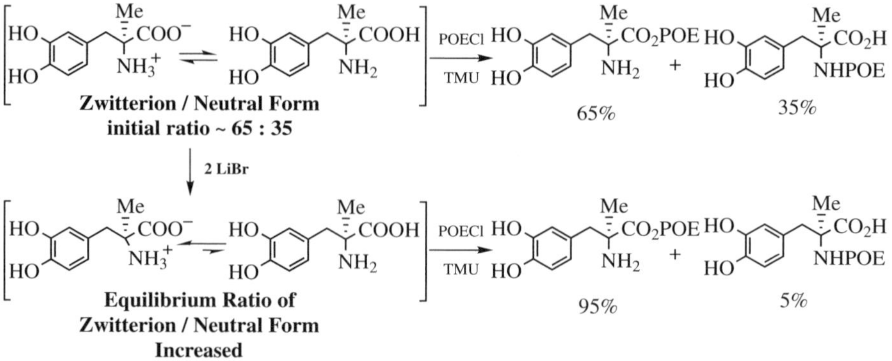

Figure 1.2. Direct alkylation of methyldopa in TMU employing Li Br addition.

his formidable background in physical organic chemistry to bear on the issue. He quickly recognized that the alkylation results represented a snapshot of the equilibrium between the neutral and zwitterionic forms of methyldopa present in polar aprotic solvents.

Alkylation of the neutral form occurred on nitrogen, while alkylation of the zwitterion occurred on oxygen affording the desired product. After a detailed comparison of the pK_a's of a variety of amino acids in water and DMSO, Dave established that the equilibrium could be favorably influenced toward the zwitterionic form by adding salts to solutions of the amino acids in polar aprotic solvents. With these results in hand, Dave redesigned the ester synthesis employing two equivalents of LiBr in TMU. In this strongly polar environment the equilibrium strongly favored the zwitterionic form of methyldopa, and a >90% yield of the desired ester resulted (Figure 1.2). Dave then developed the procedure into a general method for direct and simple amino acid ester synthesis.[3] This was a prelude to Dave's work and publications on the Mitsunobu reaction, which received worldwide recognition in its day.[4]

FLUDALANINE: ONE WE NEVER SOLVED—AT LEAST I THINK SO

The 1970's brought a remarkable, orally active, broad-spectrum antibiotic to Merck development: 2-deutero-3-fluoro-D-alanine (fludalanine). The deuterium was added to slow *in vivo* metabolic oxidation to the imine, which after hydrolysis and reduction would produce fluorolactate, a possibly toxic metabolite. This was a remarkable target—all functionality, no fat, and an isotope to boot! Immediately there were concerns about cost, but we chose to proceed forward with development. This view—that is, not letting product potential cost (or synthesis complexity for that matter) be a serious impediment at the beginning of a program—has proven to be a hallmark of the Merck Process Group. While some process groups place significant weight on synthesis complexity and possible product cost very early in a program, and often refuse to accept candidates for development based on these considerations, this has never been the case at Merck. Had it been, we probably would have never developed products such as cefoxitin, imipenem, or crixivan.

This was a case with fludalanine wherein the medicinal chemistry synthesis was simply not applicable to the preparative scale. Over the years, Janos Kollonitsch, who began his Merck career in Process Research but then moved into Medicinal Chemistry, had developed a program of photohalogenation of medicinally important substrates in strong acid media to provide unusual and new halogenated analogs of the substrates.[5] Initially, the focus was on chlorination, but even 30 years ago the potential of fluorine in bioactive molecules was recognized. Janos and his co-workers developed a photofluorination protocol wherein substrates were dissolved in liquid HF at −78°C and photofluorinated employing trifluoromethylhypofluorite (CF_3OF) and unfiltered UV light from a 1000-W high-pressure Hg/Xe lamp.[6] The fluorination was radical in nature and tended to produce mixtures of products which required separation.

Photofluorination of alanine produced a mixture of 3-fluoroalanine and 3,3-difluoroalanine and left residual alanine, but reasonable yields of 3-fluoroalanine could be obtained after careful workup and chromatographic purification. The chirality and the isotope made complex matters even more complex. At that time the Merck–Frosst Isotopes Division produced 2-deutero-D-alanine for Janos' work by labeling racemic alanine with deuteroacetic acid (AcOD) at 150°C. Not only was the proton at the chiral center exchanged under these conditions, but also all of the exchangeable hydrogens had to be exchanged to ensure ~99% deuterium incorporation at carbon. Thus, 400 moles of AcOD per mole of alanine was required in the exchange process. Thereafter, an enzymic resolution produced the 2-deutero-D-alanine for the photofluorination. In terms of deuterium efficiency, this was a terrible method, yet it sufficed to make the initial quantities of material necessary to bring the compound into development (Figure 1.3). I was asked to do the initial projections for this program, and I concluded that we would need numerous photoreactors running for weeks at a time to produce the first kilogram of drug needed to initiate safety, formulation, and Phase I studies. Given that this needed to be done in liquid HF with an explosive fluorinating agent at low temperature, we were between a rock and a hard place, as they say. We desperately needed a new approach, and having no idea what it might be, I projected that we could have the first kilogram in nine months. Clearly, we were relying on organic synthesis and its impact to get this program into development.

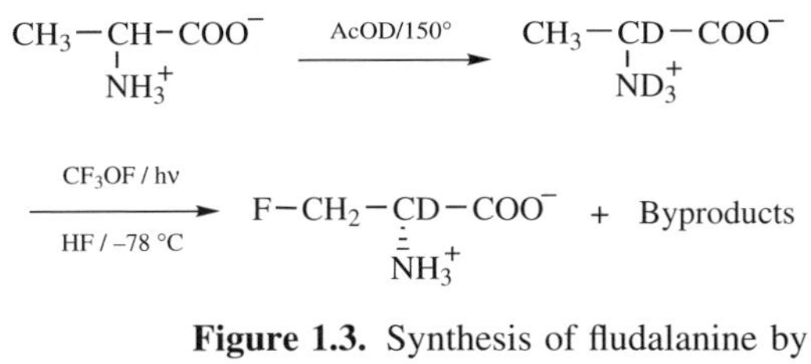

Figure 1.3. Synthesis of fludalanine by photofluorination.

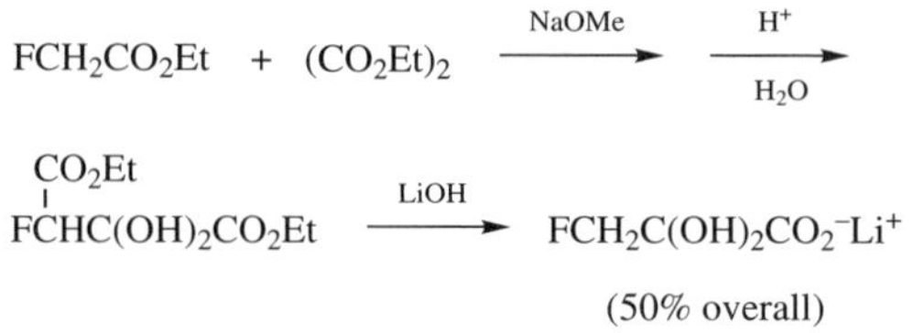

Figure 1.4. Lithium fluoropyruvate hydrate synthesis.

Working with a newly hired chemist named Ulf Dolling, who would become famous for his work on chiral, catalytic phase transfer reactions in the next decade, we decided to study the reductive amination of 3-fluoropyruvate which could be made by an ethyl fluoroacetate-ethyl oxalate condensation. For a possible source of deuterium, we asked the Frosst division if they could come up with a practical synthesis of sodium borodeuteride. They quickly showed that sodium borohydride could be exchanged with deuterium gas at 340°C, and they developed the equipment and a practical process to carry out this transformation on a kilogram scale! In approaching the fluoropyruvate reductive amination, we had three substrates to consider: esters, the acid, and salts. Of the three possibilities, we settled on the lithium fluoropyruvate hydrate, because of its relative aqueous insolubility, facile isolation, and overall stability. It was prepared by a classical condensation–decarboxylation sequence (Figure 1.4).

The reductive amination sequence on lithium fluoropyruvate hydrate proved most interesting. The amination of lithium fluoropyruvate hydrate in 13 M aqueous ammonium hydroxide at 37°C showed two FCH_2 doublets in a 95:5 ratio by proton NMR immediately after dissolution. After 1.5 hr the ratio had reversed to 5:95. Extensive NMR studies, including equilibration in $^{15}NH_4OH$, confirmed that the hydrate was immediately converted to the aminal (initial major species) upon dissolution in aqueous ammonia, and this species subsequently equilibrated to the diamine. The ketone and imine were not observed under these conditions, but the ketone and imine must be present in low concentrations to effect the observed equilibrium. The goal became to effectively reduce the imine to racemic deuterofluoroalanine and minimize back equilibrium to the ketone, which would afford deuterofluorolactate as a byproduct, keeping in mind that the borodeuteride was the cost-controlling reagent in the process. Extensive studies of the reaction kinetics, varying borodeuteride concentration, ammonia concentration, and temperature, were carried out to establish the appropriate reaction conditions. In a typical experiment, 0.2 mole of lithium fluoropyruvate hydrate was equilibrated in 300 ml of 13 M aqueous ammonia for 1.5 hr at 37°C to establish the 95:5 ratio of diamine to aminal. At this point the solution was cooled to 10°C to freeze the equilibrium ratio and then 0.085 mole (1.7 eq) of sodium borodeuteride was added (Figure 1.5). Little reduction occurred at this point because the pH of the system was too high. With the equilibrium ratio essentially frozen, the pH of the system was rapidly lowered with vacuum and nitrogen sparging to remove the excess ammonia and effect the desired reduction as the system back-equilibrated through

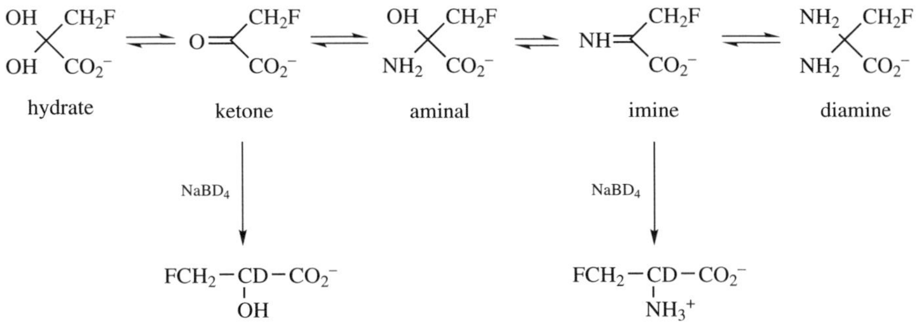

Figure 1.5. Equilibration and reduction of lithium fluoropyruvate in aqueous ammonia.

the imine. Subsequent workup employing a Dowex resin afforded a 70% crude yield (based on borodeuteride) of racemic product, which contained 99% deuterium at C-2. Using these conditions, we were able to make more than 60 kg of the racemate in the Pilot Plant over the course of the program.

I have always taken great pride in this synthesis. It illustrates the importance of understanding and controlling fundamental equilibria in a chemical process—all that stuff they taught us in freshman chemistry is actually true! At first glance it appears that this synthesis will never work, because the high pH necessary to convert the ketone hydrate to the diamine is contrary to the requirements of an efficient borodeuteride reduction. Yet the diamine:aminal ratio in the equilibrium mixture can be preserved by lowering the temperature, and the desired pH for the reduction was achieved by removing the excess ammonia. When we did the equilibration with fluoropyruvic acid hydrate, two changes to the process were noted. The equilibrium was achieved in less than 5 minutes because it was catalyzed by the protons brought into the system with the acid. But in attempting to do the reduction, the presence of these protons caused exchange of the borodeuteride under the reduction conditions, and the level of deuterium label dropped to about 90%.

Of course this process produced the racemate, and we had to face the question of resolving the product to afford the D-isomer. Given the Merck experience with the continuous resolution in the methyldopa process,[7] we opted for a continuous resolution of a conglomerate. The benzenesulfonic acid salt of 3-fluoro-2-deuteroalanine had all of the desired properties for an effective continuous resolution in *n*-propanol employing a dissolution temperature of 28°C and a crystallization temperature of 23°C which afforded a 16.8% supersaturation. It is noteworthy to mention the contributions of Dr. Mike Middler of our Chemical Engineering Department. He was one of the principals in designing the methyldopa continuous resolution, and we traded on his expertise almost daily when we were designing the conditions and system for our resolution. A summary of the details of this resolution, including a picture of the equipment necessary to perform this resolution on a laboratory scale, is included in a recently published paper.[8] With the chemistry and resolution established, we were able to make the first kilogram of drug substance in 10 months, one month more than the initial projection.[9] Employing this overall

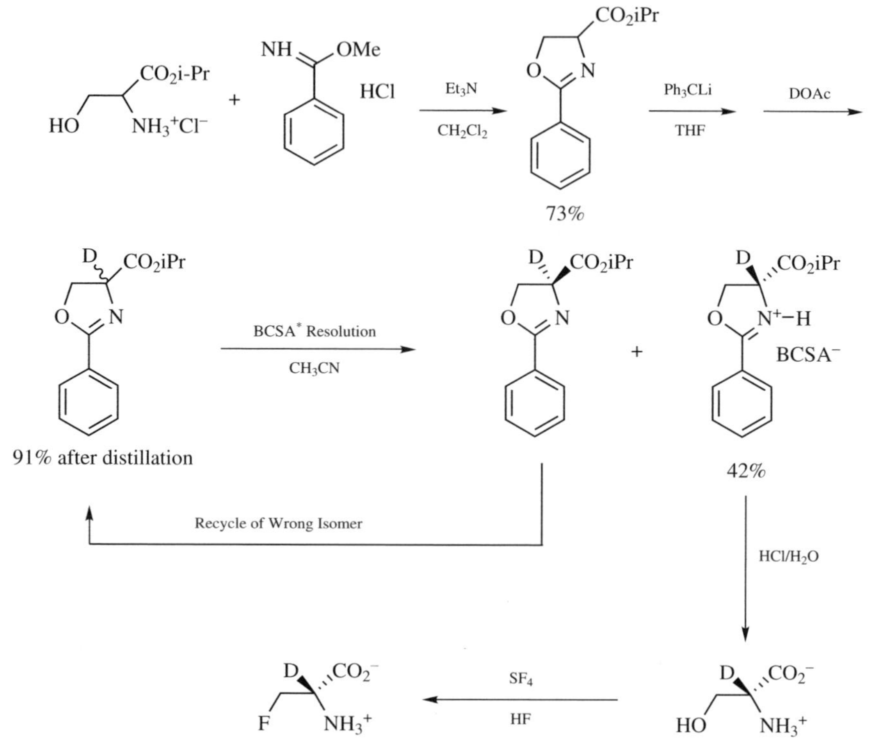

Figure 1.6. The oxazoline approach to fludalanine.

process, our engineers were able to make more than 20 kg of product to support the overall program.

In spite of the advances of this synthesis over photofluorination, it would be an inherently expensive process, relying on sodium borodeuteride for the deuterium source, making the racemate instead of the single enantiomer and employing a continuous resolution process, which would have required extensive capital investment for the program. Racemization of the free amino acid or its derivatives is not possible because of HF elimination. As we moved further into the program and began thinking of a synthesis that could produce hundreds of kilograms instead of tens of kilograms, Paul Reider, who had completed studies with Al Meyers at CSU, joined the Merck Process Group and was assigned to work on this program. Paul applied his expertise in oxazoline chemistry to propose and develope a new process for fludalanine.[10] The key in this approach was the conversion of isopropyl serinate to the 2-phenyloxazoline (Figure 1.6), followed by anion formation and deuteration. Because of concerns with exchange of the deuterium in the oxazoline, trityl lithium was used as a selective base for deprotonation and AcOD was used as the source for deuterium. This approach resulted in a dramatic improvement over the borodeuteride route because the AcOD could be readily made from acetic anhydride and D_2O, and the deuteration was essentially pinpoint, though racemic. The resulting deutero-oxazoline was resolved with D-α-bromocamphorsulfonic acid (BCSA) in acetonitrile to give the desired isomer in 42% yield and >99.8% e.e. (84% corrected for 50% availability of the desired isomer). The wrong isomer was racemized with potassium carbonate in acetonitrile. The use of the same solvent for the resolution and racemization was a bonus for the procedure. Acid hydrolysis of the freed deutero-oxazoline afforded (*R*)-2-D-serine in 92% yield. Subsequent fluorination in HF with SF_4 afforded an excellent yield of fludalanine when run at concentrations of <0.05 M. At higher concentrations, serine regeneration became a problem because of a termolecular reaction among two serines and one SF_4 which formed an $(RO)_2SF^+$ species. This subsequently afforded fludalanine and an ROS(O)F intermediate that hydrolyzed back to serine.[11]

While this procedure represented a great improvement over the borohydride route in terms of addressing the cost of the introduction of the deuterium, it was not without its liabilities. Atom economy was poor, particularly with the need for the use of trityl lithium. Resolution–racemization was required, rather than a pinpoint, enantiospecific deuteration. Finally, the use of SF_4 in HF, particularly at dilute concentrations, is inherently dangerous and expensive.

This prompted a yet another approach to fludalanine. Racemic proteo-fludalanine could be readily made via reductive amination of lithium fluoropyruvate employing sodium borohydride. The plan was to protect the amino group in the racemate as its phthalimide derivative and convert the acid to the acid chloride. Reaction with base could afford the ketene (Figure 1.7), which upon subsequent reaction with a chiral, deuterated alcohol could afford the *N*-protected deuterofludalanine ester. At this time there were the emerging reports in the literature on the diastereoselective addition of enantiopure alcohols to ketenes. The phthalimidoacid

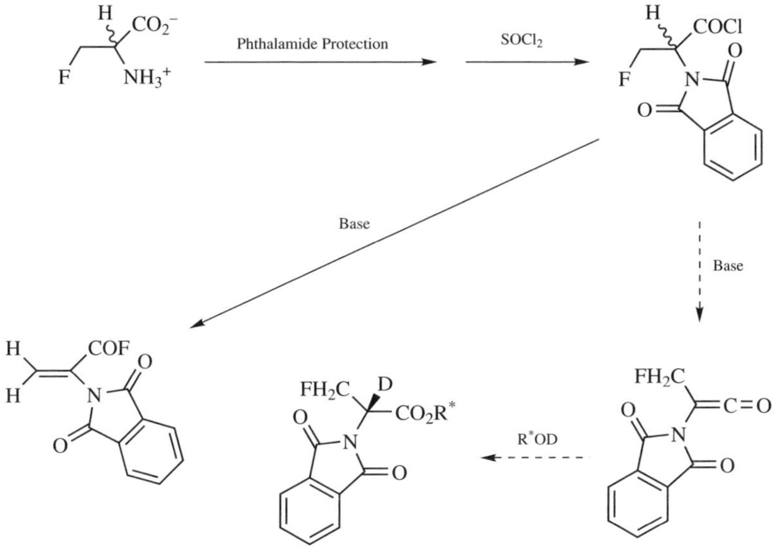

Figure 1.7. Projected ketene route to fludalanine.

chloride was readily made by standard methods. Reaction with bases ranging in strength from 4-cyanopyridine to diisopropyl ethylamine (DIPEA) afforded the same crystalline, stable product. It had IR carbonyl absorptions at $1825\,cm^{-1}$ and $1795\,cm.^{-1}$ In addition to the phthalimido group in the proton NMR spectrum, a doublet and a quartet (2*H*) were present in the 6- to 7-ppm region, with small (1 Hz) couplings. The IR should have given the structure away, but we were fascinated by the possibility of formation of cyclobutanones arising from $2 + 2$ cycloaddition of the ketene, and wandered on a misadventure. Subsequently, we recognized that we had an acid fluoride, and the solution to the structural problem was obvious. Given the choice of eliminating HCl to form a ketene (high-energy path) or HF to form an acryolyl chloride (low-energy pathway), the latter occurred. With free fluoride ion in the system, the acid chloride is converted to the acid fluoride, which is remarkably stable.[12] I have presented this problem (IR and NMR data) to many chemists over the years, and they either have failed to get the correct answer or have taken days to do so. The exception was Glen Berchtold, my MIT senior thesis advisor, who looked at the problem for less than a minute and had the correct answer. He recognized the acid fluoride carbonyl and the need for a second coupling to one of the vinyl hydrogens in the product—a W arrangement with the F atom. When I asked him how was he able to get the answer so quickly, he noted that he had been teaching the spectroscopy course for organic chemists for the prior 15 years!

The program ended due to non-chemistry issues, and our adventures on fludalanine were terminated. I do not think we came up with the ultimate process, though Paul Reider might disagree. In our ratings for that year, the lack of an enantiospecific synthesis was noted. My memory indicates that an asymmetric diastereospecific reduction of the acetamidofluoroacrylate has been reported in the literature, but that does not produce the desired target, because a new FHDC chiral center is created at C-3. Regarding the ketene approach, Rob Larsen and Ed Corely revisited that in the context of a synthesis of (*S*)-ibuprofen a few years later, and they actually solved that problem in a most elegant way which will be described later in this chapter.[13]

FINASTERIDE: CAN THE ENGINEERS ACTUALLY BE RIGHT?

In the first "Reflections..." chapter[1] I noted the unusual silylation-mediated [bistrimethylsilyltrifluoroacetamide (BSTFA)] DDQ oxidation of a lactam to form its unsaturated analog proceeding via a substrate–DDQ adduct followed by a thermal loss of the hydroquinone (Figure 1.8). This chemistry forms the nucleus of the finasteride manufacturing process,[14] but the story from the bench to the pilot plant during the early development of this product is much more complex than the scheme indicates. We have all delighted in telling this episode over the years in our finasteride presentations. It highlights the unexpected difficulties that one can encounter when trying to put a new process in place early in the drug development process, and it highlights the misplaced bravado that we process chemists are sometimes guilty of in our interactions with the chemical engineers.

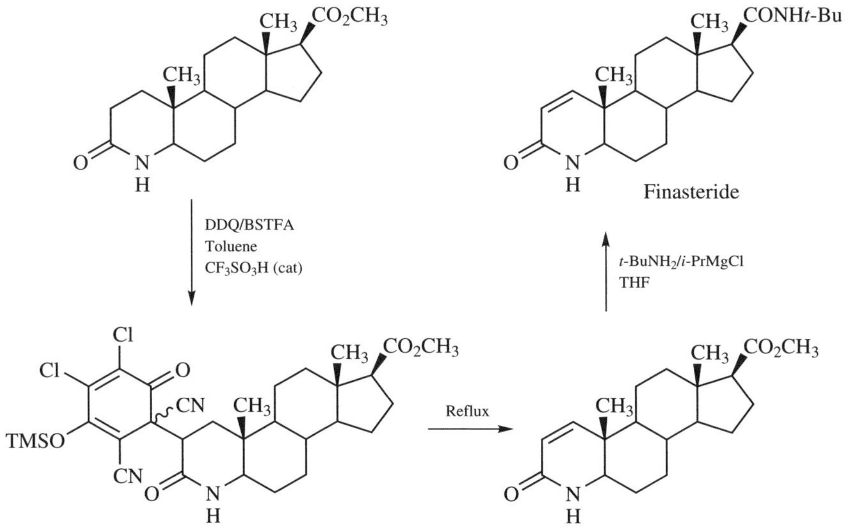

Figure 1.8. Outline of the finasteride manufacturing process.

The first indication of unexpected complexity in this chemistry occurred when we noted that the optical rotation of finasteride prepared by the silylation/quinone method was a few degrees higher than material prepared by the medicinal chemists' phenylselenic anhydride method. (We chose not to proceed with this method because of the toxicity issues associated with the handling of selenium reagents.) Of course, our first response to this situation was to simply claim that our product was purer than theirs, and the resulting rotations were the first reflection of purer final product. However, the analytical chemists would have none of this explanation, because materials prepared by both methods were >99% pure by area and weight percent. They did note, however, that we had three low-level impurities (0.1–0.3 area %) that were not produced during the selenium oxidation. Isolation and analyses of these impurities indicated that they were further oxidation products wherein one, two, and three (!!!) additional double bonds (Figure 1.9) are present in the B rings.[15] The standard optical rotations of these materials were of the order of 1000°, approximately 10 times that of finasteride. Thus, low levels of these compounds would easily increase the observed product optical rotation.

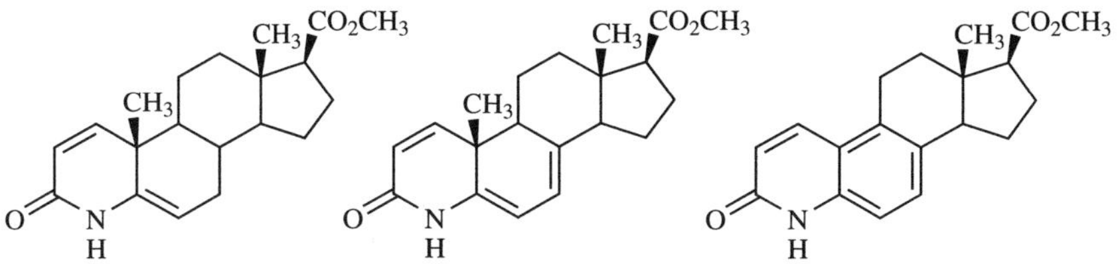

Figure 1.9. Finasteride overoxidation products produced during the silylation-mediated quinone oxidation.

Clearly the DDQ/BSTFA oxidizing system had components that we did not understand, and some sort of remedy was needed. We had already established that silylenol ethers and DDQ form substrate–quinone adducts analogous to those formed by the silyl imidates,[16] so we began to envision a stepwise protocol for conducting the finasteride oxidation. The first step would be formation of the silylimidate followed by its reaction with the quinone to form the substrate–quinone adduct. Both of these readily occurred at room temperature via what we have presumed to be an electron transfer reaction between the silylimidate and the DDQ with concurrent silyl transfer. Excess DDQ was used in the reaction to ensure complete conversion to the adduct ($<0.1\%$ starting material), because unreacted starting material could not be removed from the product mixture by crystallization. Yet, it was the excess DDQ that was the source of our overoxidation byproducts. Thus, we reasoned that addition of a ketone to the mixture that would form a silylenol ether which would subsequently react with the excess DDQ to form a new adduct and remove the DDQ from the picture.[16] Thermolysis at this point should provide product free of over oxidation products. For the ketone we selected cyclohexane-1,3-dione, which readily formed the disilylenolether under the reaction conditions. Reaction with DDQ would produce an adduct, and thermolysis would afford monosilylated 1,3-dihydroxyphenol and the hydroquinone, both of which would be subsequently silylated by the BSTFA. This would render the excess DDQ benign, and we would not create these low-level impurities (Figure 1.10).

This chemistry progressed well in laboratory runs, and we proceeded to our first Pilot Plant campaign. Appropriate charges of the azasteroid ester (1 part), BSTFA (four parts to ensure silylation of all hydroxyls to prevent them from undergoing Michael reactions with the unsaturated lactam product), and DDQ (1.2 parts) at 25°C produced the desired silylated steroid–quinone adduct with $<0.1\%$ residual starting material as expected. The cyclohexane-1,3-dione ($\sim$0.2 parts) was added and the reaction aged an hour to produce its dienol ether and subsequent adduct, and then the reaction was heated at reflux to complete the overall transformations. To our dismay, the resulting product contained $\sim$1% starting azasteroid, despite the fact that we had consumed $>99.9\%$ of it in the DDQ adduct formation. Faced with this conundrum, we did what all process chemists do: We blamed the engineers! Obviously we assumed that the reactor and/or its lines were contaminated with

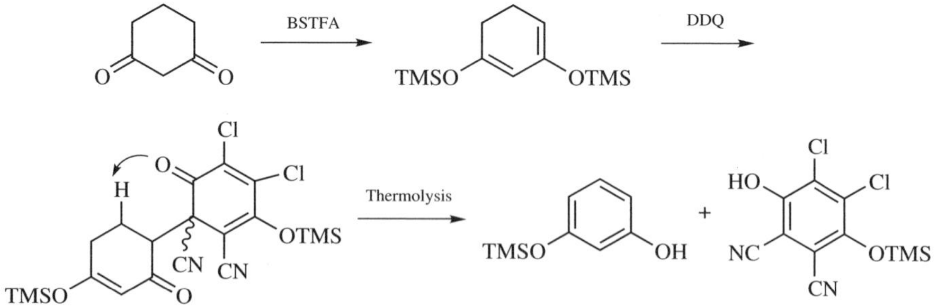

Figure 1.10. Reaction of DDQ with cyclohexane-1,3-dione and thermolysis of the adduct.

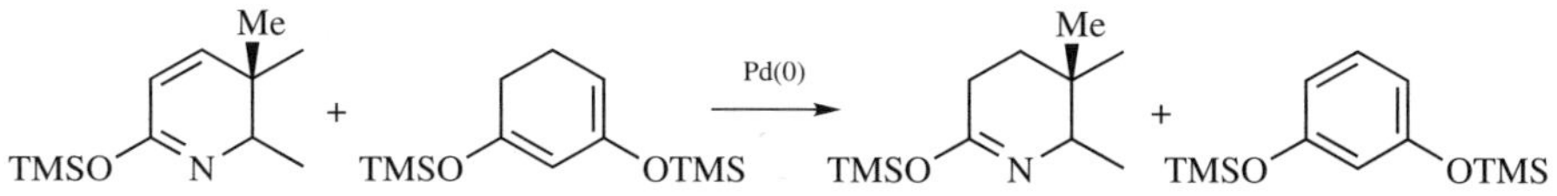

Figure 1.11. Inadvertant reduction of the oxidation product under transfer hydrogenation conditions brought about by the wrong choice of a ketone to consume the excess DDQ and the presence of trace levels of palladium in the starting lactam.

starting material, which was reintroduced into the reaction during its later stages. What we had was *not* some bizarre chemistry, but a simple contamination issue. We were steadfast in our explanation and would consider no others. Our engineering team was equally committed to the opposite thought—that is, that the reactors and their lines were *not* contaminated with starting material. To their credit, they absolutely refused to accept our explanation of the events. No matter how hard we pushed, they refused to budge—the system was clean, and something was wrong with the chemistry.

Desperate for an answer, we began to consider the engineer's views. When we ran a control experiment wherein we took the desired unsaturated product (namely, cyclohexane-1,3-dione) and BSTFA and heated the solution at reflux in toluene we saw, much to our dismay, the slow formation of *saturated* starting material. The engineers were right: There was a fatal flaw in our chemistry that we had missed. At this point we recognized that the dione forms the 1,3-disilyloxycyclohexadiene under our reaction conditions—in fact it acts as a transfer hydrogenation reagent! Could we be reducing the product back to starting material during the thermolysis reaction? As soon as we asked this question, the answer became obvious. During the preparation of the saturated azasteroid ester starting material, a double bond in the B ring of the azasteroid is reduced over Pd/C. A quick assay showed that this starting material contained ~5 ppm residual Pd from the reduction, and this Pd along with the inadvertent addition of a transfer hydrogenation reagent created the reductive environment and produced starting material from the desired product (Figure 1.11). The engineers were correct: The equipment was clean. Once the problem was understood, the solution was simple. The dione was replaced with methyl acetoacetate, which forms a silylenol ether in the reaction that reacts with DDQ to form a benign adduct under the reaction conditions.

I have delighted in telling this story in lectures over the years—it clearly shows the fallibility of process chemists, as we often think we know more than we really do. It is also an excellent example of how things can go wrong despite the best of intentions and the projections of excellent science.

KETENE ADDITIONS REVISITED

As noted during the previous discussion on fludalanine, we did not succeed in doing a diastereoselective addition of a chirally pure alcohol to a ketene. Many years after these efforts, we, along with many in the pharmaceutical industry, became interested in the single enantiomers of the common 2-arylpropionic acid nonsteroidal anti-inflammatory drugs (NSAIDs). At that time, additions of ketenes to chirally pure alcohols with d.e.'s as high as 80% had been reported, but the alcohols needed were of limited availability and expensive. Rob Larsen and Ed Corley, who were not aware of the previous work on fludalanine, rose to the challenge and defined a simple protocol based on naturally and/or commercially available chirally pure α-hydroxy esters and lactones.[13]

The plan was straightforward: generate the ketene from an NSAID such as racemic ibuprofen and define conditions for a diastereoselective addition to a chirally pure simple alcohol. (*S*)-Ethyl lactate proved to be one of the best-performing alcohols, and the reaction conditions that emerged proved to be highly specific. Nonpolar solvents such as hexane or heptane provided the best d.e.'s. Phenomenological studies indicated that the addition proceeded best in the presence of small tertiary amines (i.e., trimethylamine, dimethylethylamine, and *N*-methylpyrrolidine, for example), which were already in play to generate the ketene from the acid via the acid chloride. Without an added amine, d.e.'s were in the 60% range. Both low reaction temperatures and lower concentrations provided the maximum d.e.'s. Optimum conditions with (*S*)-ethyl lactate were as follows: −78°C; 0.02 M concentration; in heptane with trimethylamine as the base. A 98.6:1.4 SS/SR ratio of diastereomers resulted under these conditions. The d.e. dropped to 95:5 when the reaction concentration was raised to a more practical 1.0 M. A study of the structure versus d.e. for various alcohols revealed that (*R*)-pantolactone [(*R*)-dihydro-3-hydroxy-4,4-dimethyl-2(3*H*)-furanone] was by far the best affording a d.e. ratio of 0.5:99.5 for the RS/RR isomers at preparatively useful reaction conditions. The results are summarized in Figure 1.12.

The addition reaction proved to be third order overall, being first order in ketene, alcohol, and tertiary amine. A k_H/k_D of ∼4 was noted for the ROH/ROD isotope effect. These results prompted many vigorous mechanistic discussions; however, without appropriate carbon and oxygen isotope effects, it was difficult to provide a convincing picture of the mechanism of the addition. At one of the Gordon Conferences, Rob and I were able to convince Ken Houck to consider the problem from

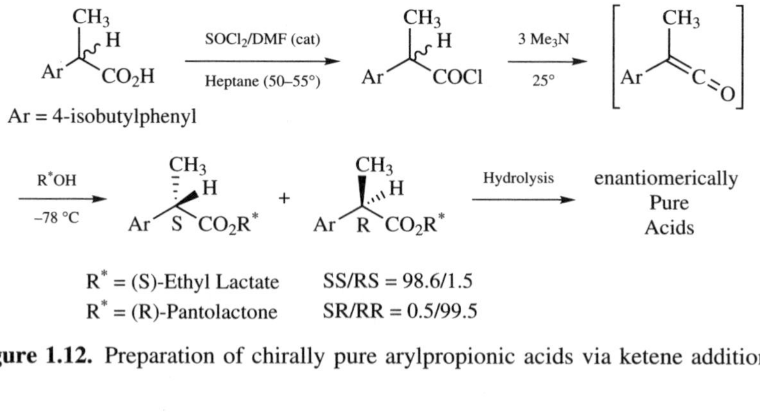

Figure 1.12. Preparation of chirally pure arylpropionic acids via ketene additions.

a theoretical view. The reader is referred to the resulting JACS papers for a discussion of the outcome of the calculations.[17]

Of course we were between a rock and a hard place on this problem. Despite publishing our results in JACS, we were far from a practical and economical process. Indeed, any process dependent on purchased racemic ibuprofen as a starting material was doomed to failure at the outset. Commercial suppliers were producing millions of kilograms of this material each year by well-established and highly efficient processes. Since the chiral center is between activating aryl and carboxyl groups, it is readily amenable to racemization, and single enantiomers can be prepared by simple resolution/racemization processes.

THE CONTINUOUS CONUNDRUM

A problem that besets all process chemists is how much effort to apply to a given program. When I began my Merck career, we took the long-range view on every compound entering development. Each was considered a likely product, and we strove to establish the best synthesis as quickly as possible. Safety Assessment typically did 14-week studies; Pharmaceutical Research began by immediately trying to develop a solid dosage form; and the early planned clinical studies were extensive. Materials requirements to initiate a program were generally high. In the final analysis, we had a lot of superb processes for candidates that died very early in the development cycle, and lots of left over material in the sample collection awaiting some unforeseen use. Obviously, we were not using our resources efficiently. Today, we recognize that of twenty compounds entering development only one is likely to reach product status. (These are my figures, and others may dispute them.) With the probability of success of a typical candidate being in the vicinity of 5%, what resources should we devote to each new candidate as it enters development? Unfortunately, the equation that determines such is not a simple one. The first consideration is where are the quick GO/NO GO points in a program? If a compound has a serious potential safety issue, the answer might be had in short safety studies. If there are questions about half-life or bioavailability of a given candidate, this information can be gotten with short safety studies and a simple Phase I study. Neither situation requires a lot of material or a synthesis resembling a manufacturing process. On the other hand, the key GO/NO GO point in a program might not come until the Phase III clinical studies, and the problem has to be approached differently. If you proceed conservatively, develop a solid process and make a large quantity of material, and the compound dies in the first few weeks of safety assessment, you have wasted a lot of effort. If, on the other hand, you prepare a small quantity of material by bulling through the medicinal chemistry synthesis to quickly answer a half-life or bioavailability question, and the results are strongly favorable and the project catches fire, you are left with a synthesis that can not support the program's needs, and drug supply becomes the program's rate-limiting issue. Companies have developed their own protocols for addressing these questions, and much more thought is now given up-front as to how to proceed on an individual program.

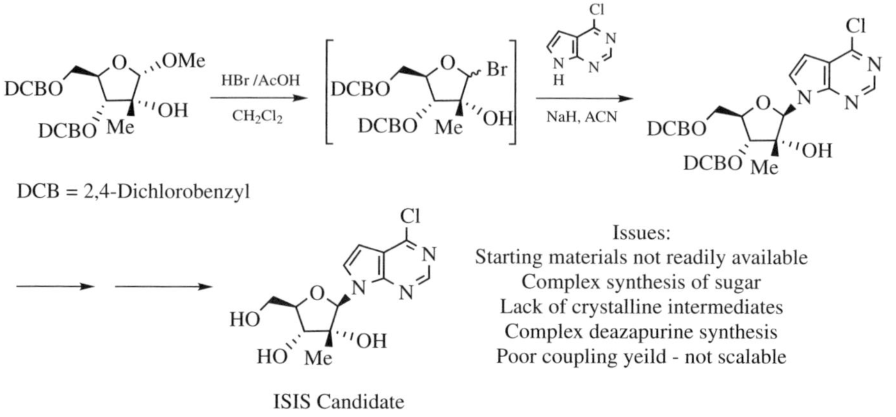

Figure 1.13. Original synthesis of ISIS candidate.

Today we might use outsourcing to prepare early material by the medicinal route to give the process chemists time to develop a preparatively viable route. Or, if the program does not have high priority, we might sequence all of the program activities in series, until we have a clear and definite GO indication. Despite all of the current planning, we still end up with imbalances. Some of our work illustrates the complexity of developing the path forward.

Recently, Merck licensed a hepatitis C candidate from ISIS (Figure 1.13), and it was brought into development. The candidate was nucleoside-like in structure, and in my entire career I had yet to work on a candidate of this structural type. I have never been able to predict which candidates will become products (we all want to have products listed on our cv's), and I have always made it a point, when possible, to choose candidates to work on based on potential chemical interest and complexity. The ISIS candidate looked like a good one from my point of view, and it was assigned to my group at my instigation. The medicinal synthesis was long and complex and was only suitable for making a few grams of product with a substantial effort. It is noted in detail in our paper on this program.[18]

Quickly, a number of issues with this synthesis were identified. The syntheses of the ribose and deazapurine portions were difficult and based on poorly available, expensive reagents. In proceeding forward with projected papyrochemical syntheses, we realized that much of the ISIS sugar and purine chemistry might not be useful for preparative work. The success of any sugar synthesis is highly dependent on selection and study of the appropriate protecting groups, which have to properly and successfully serve numerous functions during the course of the synthesis. One is to afford crystallinity and purification points, something that was lacking in the existing ribose synthesis. The synthesis of the deazapurine fragment was complex and not amenable to scale, and a more viable one was definitely needed. The most significant problem with this synthesis was the activation of the sugar and its coupling to the deazapurine anion. The intermediate bis-*O*-dichlorobenzyl-protected bromide was not stable to the conditions used to generate it. The net result was that as you scaled the synthesis, the actual amount of product made remained essentially the same! With so many complex synthetic issues on the table, I became quite conservative in approaching this problem. Any process for this compound was going to be an *a priori* complex, and we wanted stability and control at each stage of the synthesis. Our goal became the development of a scaleable synthesis of a protected ribose that would serve all program needs, particularly with regard to crystallinity and points of purification. We also needed a new and efficient synthesis of the deazapurine. Preferably both the syntheses would be amenable to outsourcing. We also needed an efficient and stable method of activation, as well as a high-yielding coupling reaction to a penultimate protected intermediate that could be readily deprotected to final product. Part of my conservativism in approaching this program was that I wanted to be sure that we could continue to strongly support it if the early safety and clinical studies proved successful. Hepatitis C is a major threat to the world, and early successes in safety and the clinic would mean a fast-track program that could create the excitement of the protease inhibitors of the 1990s.

Recognizing that achieving these goals was going to be a difficult task, we assigned two teams to the program: The first was to develop the ribose synthesis and the coupling protocol, and the second was to develop the deazapurine synthesis. The latter was far more complex than anticipated because we were uncertain about how to introduce the primary amino group in this portion of the molecule. ISIS had used the chloro analog that was displaced with ammonia to complete the synthesis. In addition to ISIS's use of the chlorodeazapurine, we examined the azido group, simple amides, and imides and finally selected the phthalimido group. This meant that the team aiming for a practical deazapurine synthesis was shooting at a moving target and had to work very closely with those working on the ribose and coupling parts of the synthesis.

The process that resulted proved to be just about everything that we wanted. A synthesis of the ribose from readily available diacetone glucose was developed. Study of a variety of protecting groups revealed that the use of the *p*-methylbenzoyl protecting groups on the ribose portion afforded stable and crystalline intermediates throughout the synthesis. The epoxide group proved to be the coupling group of choice. It was readily generated from the diol, and it was stable to its generation conditions. It was also crystalline, but its isolation was not necessary. The deazapurine was readily prepared in phthalimido-protected form via a new synthesis from bromoacetaldehyde diethyl acetal, malonitrile, thiourea, and phthalic anhydride, all commercially available and inexpensive reagents. The coupling was catalytic in sodium hydride, and, most amazing, a global deprotection could be achieved at the end with butyl amine (Figure 1.14). The overall yield for this new process was excellent, and 3 kg of drug was initially prepared to demonstrate its viability and to provide material to begin development.[18] This could be a manufacturing process for this candidate. The only issue that I see is that of atom economy, because a large portion of the molecular weight is removed in the final deprotection step. But, in our defense, the sources of the protecting groups (*p*-methylbenzoyl chloride and phthalic anhydride) are materials of commerce.

Based on the chemistry that I have presented and the amount of drug that we made, the reader might consider this a very successful program. But, in the final

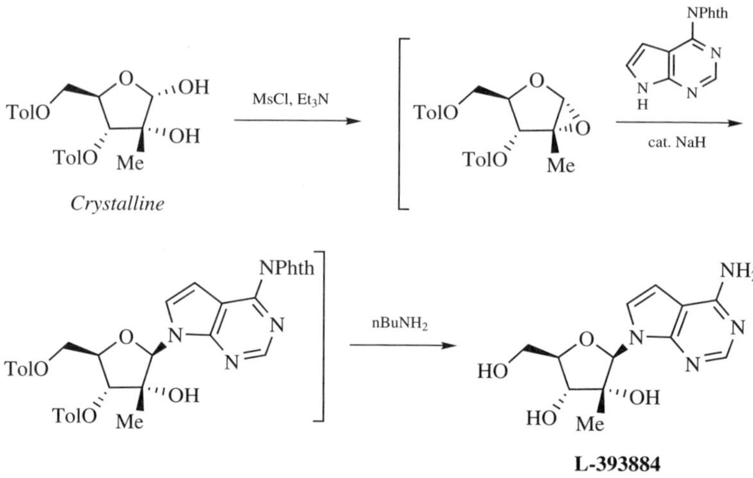

Figure 1.14. Practical synthesis of L-393884 from diacetone glucose.

analysis, the chemistry again ran well ahead of the rest of the development program. Issues arose during the safety studies. We could have achieved the same results with one-third of the material we actually made. Once we defined the outlines of the described synthesis, we could have proceeded forward without dotting most every i and crossing most every t. Generous use of preparative chromatography throughout the synthesis could have saved a lot of development time. But we did not know that issues would arise in the preclinical part of the program—all of the short-term studies done to bring the compound to development were on target. We chose to proceed conservatively and we need not have. This remains the continuous conundrum, and I do not know if anyone has a real answer. Some might suggest that we seek the help of the MBAs in planning our programs—my history indicates that this leads only to a lot of nonsense. We do the best planning that we can with the information available, and we take our kudos or lumps as they come. I could have chosen a program that illustrates success at every turn, but the readers would recognize that this is simply not a reflection of reality.

I hope that this very personal recounting of these process research wars has entertained and enlightened the reader and, hopefully, provided some useful chemistry and thoughts that will be of help to his/her future programs. Clearly, organic synthesis will continue to have a dominant role in early development. Contrary to current thinking, true process research is not just outsourcing and running lots of reactions. Serious opportunities for creative new chemistry are always present, and it is those who believe in their programs who are the ones who will really make things happen. The real question is whether process chemists will be able to continue to pull rabbits out of their hats on a timely basis to meet ever-demanding program timelines. As always, I must thank the members of the Merck Process Research Group for their superb efforts. From my position in retirement, I dearly miss them, their intellectual stimulation, and their innate drive for success.

REFERENCES

1. Grabowski, E. J. J. Reflections on Process Research, in *Chemical Process Research—The Art of Practical Organic Synthesis*, Abdel-Magid, A. F., and Ragan, J. A., Eds., ACS Symposium Series 870, American Chemical Society, Washington, DC, 2004, Chapter 1, pp. 1–21.
2. Budavari, J., and Grabowski, E. J. J. US Patent 4,670,578 (1987).
3. Hughes, D. L., Bergan, J. J., and Grabowski, E. J. J. *J. Org. Chem.* **1986**, *51*, 2579–2585.
4. Hughes, D. L., Reamer, R. A., Bergan, J. J., and Grabowski, E. J. J. *J. Am. Chem. Soc.* **1988**, *110*, 6487–6491.
5. Kollonitsch, J, Rosegay, A., and Doldouras, G. A. *J. Am. Chem. Soc.* **1966**, *86*, 1858–1859.
6. Kollonitsch, J, and Barash, L. *J. Am. Chem. Soc.* **1976**, *98*, 5591–5593.
7. Midler, Jr., M. US Patent 3,892,539 (1975).
8. Grabowski, E. J. J. *Chirality,* **2005**, *17*, S249–S259.
9. Dolling, U.-H, Douglas, A. W., Grabowski, E. J. J., Schoenewaldt, E. F., Sohar, P., and Sletzinger, M. *J. Org. Chem.* **1978**, *43*, 1634–1640.
10. Reider, P. J., Eichen Conn, R. S., Davis, P., Grenda, V. J., Zambito, A. J., and Grabowski, E. J. J. *J. Org. Chem.* **1987**, *52*, 3326–3334.
11. Reider, P. J., and Douglas, A.W. *Tetrahedron Letters* **1984**, *25*, 2851–2854.
12. Grabowski, E. J. J. Unpublished work.
13. Larsen, R. D., Corley, E. G., Davis, P., Reider, P. J., and Grabowski, E. J. J. *J. Am. Chem. Soc.* **1989**, *111*, 7650–7651.
14. Bhattacharya, A., DiMichele, L. M., Dolling, U.-H., Douglas, A. W., and Grabowski, E. J. J. *J. Am. Chem. Soc.* **1988**, *110*, 3318–3319.
15. Williams, J. M., Marchesini, G., Reamer, R. A., Dolling, U.-H., and Grabowski, E. J. J. *J. Org. Chem.* **1995**, *60*, 5337–5340.
16. Bhattacharya, A., DiMichele, L. M., Dolling, U.-H., Grabowski, E. J. J., and Grenda, V. J. *J. Org. Chem.* **1989**, *54*, 6118–6120.
17. Cannizzaro, C. E., and Houk, K. N. *J. Am. Chem. Soc.* **2004**, *126*, 10992–11008.
18. Bio, M. M., Xu, F., Waters, M., Williams, J. M., Savary, K. A., Cowden, C. J., Yang, C., Buck, E., Song, Z. J., Tschaen, D. M., Volante, R. P., Reamer, R. A., and Grabowski, E. J. J. *J. Org. Chem.* **2004**, *69*, 6257–6266.

2

DEVELOPMENT AND SCALE-UP OF A HETEROCYCLIC CROSS-COUPLING FOR THE SYNTHESIS OF 5-[2-(3-METHYL-3*H*-IMIDAZOL-4-YL)-THIENO[3,2-*b*]PYRIDINE-7-YL]AMINO-2-METHYL-1*H*-INDOLE

John A. Ragan
Chemical Research and Development, Pfizer Gobal Research & Development, Eastern Point Road, Groton, CT 06340

INTRODUCTION

Angiogenesis is a requirement for tumor growth and metastasis and occurs through several discrete biochemical signaling pathways. One key pathway that initiates proliferation and migration of endothelial cells is signaling through the vascular endothelial growth factor receptor-2 (VEGFR-2).[1] Therefore, small molecules that block this signaling pathway through inhibition of VEGFR kinase activity could potentially inhibit angiogenesis and tumor growth. 5-[2-(3-Methyl-3*H*-imidazol-4-yl)-thieno-[3,2-*b*]pyridine-7-yl]amino-2-methyl-1*H*-indole (**1**, Figure 2.1) recently emerged as a promising VEGFR kinase inhibitor (7 nM IC_{50} against VEGFR-2) and was thus of interest for clinical evaluation in the treatment of cancer.[2] This prompted us to seek a practical synthesis of this imidazolyl-thienopyridine to provide the kilogram quantities required for regulatory toxicology studies and clinical evaluation.

Fundamentals of Early Clinical Drug Development: From Synthesis Design to Formulation, edited by Ahmed F. Abdel-Magid and Stéphane Caron

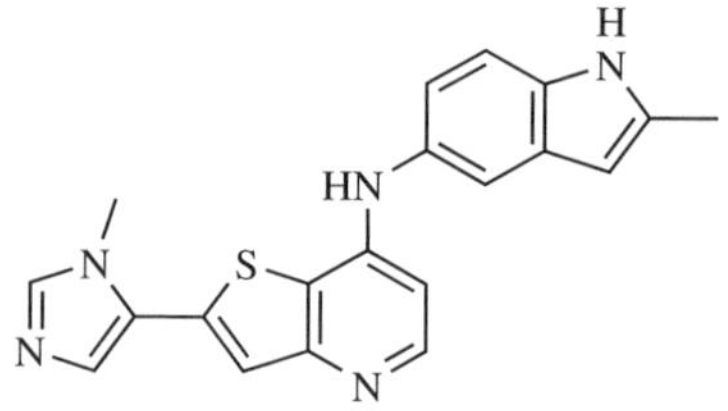

1

Figure 2.1. Structure of 5-[2-(3-methyl-3*H*-imidazol-4-yl)-thieno-[3,2-*b*] pyridine-7-yl] amino-2-methyl-1*H*-indole.

RETROSYNTHESIS

The retrosynthetic analysis of thienopyridine **1** is shown in Figure 2.2. The modular nature of the molecule lends itself to disconnection to subunits **2**, **3**, and **4**. Aminoindole **2** is commercially available. Iodothiophene **3** is available in a few steps from 3-amino-2-carboxythiophene (**5**). The exact nature of the imidazole organometallic reagent **4** took on several forms during the course of this work, and it constitutes the major focus of research; variants examined include organozinc, organomagnesium, organoboron, and organostannane reagents, as will be discussed throughout this chapter. However, these basic bond disconnections were maintained

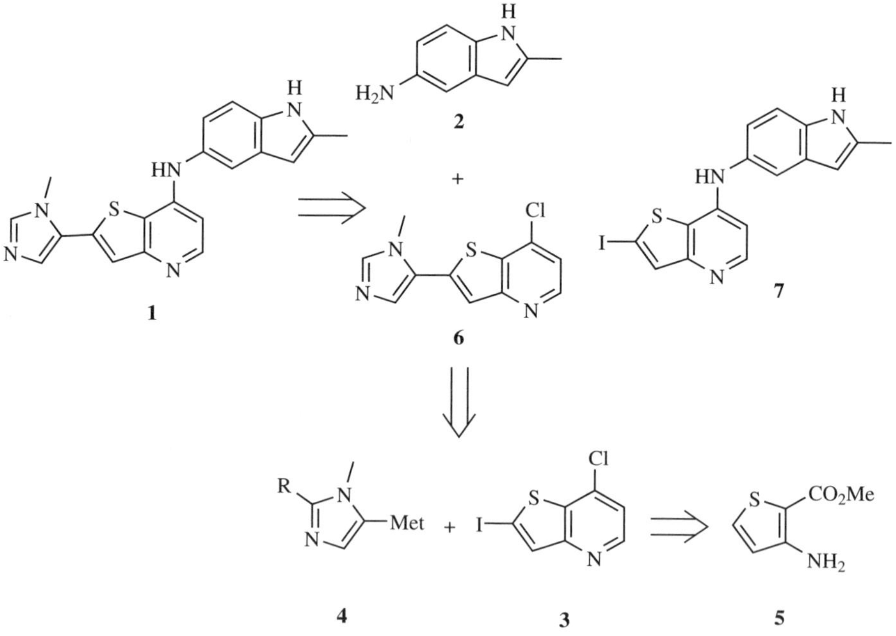

Figure 2.2. Retrosynthetic analysis.

from the initial Discovery synthesis through the subsequent cGMP (current Good Manufacturing Practice) bulk campaign.

This approach installs the aminoindole moiety in the final bond-forming step, and it is reasonable to consider the alternative ordering of events—that is, aminoindole coupling with **3** to generate an intermediate such as **7**. Although one of the early synthetic routes in Discovery showed that this route was valid for a Stille coupling (Figure 2.6), we moved away from this approach primarily for two reasons: First, for routes seeking alternatives to the organostannane reagent, more basic organometallic reagents (e.g., organozinc, organomagnesium) were anticipated to be incompatible with the acidic NH groups on the aminoindole side chain. Second, when we did ultimately return to the Stille coupling, we preferred to have this reaction as early in the synthesis as possible in order to maximize our chances of purging residual organostannane impurities by crystallization of downstream intermediates.

PREPARATION OF THE 7-CHLOROTHIENO[3,2-*b*]-PYRIDINE: A HIGH TEMPERATURE DECARBOXYLATIVE PYRIDINONE ANNULATION

Preparation of the known[3] thienopyridine **11** is shown in Figure 2.3; we made only minor adjustments to this chemistry, executing it essentially as described in the literature. Additionally, several fine chemical manufacturers were found to provide kilogram quantities of this intermediate. Saponification and oxalic acid-mediated decarboxylation of methyl 3-amino-thiophene-2-carboxylate provides 3-aminothiophene (**8**) as its oxalate salt. Pyridine annulation was then effected by condensation of Meldrum's acid, triethylorthoformate, and the aminothiophene to generate vinylogous carbamate (**9**). Thermolysis of this intermediate at 240–250°C in Dowtherm® provided thieno[3,2-*b*]pyridin-7-one (**10**). Treatment with oxalyl

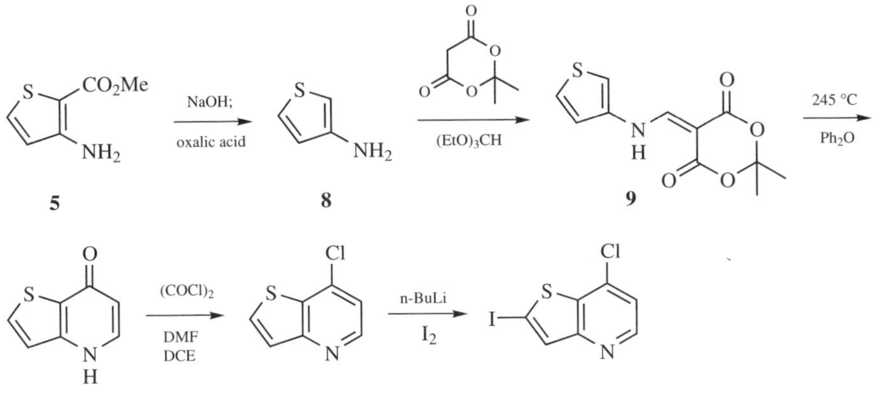

Figure 2.3. Preparation of the iodothienopyridine.

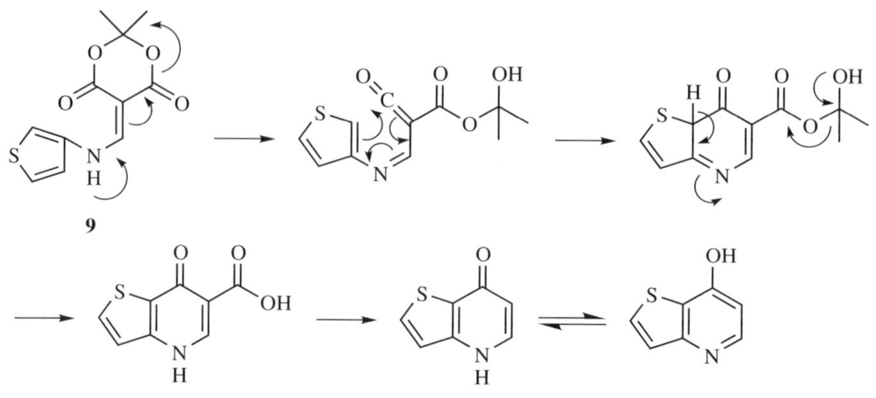

Figure 2.4. Possible mechanism for the pyridine annulation reaction cascade.

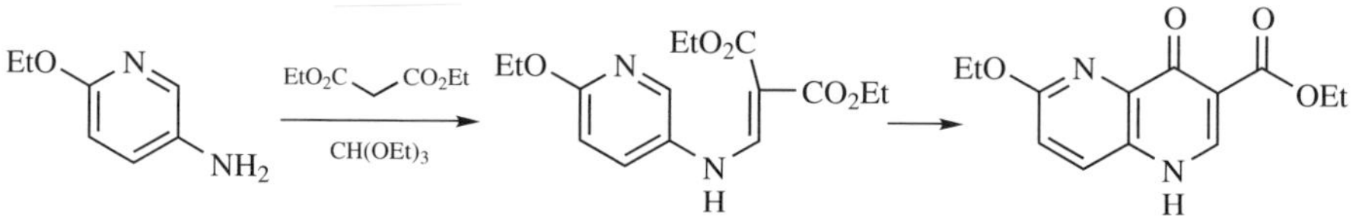

Figure 2.5. Synthesis of a carboethoxypyridone.

chloride and catalytic dimethylformamide in dichloroethane provided 7-chlorothieno [3,2-*b*]pyridine (**11**). *Alpha*-lithiation of the thiophene and trapping with iodine provided 2-iodo-7-chlorothieno[3,2-*b*]pyridine (**3**).

A possible mechanism for the pyridine annulation reaction cascade is shown in Figure 2.4. The decarboxylation is a direct result of the use of Meldrum's acid; thermal rearrangement to the ketene generates a hemi-acetal, which liberates acetone to generate carboxypyridone **12**. This species decarboxylates under the thermal reaction conditions to provide thienopyridinone **10**. This is in contrast to the analogous pyridinone annulation with dialkylmalonates in place of Meldrum's acid, in which case the ester carbon is retained, as shown for the synthesis of carboethoxypyridone **13** (Figure 2.5).[4]

FIRST- AND SECOND-GENERATION DISCOVERY SYNTHESES: STILLE COUPLING AND DEVELOPMENT OF AN ORGANOZINC CROSS-COUPLING

As discussed earlier, the first Discovery synthesis of **1** utilized a Stille cross-coupling[5] of iodothiophene **7**, in which the aminoindole moiety was installed prior to cross-coupling (Figure 2.6).

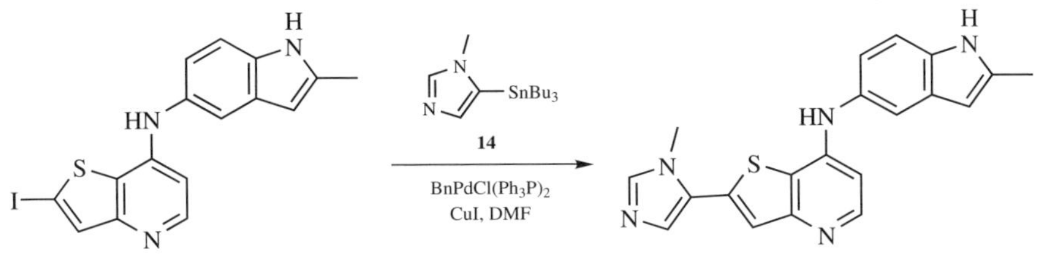

Figure 2.6. First Discovery synthesis of 5-[2-(3-methyl-3*H*-imidazol-4-yl)-thieno-[3,2-*b*] pyridine-7-yl]amino-2-methyl-1*H*-indole.

While this chemistry was effective in forming the requisite imidazole-thiophene and supported early SAR efforts, Discovery appreciated the scale-up and analytical challenges associated with a stoichiometric organostannane coupling, and it sought an alternative approach well in advance of the compound's nomination for development.

This led them to the organozinc imidazole reagent **15**, formed from 2-TBS-1-methylimidazole by lithiation at the 5-position and transmetalation with $ZnCl_2$ in THF (Figure 2.7). On laboratory scale, this palladium-mediated organozinc coupling (Negishi reaction) was effective in providing imidazolyl-thiophene **16**. Desilylation and aminoindole coupling then provided **1**.

The Negishi coupling was viewed as a significant synthetic breakthrough, because it worked well on small laboratory scale (5–10 g) and avoided the handling and analytical issues associated with an organostannane reagent. However, it became clear as the chemistry was scaled up further (50–100 g) there were factors influencing the success or failure of this reaction for which we did not have a solid understanding or ability to control. Certain runs would fail to deliver useful yields of the desired product, and these failures occurred in an essentially unpredictable fashion. Numerous factors were considered and analyzed, including:

- *Catalyst Source*: Certain batches of $Pd(Ph_3P)_4$ provided more consistent results than others. Some empirical correlations with particular manufacturers

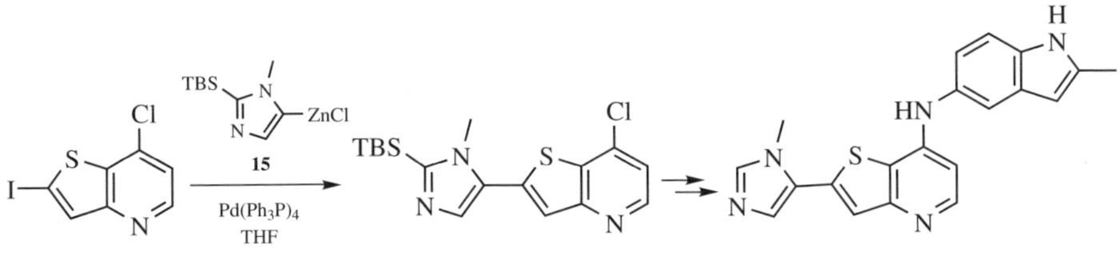

Figure 2.7. Use of the Negishi reaction, desilylation, and aminoindole coupling to synthesize 5-[2-(3-methyl-3*H*-imidazol-4-yl)-thieno-[3,2-*b*]-pyridine-7-yl]amino-2-methyl-1*H*-indole.

were seen, but no analytical correlation could be identified (e.g., melting point, combustion analysis, trace metal analysis). Use of other homogeneous palladium catalysts offered no significant improvement, nor did use of freshly recrystallized catalyst. This did not bode well for our ability to set reliable specifications for purchase of this catalyst in the future.

- *Purity of Iodothiophene*: Even more so than with different batches of catalyst, both Discovery and Process Chemistry labs observed empirical variations in reaction outcome with different batches of iodothiophene **3**. Successful outcomes were frequently seen with iodothiophene that had been freshly triturated with solvents such as methyl *tert*-butyl ether (MTBE) or isopropyl ether (IPE). As with the catalyst, however, no analytical correlation could be made between "good" and "bad" batches (e.g., HPLC, ^{1}H or ^{13}C NMR, combustion analysis, melting point, or trace metal analysis). Reasoning that very low levels of an unidentified impurity might be poisoning the catalyst, and that this impurity might have an inordinate affinity for the palladium catalyst, we tried recrystallizing a batch of "bad" iodothiophene from dimethylacetamide (DMAC) in the presence of 5 mol% $Pd(Ph_3P)_4$.[6] The hope was that the palladium catalyst would bind selectively to the impurity and solubilize it. While this procedure did provide beautiful white crystals of iodothiophene, the purified material did not perform more reliably in the Negishi coupling.
- *Regioselectivity in the TBS-Imidazole Lithiation*: The purity of the *in situ*-generated organozinc reagent was also recognized as a key parameter. Purification of this reagent was less straightforward than for either catalyst or iodothiophene, because the organozinc reagent is not an isolable species. However, we did observe byproducts in some of the failed cross-couplings that suggested several possible competing pathways (Figure 2.8). Silanol **22**

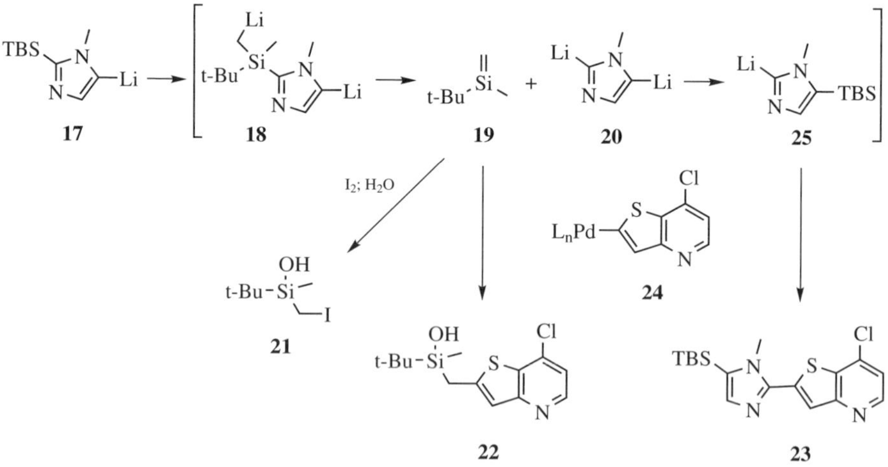

Figure 2.8. Regioselectivity in TBS-imidazole lithiation.

was observed as a low-level byproduct in several failed organozinc coupling reactions. In some of these experiments, the regioisomer (**23**) was also observed, suggesting that silyl migration from the 2- to the 5-position of the imidazole had occurred prior to coupling (this TBS migration has been reported as the exclusive reaction pathway for the same 5-lithio-2-TBS-1-methylimidazole anion prepared via halogen-metal exchange from 5-bromo-2-TBS-*N*-methylimidazole; the fact that it does not occur as readily in our system may suggest that it is caused by the presence of LiBr in the halogen-metal exchange reaction, which is not present in our system).[7,8] Although an exact mechanism was not ascertained, silanol **22** appears to arise from oxidation of one of the two methyl groups on the *tert*-butyldimethylsilyl group. This led us to consider the possible competitive lithiation of the TBS group to generate a species such as **18**, which might disproportionate to form 2,5-dilithio-1-methylimidazole (**20**) and silylene **19**. Silylation of **20** at the more reactive 5-position could then form 2-lithio-5-TBS-imidazole (**25**), which would lead to regioisomer **23**. Dilithioimidazole **20** has been reported previously (LiBr serves to solubilize this species, and its absence in our system may help minimize this pathway).[7] The silylene would clearly be a short-lived species (there is literature precedent for this compound, formed by photolysis of a silacyclobutane in MeOH).[9] It might then react with organopalladium intermediate **24** to form the observed byproduct (**22**). This mechanism is consistent with other experiments in which the lithiated TBS-imidazole was trapped with I_2; following aqueous workup, small quantities of iodosilanol **21** were isolated, which also might arise from silylene **19**.

While the above observations were frustrating and led to no clear conclusions regarding the irreproducibility of the organozinc coupling, it was clear that this approach would not be viable for further scale-up. In parallel with the ongoing experiments to improve our understanding of the organozinc coupling, we began to seek an alternative cross-coupling approach.

EVALUATION OF ALTERNATIVE CROSS-COUPLING PROCEDURES

As summarized in Table 2.1, a variety of transition metal-catalyzed cross-couplings were examined, with limited success. The only two methods that displayed any degree of reliability on even laboratory scale were the aforementioned organozinc (Negishi) coupling (entry 1) and the organostannane (Stille) coupling (entry 12). Only the latter method performed consistently on scales >50 g. Although iodothiophene **3** was the electrophile in the majority of cases, we did explore some couplings with reversed polarity (e.g., nucleophilic thiophene plus electrophilic imidazole): entries 2, 6, 8, and 9. In some cases these worked with model electrophiles (e.g., iodobenzene, entry 2), but none worked with the requisite haloimidazoles.

TABLE 2.1. Cross-Couplings Examined for Formation of Thiophene-Imidazole

Entry	Reference	Reaction/Comments
1	Negishi	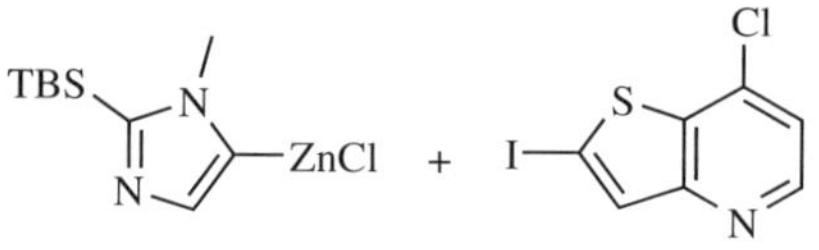 **15** **3** $Pd(Ph_3P)_4$ (numerous other Pd sources also examined)
		Effective on laboratory scale (<50 g), but capricious upon scale-up (50–250 g).
2	Negishi	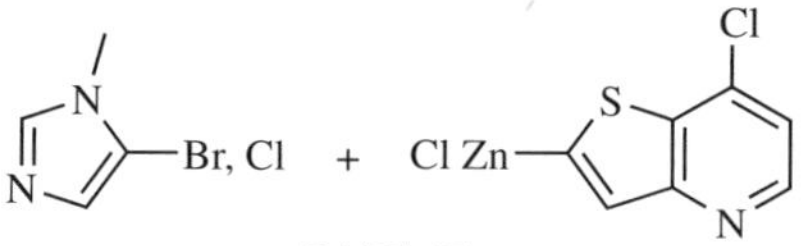 $Pd(Ph_3P)_4$
		Reverse polarity from entry 1. Worked with iodobenzene as electrophile, but failed with haloimidazoles.
3	Heck	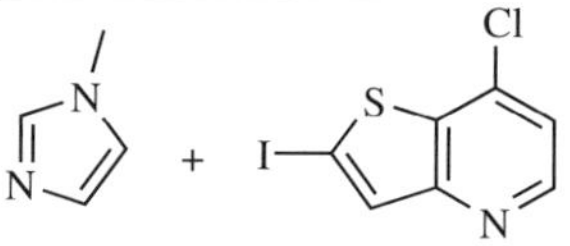 $Pd(OAc)_2/Ph_3P$ or $Pd(OAc)_2/Cy_2NMe$
		No reaction under standard Heck conditions. In a phosphine-free system, 16% conversion was observed after 24 h at 110°C, but a mixture of regioisomers was observed.
4	Hiyama	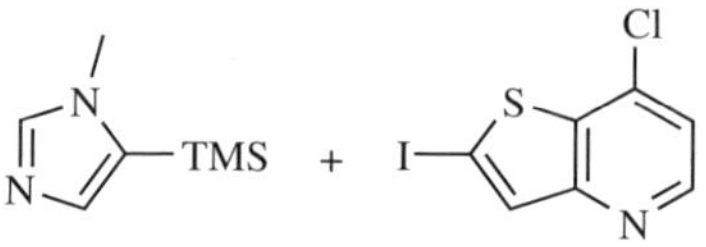 Pd(0) or Pd(II), Ph_3P or BINAP
		No coupling observed, reduction of iodide was seen. Added CsF led to desilylation.
5	Suzuki–Miyaura	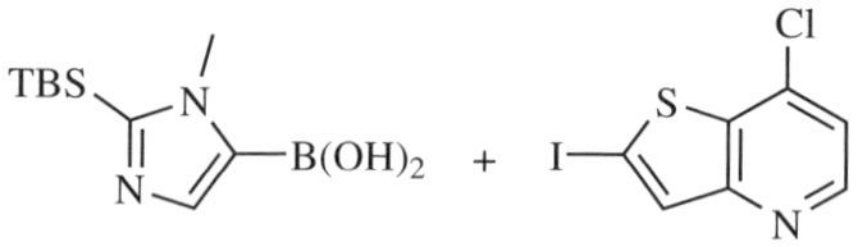 **26** $Pd(Ph_3P)_4$ or $Pd(OAc)_2$ / dppf, dppb
		Isolation of the requisite boronic acid was problematic. Competitive deborylation was observed to form 2-TBS-*N*-methylimidazole (see discussion in text).

TABLE 2.1. ***(Continued)***

Entry	Reference	Reaction/Comments
6	Suzuki–Miyaura	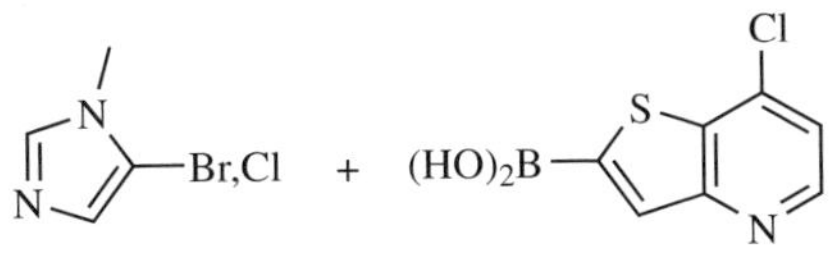 $Pd(Ph_3P)_4$ or $Pd(OAc)_2$ / dppf, dppb
		The thienopyridine-boronic acid was isolated as a stable, well-behaved solid, but couplings with haloimidazoles were unsuccessful. Deborylation was observed in some cases.
7	Kumada–Tamao	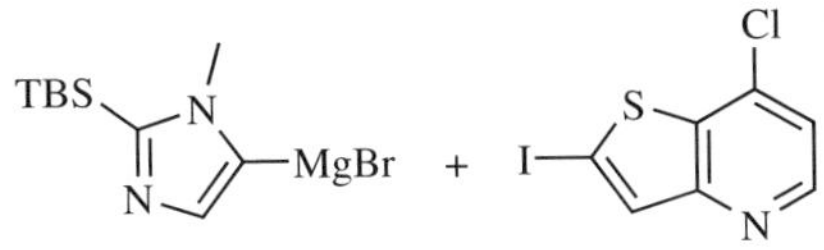 Pd(0) or Ni(0) catalysts
		No coupling product observed. A model study with PhMgBr and iodothienopyridine was successful.
8	Kumada–Tamao	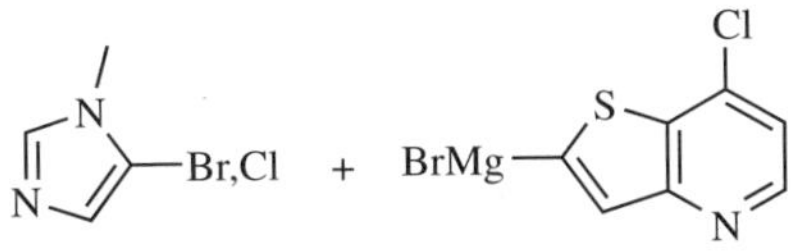 Pd(0) or Ni(0) catalysts
		No coupling product was observed. Reaction also failed with a model electrophile (PhI).
9	Kobayashi	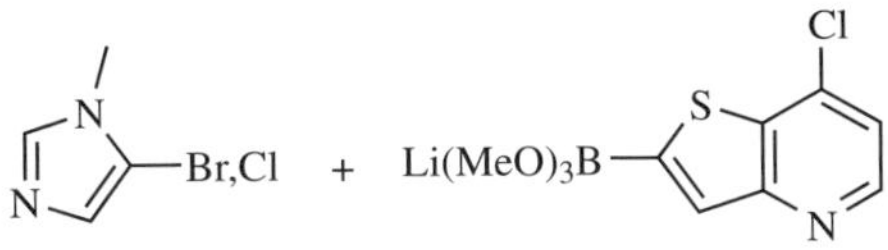 $NiCl_2(Ph_3P)_2$ or $Pd(Ph_3P)_4$
		No coupling product was observed, although a model study with iodobenzene was successful.
10	Kobayashi	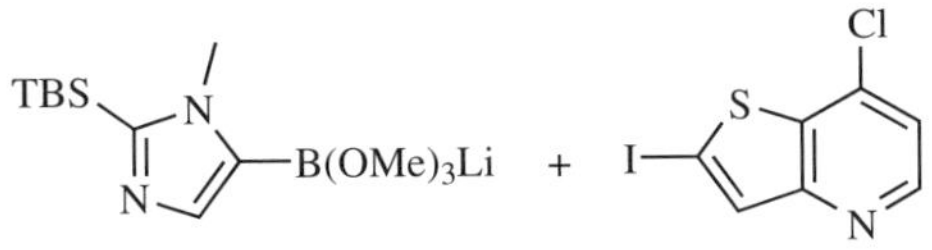 $NiCl_2(Ph_3P)_2$ or $Pd(Ph_3P)_4$
		Poor conversion, several byproducts observed.

TABLE 2.1. (***Continued***)

Entry	Reference	Reaction/Comments
11	DeShong	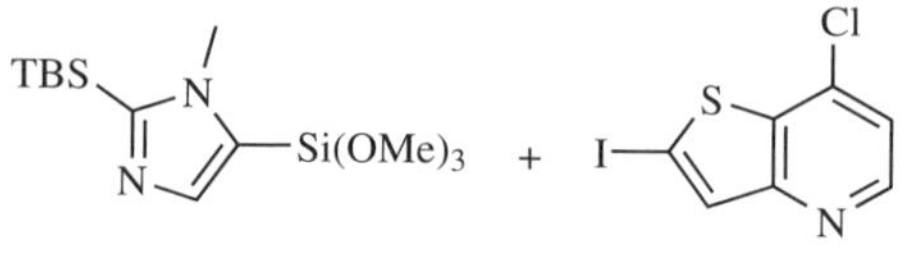 $Pd(dba)_2$ / Bu_4NF Desilylation was observed, no coupling product formed. A model study with $PhSi(OMe)_3$ also failed.
12	Stille	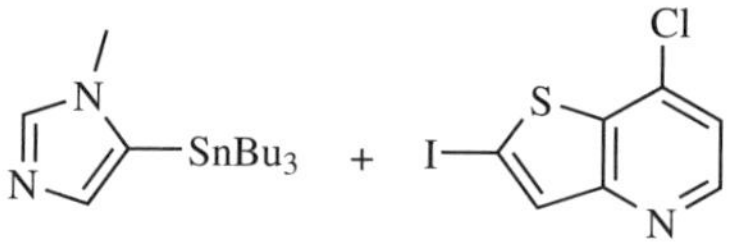 $Pd(Ph_3P)_4$ Robust and scalable; only method that worked reliably on >50-g scale.

The Suzuki coupling (entries 5 and 6) warrants further discussion, because this approach was the most extensively examined behind the Negishi and Stille couplings. There were several appealing aspects to the Suzuki coupling, particularly the avoidance of toxic heavy metals (such as organostannanes) and the use of an isolable nucleophilic coupling partner (i.e. an organoboronic acid), as opposed to the *in situ*-generated organozinc reagent of the Negishi coupling. Additionally, there is a compelling literature precedent for the use of imidazole-boronic acid **29** (Figure 2.9).

In the literature report,[10] boronic acid **29** was prepared via 1-(2-trimethylsilylethyloxymethyl)-2-trimethylsilylimidazole (**28**). This intermediate differs from our substrate (1-methyl-2-TBS-imidazole) in two important respects: the coordinating ability of the *N*-alkyl substituent and the migratory aptitude of the 2-silyl substituent. We suspect that the presence of the SEM group on the imidazole nitrogen plays a critical role, by facilitating lithiation at –78°C via coordination of the

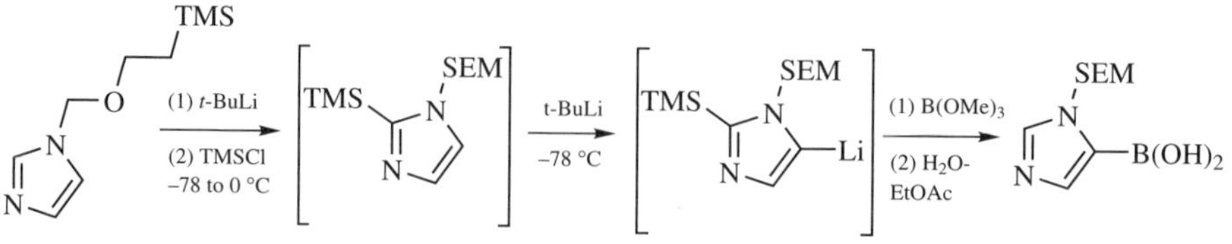

27 **28** **29**

Figure 2.9. Suzuki coupling.

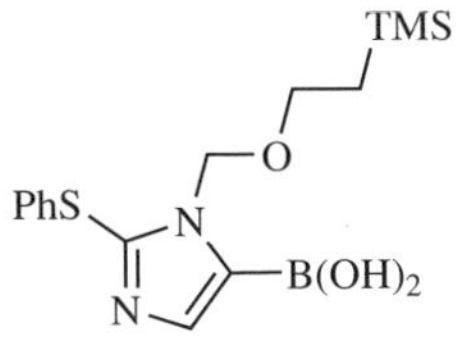

30

Figure 2.10. A SEM-protected imidazoylboronic acid.

alkyllithium reagent. At that temperature, the TMS group does not migrate from the 2- to 5-position, and of course the competitive TBS lithiation discussed earlier (Figure 2.8) does not come into play. In general, *N*-protecting groups that have Lewis basic coordinating sites (e.g., CH_2OR, SO_2NMe_2) minimize the potential for silyl migration from the 2- to 5-position in 5-lithiated imidazoles.[11] In our system, higher temperatures are required to effect lithiation of the 5-position on the unactivated imidazole; this is most likely due to the noncoordinating nature of the *N*-methyl substituent on the imidazole ring (versus the coordinating *N*-SEM substituent). Thus, we cannot use TMS as a blocking group, because it migrates readily from the 2- to 5-position at that temperature. Silyl migration can be avoided by use of a bulkier silyl group such as TBS and a higher lithiation temperature. However, the higher temperature leads to many of the problems we encountered and our ultimate inability to prepare useful quantities of boronic acid **26**.

In addition to the above analysis, it should be noted that deborylation during attempted isolation and couplings appeared to be problematic. This suggests that the carbon–boron bond of the 5-imidazoylboronic acid moiety is somewhat labile, which may also contribute to the modest coupling yield reported for boronic acid (**29**) (37% in a bis-coupling with an aryliodide substrate).[10] Another report of a similar boronic acid (**30**, Figure 2.10)[12] also encountered low coupling yields with a 2-bromoindole electrophile, and this might also reflect lability of the C-5 carbon–boron bond.

BACK TO THE STILLE COUPLING: SCALE-UP AND ANALYTICAL CHALLENGES FOR A STOICHIOMETRIC ORGANOSTANNANE COUPLING

As the studies discussed above were nearing completion, it became clear that the Stille coupling was the only consistently reliable method for preparation of **1**. While we realized that an organostannane coupling would likely not be a commercially viable synthetic route, the near-term time pressures associated with impending regulatory toxicology studies in support of the planned IND filing required kilogram quantities of material. Thus, we decided to utilize the Stille coupling for preparation of the initial cGMP bulk lot, planning to introduce a new, non-stannane route in future bulk campaigns if the compound moved forward.

This tension between the need for bulk API (active pharmaceutical ingredient) as quickly as possible and the need to make changes and improvements in the Discovery synthesis is a fundamental issue within exploratory development. The fact that the rate of compound attrition is generally higher in early development than in mid- or late development argues for "enabling" the Discovery route to make just enough bulk to run the regulatory toxicology program and initiate Phase 1. Since there is a significant probability that the compound will not survive this stage of development, this strategy potentially saves the investment of developing a new synthetic route. But balancing this strategy is the fact that if the compound does survive preclinical and Phase 1 studies, then bulk demands will increase dramatically, and a more robust synthesis to prepare material for Phase 2 (and ultimately for registration) will be needed. If new route identification work is deferred until the compound has shown clinical efficacy, you will possibly have to delay the clinical program (or invest significant resources from other projects) to play "catch up" on identifying a viable commercial synthesis. Many of the challenges of operating in exploratory development (i.e., preclinical to initiation of Phase 3) come from managing this tension in a manner that is both resource-sparing and developmentally and strategically sound.

The major issues we faced with the Stille coupling were equipment contamination and analysis and control of residual organostannane byproducts in drug substance. The former issue was addressed by limiting our runs to 22-liter glassware, so that we could simply dispose of the reaction vessel upon completion of the campaign. While obviously not a long-term solution, this strategy was acceptable for an initial campaign targeting 1–2 kg of final product (fortunately, the Stille coupling works well at high concentrations: ~4 liter solvent per kilogram of substrate). For the latter issue, inductively coupled plasma emission spectroscopy (ICP) analysis was used to determine total stannane and palladium content with a lower limit of quantification (LLOQ) of 2 ppm. Our target for both tin and palladium was <20 ppm, based on the proposed clinical doses. Fortunately, both final product **1** and penultimate intermediate **6** are solids that provide the option of purging residual impurities by crystallization (Figure 2.11).

Table 2.2 summarizes tin and palladium levels from several laboratory-scale experiments. Entry 1 shows that a simple reslurry of the crude Stille product removes the bulk of the residual stannane (154 ppm versus an initial level of

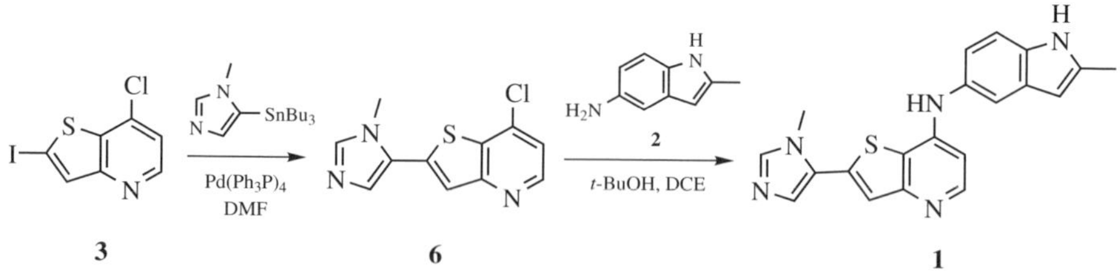

Figure 2.11. Synthesis of 5-[2-(3-methyl-3*H*-imidazol-4-yl)-thieno-[3,2-*b*]pyridine-7-yl] amino-2-methyl-1*H*-indole using stille coupling.

TABLE 2.2. Tin and Palladium Levels in Laboratory Pilots

Entry Compound	Purification Method	Sn Level (ppm)	Pd Level (ppm)
1	Aqueous pH workup, then reslurry with MTBE (57% yield).	154	52
2	Recrystallization of entry 1 material from ethyl acetate (73% recovery).	19	38
3	Aminoindole coupling of **6** (from 5 entry 1), followed by crystallization from MeOH (isolated HCl salt, 37% yield).	5	7
4	Aminoindole coupling of **6** (from 2 entry 1), followed by silica gel chromatography (isolated free base, 50% yield).	2	<1

[a] For clarification, note that entries 2–4 all involve further processing of the material form entry 1: recrystallization (entry 2), aminoindole coupling followed by recrystallization (entry 3), or aminoindole coupling followed by silica gel chromatorgraphy (entry 4).

~170,000 ppm).[13] This material was further purified by recrystallization to just below our target tin level (entry 2, 19 ppm). We hoped that we could bypass recrystallization of **6** and proceed directly into the final aminoindole coupling, with isolation of this product providing further reduction in tin levels. This hope was borne out by entries 3 and 4, which show that utilization of the 154 ppm lot of **6** in the aminoindole coupling provides acceptable purity by either crystallization of the HCl salt (entry 3, 5 ppm) or silica gel chromatography of the free base (entry 4, 2 ppm). On the basis of these results, we were confident that we would be able to purify our bulk campaign material to <20 ppm organostannane residues, and this prediction was confirmed in the cGMP campaign.

The organostannane component of the Stille coupling, 5-tributylstannyl-1-methylimidazole (**14**), was prepared from 1-methylimidazole, per the literature (Figure 2.12).[14] 2,5-Dilithioimidazole (**20**) was formed by metalation in TMEDA-hexane at –20°C, and quenched with Bu_3SnCl to form bis-stannane **31**. Aqueous workup cleaved the more labile 2-stannyl moiety to provide **14**. This material could be used directly on laboratory scale, but on scale-up it was deemed prudent to remove Bu_3SnCl-derived impurities prior to coupling. This was achieved by a hexane–acetonitrile partition, in which the nonpolar stannane impurities were

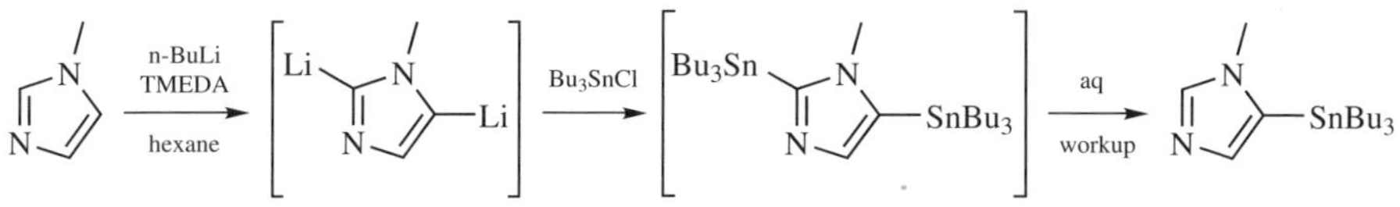

20 31 14

Figure 2.12. Preparation of the organostannane component of the Stille coupling, 5-tributyl stannyl-1-methylimidazole.

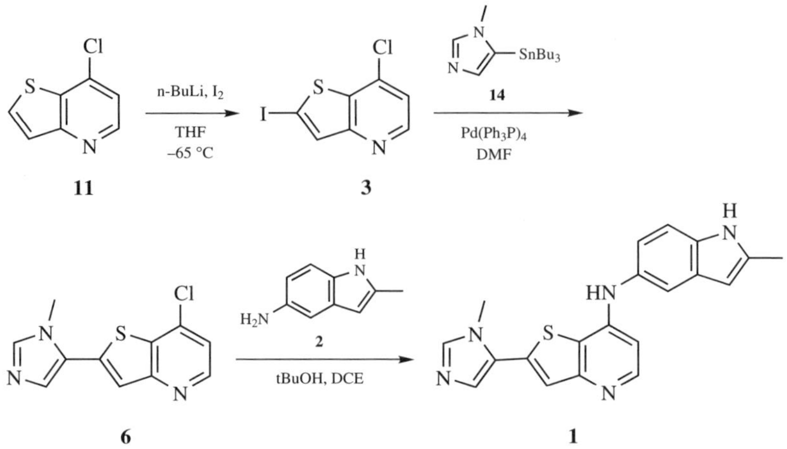

Figure 2.13. Synthesis of 5-[2-(3-methyl-3*H*-imidazol-4-yl)-thieno-[3,2-*b*]pyridine-7-yl] amino-2-methyl-1*H*-indole using the cGMP campaign.

selectively partitioned into the hexane phase, while the more polar imidazole reagent remained in the acetonitrile phase. This method was used to prepare 1.7 kg of **14**.

For the cGMP campaign (Figure 2.13), we began with 3 kg of 7-chlorothienopyridine (**11**). Lithiation was achieved by treatment of 1-methylimidazole with *n*-BuLi (1.6 eq) in THF-hexane at −70°C for 60 min (D_2O quench of an aliquot showed >95% deuterium incorporation by ^{1}H NMR analysis), followed by addition of iodine (1.6 eq) in THF at such a rate that the internal temperature remained below −65°C. Addition of water precipitated iodide **3** as a light brown solid, which was then washed with water and hexanes to provide an 84% yield of **3** with >99% HPLC purity. Running the lithiation sequence at −20°C led to increased levels (5–10%) of a bis-iodide impurity as determined by mass spectrometry analysis. The sequential triturations were found to provide a more efficient isolation and purification than an aqueous workup (e.g., partitioning between ethyl acetate and aqueous $Na_2S_2O_3$).

The Stille coupling was effected with 5 mol% $Pd(Ph_3P)_4$ in DMF at 90°C. Unlike other coupling methods investigated, this reaction was robust and nondiscriminating to catalyst source; it was scaled up from 10 to 500 g with no significant change in isolated yield (63–67% on 530-g scale). Considering the extensive efforts to identify alternative cross-coupling methods (Table 2.1), this is a compelling testimony to the reliability of Stille's methodology.

AMINOINDOLE COUPLING AND SALT FORMATION

Coupling of 5-amino-2-methylindole (**2**) with chloropyridine **6** was accomplished under moderate pressure in a Parr reactor in *tert*-butyl alcohol and dichloroethane.

High concentration (2 M in 1:1 *t*-BuOH/DCE), excess aminoindole (2 eq), and high temperature (100°C, 17 psi) were critical to drive this reaction to reasonable (>90%) conversion. A silica gel column was utilized to purify this reaction mixture (11:1 weight ratio of silica gel to starting material **6**). Following a reslurry in 2-propanol, a 65% yield of **1** was isolated (500-g scale). Conversion to the (−)-camphorsulfonic acid salt then provided the desired drug substance, ICP analysis of which showed 3 ppm stannane and 3 ppm palladium, consistent with the lab pilots described in Table 2.2.

Subsequent to the cGMP campaign, it was found that running the aminoindole coupling in EtOH provided much cleaner product and a simpler isolation. This simple solvent change was not tried earlier based on literature precedent which suggested that ethoxide would displace the 4-chloropyridine.[15] Thus, combining equimolar amounts of chloropyridine **6** and the aminoindole **2** in refluxing EtOH for 48hr provided the desired product, which crystallized (as the HCl salt) from the reaction mixture upon cooling in 87% yield (99% HPLC purity) on 25-g scale.

CONCLUSIONS

The development of an efficient synthesis of **1** required investigation of a wide variety of synthetic issues. We have demonstrated that the preparation of clinical bulk supplies under cGMP conditions poses a wide array of challenges to our skills as synthetic chemists. Many of these challenges arise from the more stringent analytical requirements in preparing cGMP bulk material. For example, while chromatographic purification is frequently acceptable in the synthesis of milligrams or even grams of material, this purification technique is impractical on a multi-kilogram scale. Likewise, the use of toxic intermediates such as heavy metal organometallic reagents can be dealt with on a laboratory scale, but poses much more serious challenges on a kilogram scale when preparing material slated for use in a clinical setting. These observations suggest that there remain many challenging areas in the arena of organic synthesis where the development of selective, low-cost, and environmentally friendly methods will be of tremendous benefit to chemists in both academic and industrial fields.

REFERENCES AND NOTES

1. (a) Fan, T. P. D., Jaggar, R., and Bicknell, R. *Trends Pharmacol. Sci.* **1995**, *16*, 57–66. (b) Folkman, *J. Nature Medicine* **1995**, *1*, 27–31.
2. Ragan, J. A., Raggon, J. W., Hill, P. D., Jones, B. P., McDermott, R. E., Munchhof, M. J., Marx, M. A., Casavant, J. M., Cooper, B. A., Doty, J. L., and Lu, Y. *Org. Proc. Res. & Dev.* **2003**, *7*, 676–683.
3. (a) Barker, J. M., Huddleston, P. R., Holmes, D., Keenan, G. J., and Wright, B. *J. Chem. Res. Miniprint* **1984**, 771–795. (b) Barker, J. M., Huddleston, P. R., Holmes, D., and Keenan, G. J. *J. Chem. Res. Miniprint* **1982**, 1726–1746. (c) Barker, J. M., Huddleston, P. R., and Wood, M. L. *Synth. Commun.* **1995**, *25*, 3729–3734.

4. Beaudin, J., Bourassa, D. E., Bowles, P., Castaldi, M. J., Clay, R., Couturier, M. A., Karrick, G., Makowski, T. W., McDermott, R. E., Meltz, C. N., Meltz, M., Phillips, J. E., Ragan, J. A., Ripin, D. H. B., Singer, R. A., Tucker, J. L., and Wei, L. *Org. Process Res. & Dev.* **2003**, *7*, 873–878.

5. (a) Mitchell, T. N. In *Metal-Catalyzed Cross-Coupling Reactions*, Diederich, F., and Stang, P. J., Eds., Wiley-VCH Verlag, Weinheim, Germany, 1998; Chapter 4. (b) Farina, V., Krishnamurthy, V., and Scott, W. J. *Org. React.* **1997**, *50*, 1–652. (c) Stille, J. K. *Angew. Chem., Int. Ed. Engl.* **1986**, *25*, 508–524.

6. The author would like to thank Professor Dan Kemp (MIT) for this suggestion.

7. Shapiro, G., and Marzi, M. *Tetrahedron Lett.* **1993**, *34*, 3401–3404. These authors describe the role of LiBr in solubilizing 2,5-dilithio-*N*-methylimiazole; they also note differing 2- to 5-migration propensity for the 5-lithio-2-thiophenyl-*N*-methylimidazole anion prepared *via* halogen metal exchange of the 5-bromo precursor versus deprotonation of 2-thiophenyl-*N*-methylimidazole.

8. For a comprehensive review of lithiated imidazoles, see Iddon, B., and Ngochindo, R. I. *Heterocycles*, **1994**, *38*, 2487–2568.

9. Leigh, W. J., and Boukherroub, R., Kerst, C. *J. Am. Chem. Soc.* **1998**, *120*, 9504–9512.

10. Han, Q., Chang, C.-H., Li, R., Ru, Y., Jadhav, P. K., and Lam, P. Y. S. *J. Med. Chem.* **1998**, *41*, 2019–2028.

11. For example, see: Carpenter, A. J., and Chadwick, D. J. *Tetrahedron* **1986**, *42*, 2351–2358.

12. Kawasaki, I., Katsuma, H., Nakayama, Y., Yamashita, M., and Ohta, S. *Heterocycles* **1998**, *48*, 1887–1901.

13. 170,000 ppm comes from the atomic weight of tin, the molecular weights of the reactants, and the stoichiometry of the reaction in the following equation: (at wt Sn) $\times$ (eq stannane reagent)/[(mol wt iodothiophene) $\times$ (eq iodothiophene) + (mol wt stannane reagent) $\times$ (eq stannane reagent) + (mol wt Pd tetrakis) $\times$ (eq cat.)] $= (118 \times 1.1)/(295 \times 1.0) + (370 \times 1.1) + (1155 \times 0.05) = 0.17$ (17%) = 170,000 ppm.

14. Gaare, K., Repstad, T., Benneche, T., and Undheim, K. *Acta Chem. Scand.* **1993**, *47*, 57–62.

15. (a) Barker, J. M., Huddleston, P. R., Holmes, D., Keenan, G. J., and Wright, B. *J. Chem. Res. Miniprint* **1984**, 771–795. (b) Cutler, R. A., and Surrey, A. R. *J. Am. Chem. Soc.* **1955**, *77*, 2441–2444.

3

LARGE-SCALE ENANTIOSELECTIVE PREPARATION OF 2*E*,7*E*,5*S*,6*R*, 5-HYDROXY-6-METHYL-8-PHENYL–OCTA-2,7-DIENOIC ACID, A KEY FRAGMENT FOR THE FORMAL TOTAL SYNTHESIS OF THE ANTI-TUMOR AGENT CRYPTOPHYCIN 52

JAMES AIKINS, TONY Y. ZHANG, MILTON ZMIJEWSKI, AND BARBARA BRIGGS
Chemical Product Research and Development, Lilly Research Laboratories, A Division of Eli Lilly and Company, Indianapolis IN 46285

INTRODUCTION

Cryptophycins are a new class of marine natural products isolated from blue-green algae (cyanobacterium) with potent fungicides[1] and anti-mitotic agents.[2,3] This has spurred an intensive interest in both industry and academia.[4,5] A collaborative effort between scientists from Eli Lilly and Company and the University of Hawaii led to the identification of Cryptophycin 52 (**1**) (Figure 3.1) as a clinical candidate with exceptional *in vivo* activity and tumor-selective cytotoxicity. The original convergent synthetic strategy for the synthesis of **1** featured the four distinct building fragments **2, 3, 4**, and **5** as depicted in Figure 3.2. 2*E*, 7*E*, 5*S*, 6*R*, 5-hydroxy-6-methyl-8-phenyl-octa-2, 7-dienoic acid. The original synthesis of (**2**) (Scheme 3.1) took

Fundamentals of Early Clinical Drug Development: From Synthesis Design to Formulation, edited by Ahmed F. Abdel-Magid and Stéphane Caron

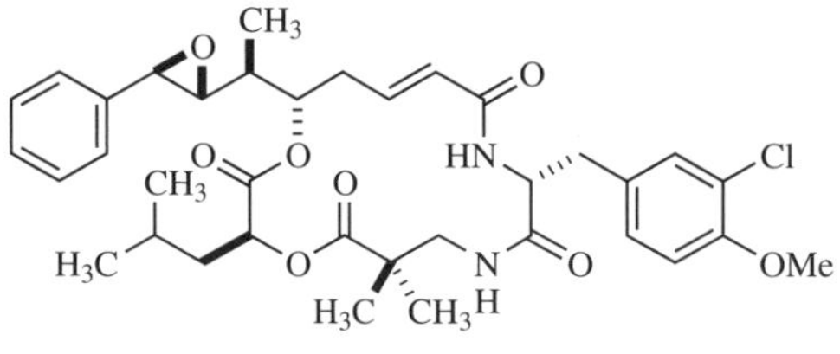

1

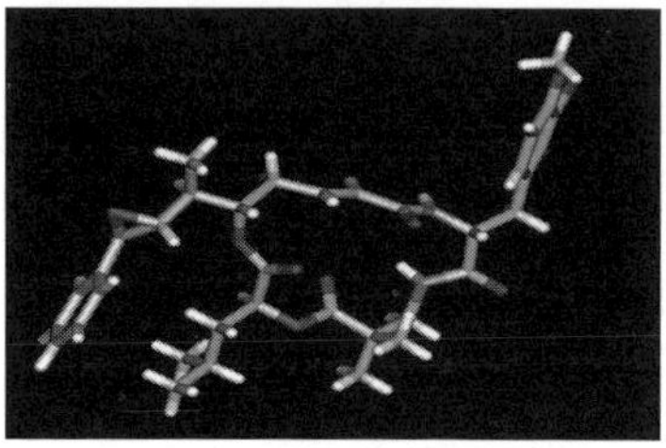

X-RAY of 1

Figure 3.1.

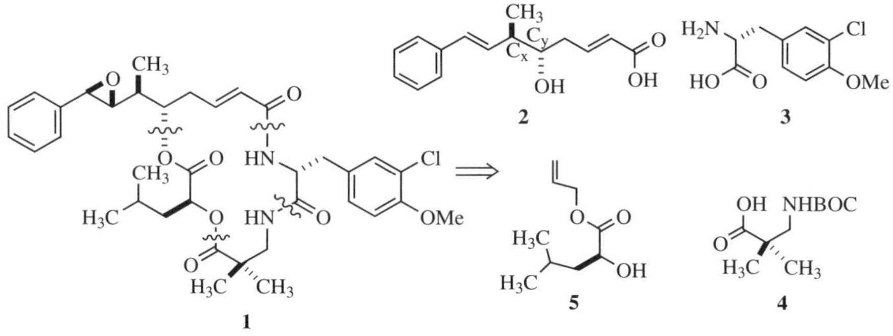

Figure 3.2.

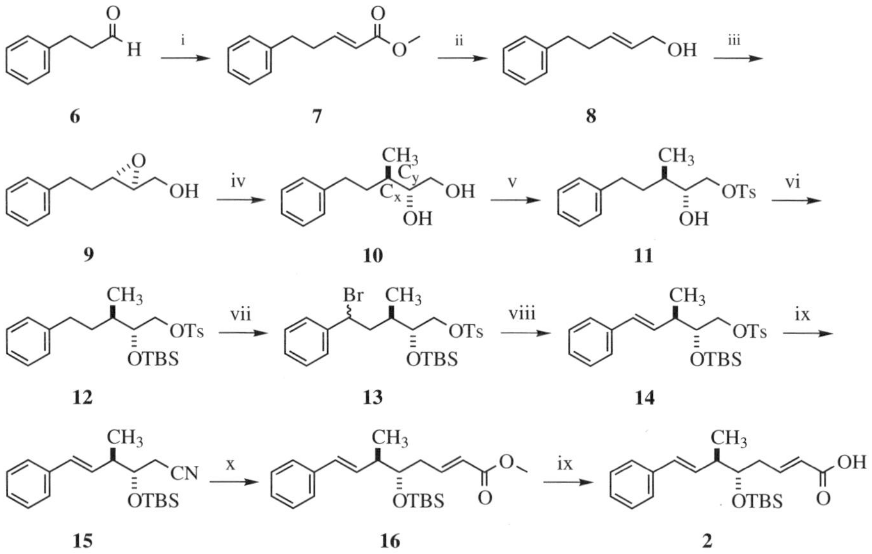

Scheme 3.1. (i) $PhCH_3$, trimethyl phosphonoacetate, TMG; (ii) DIBAL, $PhCH_3$, (iii) SAE; (iv) Heptane/$PhCH_3$, Me_3Al; (v) 1% Bu_2SnO, TsCl; (vi) TBSOTF, Et_3N; (vii) NBS/AIBN, heptane; (viii) MeCN/DBU, heptane; (ix) KCN/DMSO, $NaHCO_3$; (x) DIBAL/$PhCH_3$; $(MeO)_2P(=O)CH_2CO_2Me$ (HEW); (xi) KOH/dioxane.

advantage of Sharpless Asymmetric Epoxidation (SAE) of alcohol **8** and relied on subsequent reaction of the epoxide **9** with trimethylaluminum for delivery of the two stereocenters C_x and C_y. Tin-mediated tosylation of the primary alcohol followed by reaction of the secondary alcohol with TBSOTf in the presence of triethylamine afforded **12**. Benzylic bromination and subsequent reaction with DBU afforded the styryl compound **14**. Nitrile displacement of the tosylate followed by reduction/HEW sequence afforded the precursor **16**. The key intermediate **2** was isolated after hydrolysis of **16**. Although this approach was adequate for supplying material for preclinical toxicology and Phase I studies, there were several issues that rendered this approach problematic for larger-scale synthesis. Most of the intermediates were oils and required chromatographic purification on scale. In addition, the 11-step synthesis had a free radical bromination/debromination sequence that had associated safety concerns. The synthesis also included trimethylaluminum, a known pyrophoric reagent, and the use of toxic sodium cyanide. With these challenges, a core group of process chemists was assembled and charged with developing efficient, cost-effective, and safe technology for the large-scale synthesis of **2**.

This report details one of many efficient enantioselective syntheses of **2** from readily available ethyl acetoacetate. Proof of concept and subsequent process development of key steps (Scheme 3.2) leading to **2** from bio-reduction of the pro-chiral pyranone **19**[6] or its potassium salt **18** is discussed. A unique one-pot approach to

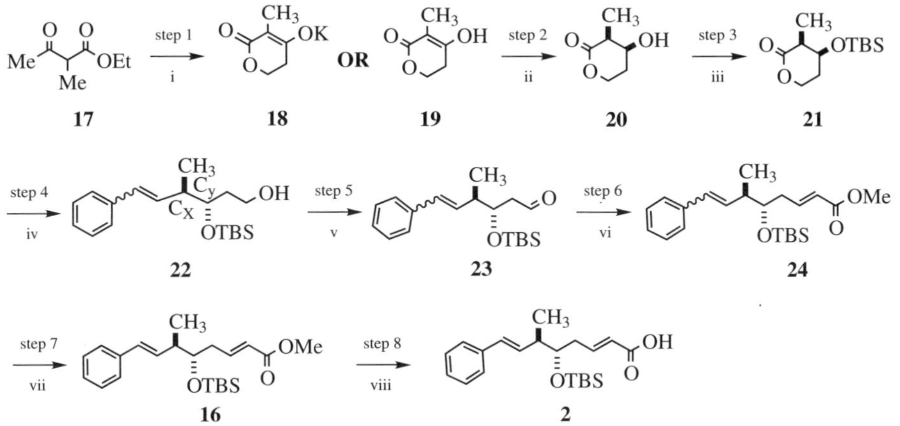

Scheme 3.2. (i) *i*-PrOH, *t*-BuOK, paraformaldehyde, 70%; (ii) pH 5.0 buffer, *Mortierella Isabellina reductase*, 70% (99%de, 99%ee); (iii) DMF, TBSCI (2.0 eq), imidazole (4.0 eq), r.t., 86%; (iv) DIBAL; Wittig, 70% (1:1) *E*-to-*Z* ratio; (v) CH_2Cl_2, Swern, 90%; (vi) CH_2Cl_2, $(MeO)_2P(=O)CH_2CO_2Me$ (HEW), 85%, (vii) Toluene, 1,1-azo bis(cyclohexanecarbonitrile) (VAZO), 1% thiophenol, reflux 60%; (viii) dioxane, KOH (4.0 eq), reflux, 90%.

synthesize ***E***-styrenoid product **22** exclusively, along with a plausible mechanism is discussed. The formal total synthesis of **1** is also discussed.

PROOF OF CONCEPT

The approach we envisioned involved accessing pyranone **19** with the appropriate carbon framework and functionality for an asymmetric hydrogenation or biocatalysis to effect reduction to **20**. The reduced product **20** can then be elaborated on to the target compound **2** as shown in the outline in Scheme 3.2. Compound **19** was prepared in 65% yield by subjecting a mixture of ethyl 2-methylacetoacetate (**17**) and formaldehyde to potassium *tert*-butoxide at ambient temperature. This process developed in-house was scaled up in the pilot plant to provide roughly 80 kg of **19** to fuel developmental efforts of subsequent steps.

To guide us in our stereochemistry analysis of later steps, we opted to initially prepare racemic samples via hydrogenolysis of **19**. Preliminary hydrogenolysis of **19** with different catalysts to effect reduction is shown in Table 3.1. Subjection of **19** to PtO_2 in ethyl acetate afforded diastereomeric mixtures of hydroxylactone **20** and **20a** (*cis* and *trans* ratio of 6:1). An improved ratio (10:1) was achieved with rhodium on alumina. Based on this encouraging data, we explored derivatization of **19** with *t*-butyldimethylsilyl chloride (TBSCl) and subject the silyl-enol ether to rhodium on alumina hydrogenation. Unfortunately, no reaction was observed. While pursuing the hydrogenation, we were simultaneously exploring biocatalysis[7] as an alternative means to produce **20** in an asymmetric fashion. Various

TABLE 3.1. Preliminary Hydrogenolysis of 19 with Different Catalysis to Effect Reduction

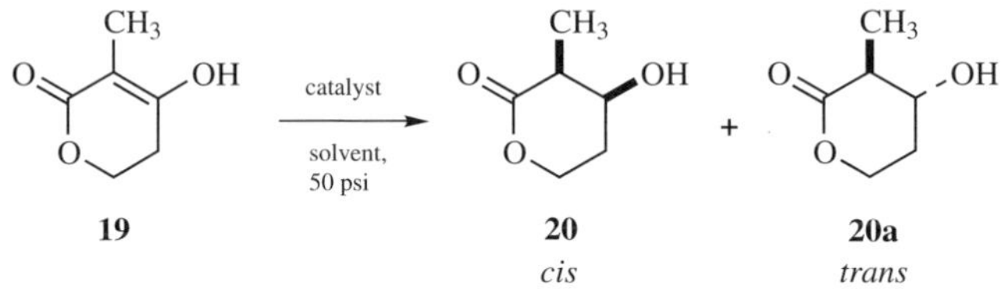

Catalyst	Percent Catalysis Load by weight	19	Solvent	Percent Yield **20 + 20a**	Temperature	Ratio **20/20a**
PtO_2	100	1.5	IPA	93.0	25°C	6:1
Rh/Al	100	1.5	IPA	95.3	25°C	10:1
Rh/Al	100	1.0	iPrOH/*t*-BuOH	—	25°C	—
PtO_2	100	1.0	EtOAc	—	25°C	—
Rh/Al	100	1.0	EtOAc	—	25°C	—
Raney Ni	100	1.0	IPA	—	25°C	—

reductases[8] were investigated with no success. However, a *Mortierella Isabellina* reductase produced a single diastereomeric product by chiral GC analysis. None of the other three possible stereo products were observed.

The unprecedented diastereoselectivity observed in the bioreduction of **19** directed our research efforts toward isolation, and subsequent elaboration of **20** led to a known intermediate in the SAE route for unequivocal stereoisomeric identification.

Isolation of bioreduction products from aqueous reaction mediums has always been problematic, especially when the product and the starting material have similar solubility profiles. The extreme solubility of **19** and **20** in the aqueous reaction medium required an organic solvent or combinations thereof that not only excluded the starting material but rejected other impurities as well. Preliminary attempts to extract the product with various organic solvents after supersaturation of the reaction broth with brine were discouraging. To facilitate development of subsequent steps, the solution from the bioreduction broth was concentrated to dryness. In the absence of water, extremely pure **20** can be extracted with *t*-butylmethyl ether (MTBE). Typical yields obtained with MTBE extract ranged from 68% to 70%.

Although the isolation protocol discussed above was not an ideal process, sufficient material was obtained to move forward. Derivatization of **20** with TBSOTf and triethylamine afforded a 1:1 mixture of the expected product **21** and the dehydration product. An alternative approach involving TBDSCl and imidazole in dimethylformamide afforded **21** in near quantitative yield. DIBAL reduction of **21** and subsequent reaction with a Wittig reagent proceeded in good yield with a 1:1 mixture of ***E*** and ***Z***-isomers **22A** and **22B**. The Wittig reagents utilized with moderate success in isomer ratios are shown in Table 3.2. The highest isomeric

TABLE 3.2. Wittig Reagents Utilized with Moderate Success in Isomer Ratios

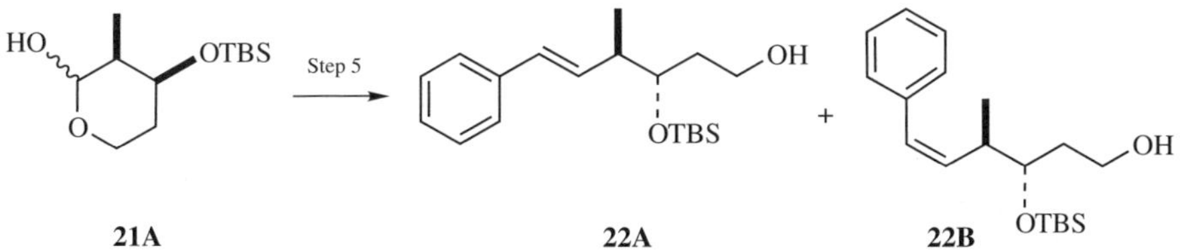

21A **22A** **22B**

Reagents	22A		22B
$PhCH_2PPh_3Cl$ / *t*-BuOK	1	:	1
$PhCH_2PO(OEt)_3$ / *n*-BuLi, –78 °C	7	:	1
$PhCH_2P(n\text{-}Bu)_3Br$ / *n*-BuLi, –78 °C	6	:	1

ratio (7:1 in favor of the ***E*** isomer) was obtained with benzyldiethylphosphonate and *n*-butyl lithium at cryogenic condition. Swern oxidation followed by Horner–Emmons olefination afforded **24** in greater than 85% yield.

Improvement of the isomeric ratio to the exclusivity of the ***E***-isomer was achieved with 1,1-azo-bis(cyclohexanecarbonitrile) (VAZO) and catalytic amount of thiophenol in toluene at 110°C. Desilylation and hydrolysis of **16** afforded the target compound **2**. The analytical data and stereochemical purity of **22A** from the bioreduction route compared favorably with the sample from the original Sharpless Asymmetric Epoxidation route.[9]

With stereochemical identity of **20** demonstrated, attention was then focused on developing the individual steps depicted in Scheme 3.2 to provide a viable process for larger-scale production of **2**. Unlike steps 1 and 2, which were in good process condition to be scaled up, the critical challenges posed in later steps were twofold: (a) telescoping of steps 2 and 3 to isolate **21** from aqueous reaction broth and (b) the selective ***E***-isomer formation of the pivotal intermediate **22A**.

DEVELOPMENT OF CRITICAL STEPS AND SCALE-UP

Telescoping of Steps 2 and 3 (Isolation of 21)

The initial isolation of **20** from reaction broth, as mentioned earlier, needed development before implementation on larger scale. An alternative isolation protocol, eventually developed, involved saturation of the reaction broth with 20% brine followed by multiple acetonitrile extractions. Careful azeotropic[10] removal of residual water from the acetonitrile/water extract is achieved under reduced pressure at 35°C. The acetonitrile solution is carried through subsequent silylation without any ramifications. The rationale behind the choice of acetonitrile as the solvent for extraction is threefold. First, silylation reaction performs well in this solvent which allows for a streamlined process. Second, insolubility of the silylated hydroxylactone in hexane or heptane at −20°C allows for an easy recrystallization to reject impurities. Finally, an aliphatic acid impurity[11] which we speculate originates from degradation of the cell wall of the yeast can be extricated from the acetonitrile with multiple hexane or heptane washes at ambient temperatures. This process was scaled up at 12 g/liter in a 1000-gallon tank to provide an acetonitrile solution of **20** (70% *in situ* yield).

As previously discussed the initial silylation of **20** with *t*-butyldimethylsilyl triflate and triethylamine in methylene chloride afforded product **21** with substantial amount of impurity **25**[12] (Figure 3.3) at ambient temperatures. Although silylation with trimethylsilyl chloride seemed extremely attractive due to the cost, repeated attempts to carry out this reaction failed due to labile trimethylsilyl moiety upon workup. Silylation eventually proceeded in greater than 95% yield with *t*-butyldimethylsilyl chloride and imidazole[13] in acetonitrile. Forward processing of the isolated solution of **20** in acetonitrile under these silylation conditions afforded

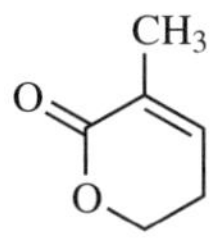

25

Figure 3.3.

21 as an off-white low-melting[14] solid in 85% isolated weight yield with an excellent purity profile. The telescoping of steps 2 and 3 provided roughly 15 kg of **21** in the pilot plant.

Step 4 (Preparation of 22A)

A key transformation in the development of this route involved reaction of hemiacetal **21A** from DIBAL reduction of **21** (Scheme 3.2) with an appropriate styrenoid-forming reagent to form the *E* geometric product. The Wittig reaction has been used extensively to prepare styrene derivatives.[15] In most cases, a mixture of *cis/trans* geometric isomer is obtained with much work devoted to stereoselective synthesis of one isomer over the other.[16] While the *cis* isomer can be obtained under a variety of conditions, the *trans* isomer generally presents a significant challenge. In an attempt to eliminate the silylation protection and deprotection sequence, we investigated the possibility of directly accessing nonsilylated derivative **26** (Figure 3.4) from the isolated bioreduced pyranone. Selective oxidation of a primary alcohol in the presence of a secondary alcohol had previously been demonstrated.[17] This information played a critical role in steering our initial research efforts toward an olefination methodology that would transform **20** to the styrenoid diol **26**. Preliminary reaction with Wittig reagent proceeded in high yield with a 1:1 ***E***-to-***Z*** ratio.

We therefore, reversed course and turned our attention toward styrenoid formation of the silylated product **21**. Initial attempts to form **22A** from isolated hemiacetal **21A**[18] with a variety of known styrenoid-forming reagents were quite discouraging. The conventional Wittig reaction with the latent aldehyde proceeded at room temperature to give 70% yield of product which contained a 1:1 ***E***-to-***Z*** geometric isomer ratio. No reaction was observed with the Horner–Emmons reagent using a variety of bases including *n*-BuLi, *t*-BuOK, *t*-BuONa,

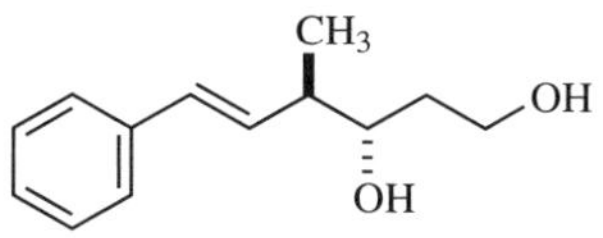

26

Figure 3.4.

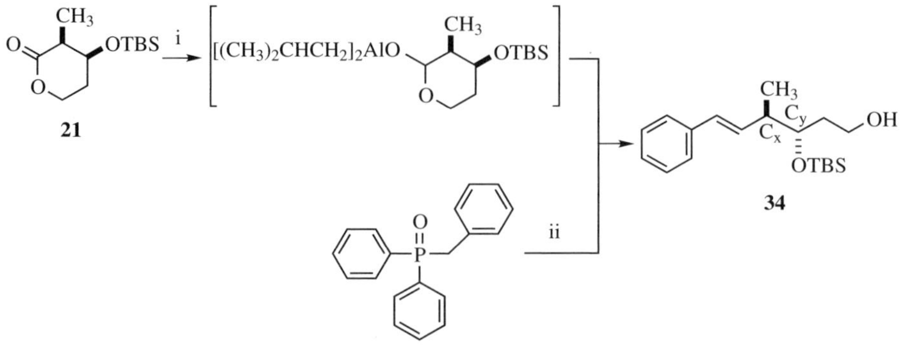

21 **34** **33**

Scheme 3.3. (i) Heptane, 1.2 eq DIBAL, −10°C; (ii) THF, $NaN(TMS)_2$, 50°C.

and $NaN(TMS)_2$. However, with $KN(TMS)_2$ and catalytic amount of crown ether (18-crown-6), 86% yield of the styrenoid product (6:1 ***E***-to-***Z*** isomer ratio) was formed, contaminated with 6% racemized product. The reaction appeared to be slower without the crown ether. Although a step in the right direction, we needed conditions to improve the ***E***-to-***Z*** ratio and also avoid racemization.

Reaction of the Horner–Wittig variant, benzyl diphenylphosphine oxide, with aldehyde is known to afford styrene with good ***E***-to-***Z*** selectivity.[19] There have also been literature reports involving reduction of lactone with an equivalent of DIBAL and subsequent *in situ* reaction with Wittig reagents to afford styrenoid derivatives in decent yields.[20] These precedents paved the way for the implementation of a facile and efficient entry into *trans* styrenoid moieties and derivatives thereof. We envisioned a convergent approach (Scheme 3.3) involving DIBAL reduction of **21** and subsequent *in situ* reaction with the appropriate Wittig variant.

Thus Horner–Wittig reagent **33** was prepared in two steps either following a literature protocol[16] (78% overall yield) or via an alternative procedure (65% overall yield) as depicted in Scheme 3.4.

A pilot reaction of isolated hemiacetal **21A** with this reagent under a nitrogen blanket resulted in the formation of stilbene and starting hemiacetal but no product formation. This result prompted us to probe the mechanism of stilbene formation. A control experiment with the Horner–Wittig reagent and base devoid of hemiacetal and nitrogen blanket gave essentially stilbene and no starting Horner–Wittig reagent; a phenomenon documented by others.[21] Incorporation of a freeze–thaw cycle prior to execution of reaction resulted in avoiding the formation of stilbene, thus confirming reaction sensitivity to dissolved residual oxygen. Alternatively, an argon purge can be substituted for the freeze–thaw cycle to avoid cryogenic conditions. After these changes were incorporated, the reaction proceeded with greater than 90% efficiency using 1.8 equivalents of Horner– Wittig reagent and 1.6 equivalents of base.[22] A subcess of base is needed to prevent possible racemization of stereocenters C_x or C_y. In preparing the hemiacetal **21A** for the Horner–Wittig

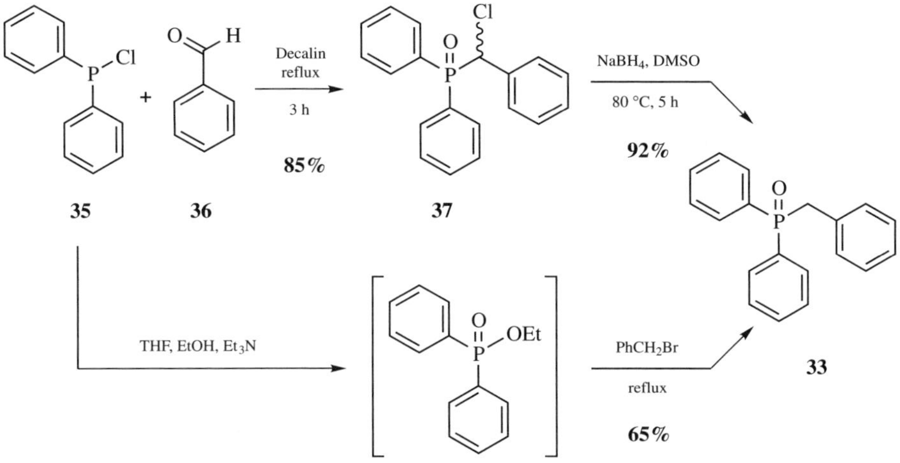

Scheme 3.4.

reaction, a slight excess of DIBAL was required for efficient hemiacetal formation. No over reduction was observed under these conditions. At the end of the Horner–Wittig condensation reaction, careful quenching was exercised to prevent foaming due to excess DIBAL. Following the completion of the Horner–Wittig reaction, it is necessary to remove any residual hexamethyldisilylazane from the base[23] to prevent silylation of primary alcohol **22A** to **39** during evaporation of organic solvent under reduced pressure. The bis-silylated compound **39** (Figure 3.5) has been isolated and identified as an impurity which is formed during workup. Formation of **39** interferes with the subsequent oxidation step by depressing the yield of **23**. This can be avoided[24] by quenching the reaction mixture with vigorous stirring in the presence of excess 1 N hydrochloric acid solution.

Plausible Mechanism for E-Isomer Formation. A plausible mechanism for exclusive ***E***-isomer formation[25] is illustrated in Scheme 3.5. Initial reaction of **32** with deprotonated **33** establishes equilibrium between starting materials and the *erythro* and *threo* intermediate phosphinate adducts (**40** and **41**). We believe that at this

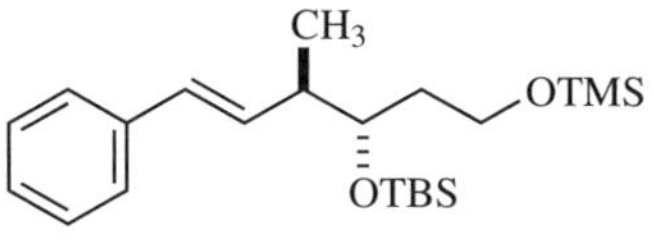

39

Figure 3.5.

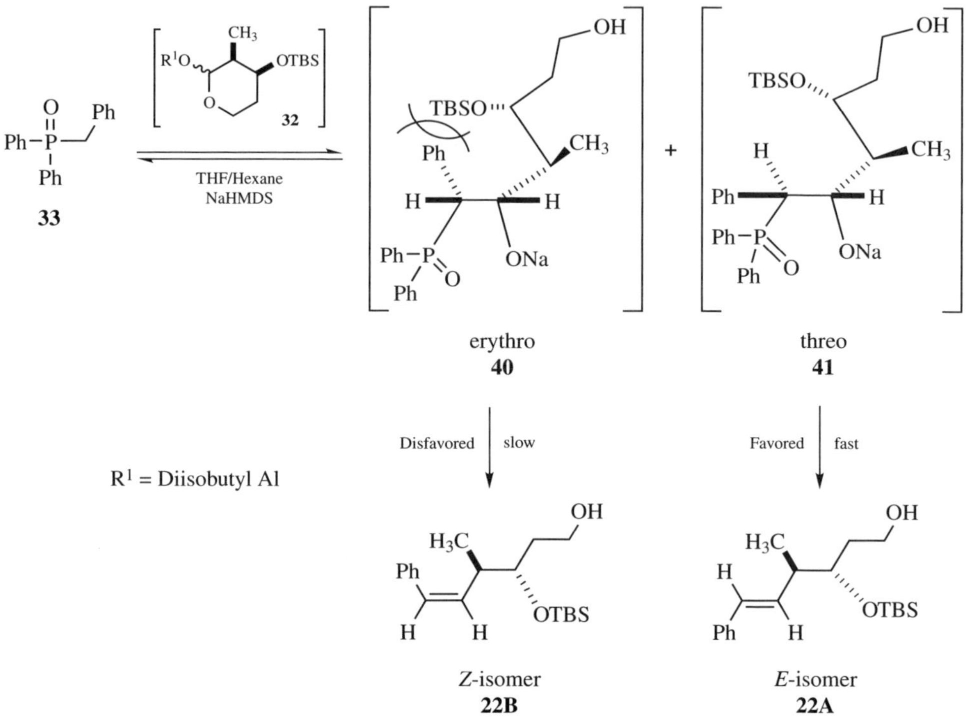

Scheme 3.5.

juncture, in the following elimination step, transition state interactions between the bulky phenyl and the TBSO groups in the *erythro* phosphinate adduct disfavor *syn* elimination to the Z-alkene **22B**. On the other hand, the transition state originating from the *threo* adduct exhibits less steric interactions. It is therefore favored and proceeds through the phosphinate *syn*-elimination to form the ***E***-isomeric product **22A**. The slow leakage of the *threo* adduct over time to ***E*** styrenoid product at the expense of the *erythro* pathway to **Z**-isomer accounts for the observed selectivity. As previoudly mentioned, the reaction appears to be highly sensitive to oxygen; hence a strict adherence to oxygen-free environment is essential to prevent auto-oxidation, which leads to *trans*-stilbene formation.

In a control experiment with hemiacetal **32** at −78°C (Scheme 3.5), no product was isolated. Proton NMR analysis after quench indicated hemiacetal and starting Horner–Wittig reagent. However, the reaction seems to proceed to product when carried out at ambient temperatures. At −78°C, it appears that the DIBAL adduct **32** (Scheme 3.6) in form of aluminum protected hemiacetal prevents the reaction from proceeding. However, at higher temperatures (−10°C to 25°C), equilibrium shifts toward unmasking of the aldehyde **32A**, hence the formation of styrenoid

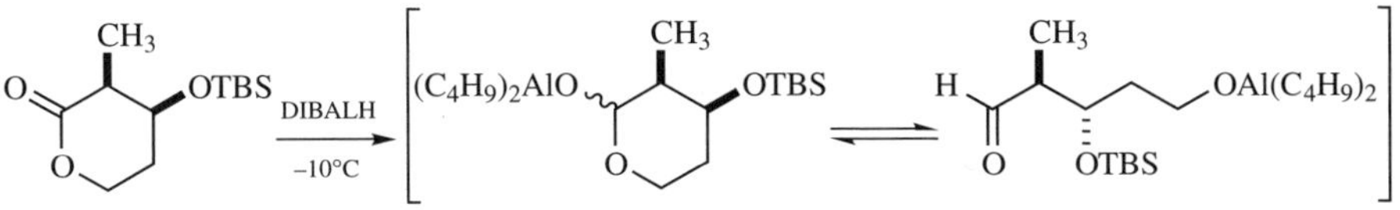

21 **32** **32A**

Scheme 3.6.

product. This phenomenon is well-documented in the literature.[26] Treatment of isolated hemiacetal failed to give any product at ambient temperature, thus confirming the need to unmask the aldehyde.

Step 5 (Preparation of 23)

Our selection of reagent for transformation of **22A** to the aldehyde **23** was based on robustness and the possibility of telescoping later synthetic steps in the sequence. Catalytic TEMPO (2,2,6,6-tetramethyl-1-piperidinyloxy radical) and sodium hypochlorite solution as an oxidant seemed appropriate based on an in-house technology with a similar and highly sensitive scaffold. Initial pilot scale attempts to oxidize **22A** to **23** was quite successful. However, upon scale-up, the reaction seemed to stall at 50% conversion. Even more discouraging was the fact that the hypochlorite began to react with the styrenoid moiety.

Recently, Einhorn et al.[27] applied catalytic TEMPO and *N*-chlorosuccinimide (NCS) as an oxidant to successfully carry out oxidations of simple primary and secondary alcohols to aldehydes and ketones under phase transfer catalysis. Although application of this procedure toward the oxidation of **22** was successful, reaction times were extremely long. However, the reaction proceeded at a faster rate when *N*-chlorophthalimide (NCP) was substituted. No detectable overoxidation or other side reactions were observed. Due to lack of bulk quantities of NCP, an alternative Swern oxidation method was implemented to enable progress of the overall synthesis. Roughly a kilogram of **23** was prepared by the Swern approach.

Following the complete development of critical steps, attention was then focused on completing the synthesis to target compound **2** (Scheme 3.7). The enantioselective synthesis was completed by olefination of aldehyde **23** to afford the ester **24**. The target compound **2** was isolated in an overall yield of 28% (96%de, 99%ee) from **17** after hydrolysis of **24**.

FORMAL TOTAL SYNTHESIS OF 1 (CRYPTOPHYCIN 52)

The synthesis of **1** (Scheme 3.8) was accomplished in a convergent fashion. The enantiopure compound **2** was coupled with fragment **3** to give **43**.

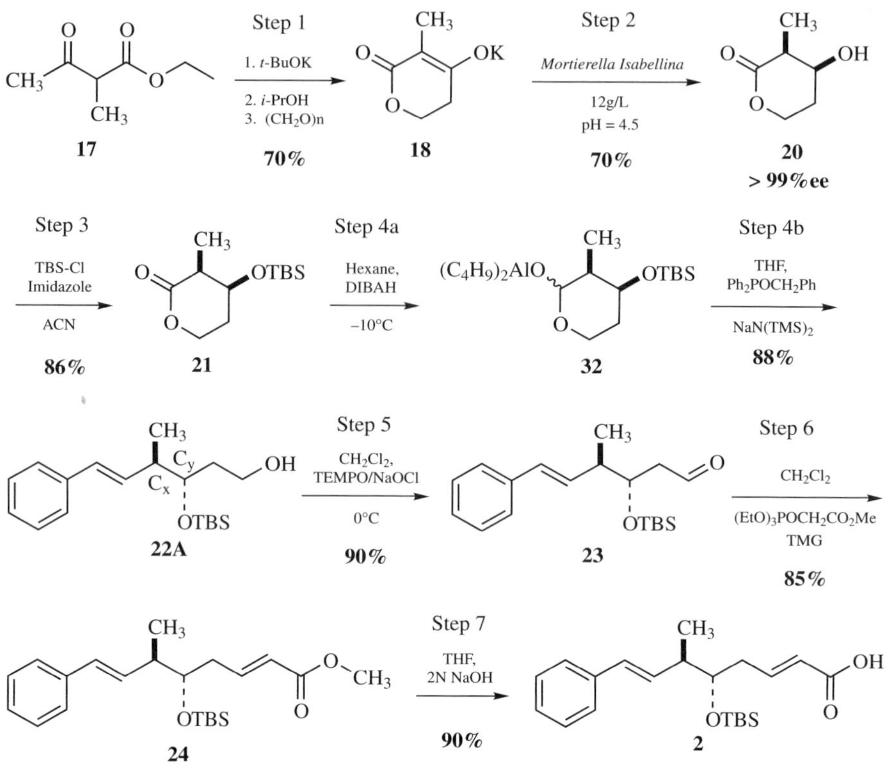

Scheme 3.7.

Simultaneously, fragment **4** was coupled with **5** to afford **44**. Removal of the allylic protective group in **44** provided **45** and subsequent coupling with **43** yielded **seco-46**. Macrolactamization of **seco-46** afforded the penultimate compound **47**. The epoxidation of **47** with *m*-chloroperoxybenzoic acid completed the formal total synthesis of the target cryptophycin compound **1** (96%de) as shown in Scheme 3.8.

CONCLUSION

In summary, a short and efficient seven-step enantioselective synthesis of a key cryptophycin 52 fragment **2** employing a novel bio-reduction of the prochiral pyranone **19** is described. The synthesis takes advantage of a sequential silylation, DIBAL reduction followed by a key ***E***-styryl homologation protocol with **33**. Swern oxidation of the styryl product **22A**, and further reaction with methyl diethylphosphonoacetate affords the penultimate compound **24** in high yield and purity. The requisite compound **2** is isolated in an overall yield of 28% (96%de)

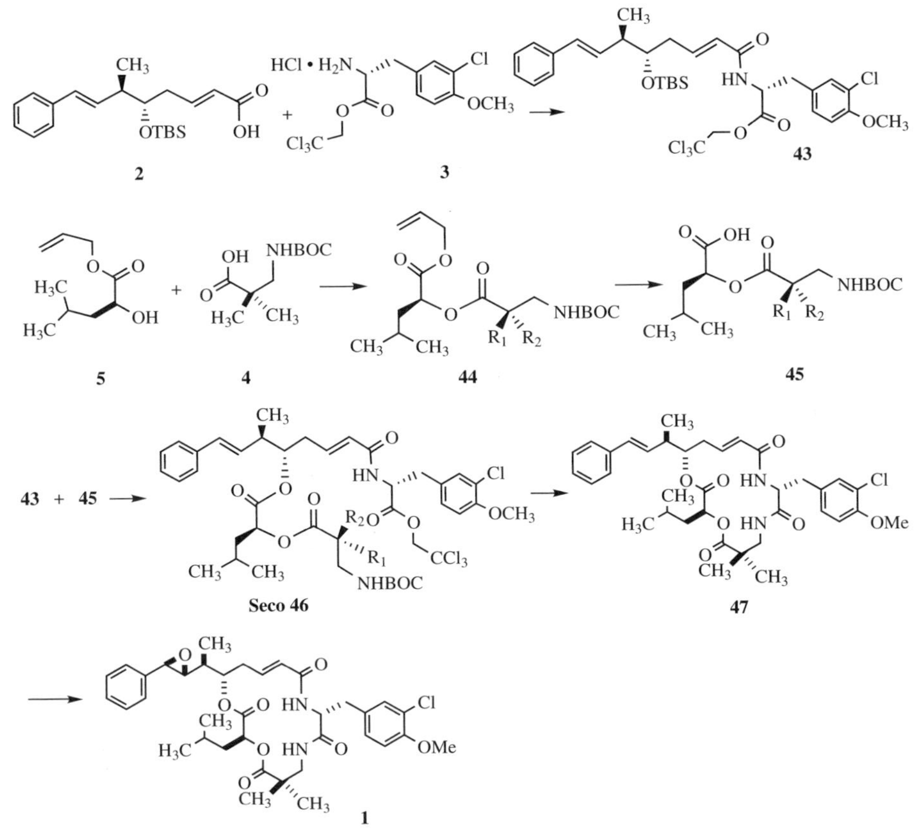

Scheme 3.8.

from the pyranone after hydrolysis of the penultimate **24**. The successful enantioselective synthesis of **2** allows for a convergent formal total synthesis of cryptophycin 52 (**1**) in five steps as shown in Scheme 3.8.

ACKNOWLEDGMENTS

The author would like to thank Dr. Keith Michael Devries, Dr. Dawn Brooks, and Dr. David Mitchell for their critical evaluation of this publication, CPR&D management for their support, as well as the following colleagues for contributions to this project:

Process Chemists—John Brennan, David Hay, William Miller, David Varie
Pilot Plant—Jason Cronin, Richard Miller

Analytical—Brad Held, Joe Kennedy, David Robbins, Joseph Turpin
X-Ray Analysis—Jennifer Olkowski, Gregory Stevenson

REFERENCES AND NOTES

1. Schwartz, R. E., Hirsch, C. F., Sesin, D. F., Flor, J. E., Chartrain, M., Fromtling, R. E., Harris, G. H., Salvatore, M. J., and Yudin, K. *J. Ind. Microbiol.* **1990**, *5*, 113–124.
2. Smith, C. D., Zhang, X., Moobery, S. L., Patterson, G. M. L., and Moore, R. E. *Cancer Res.* **1994**, *54*, 3779.
3. Barrow, R. A., Hemcheidt, T., Liang, J., Paik, S., Moore, R. E., and Tius, M. A. *J. Am. Chem. Soc.* **1995**, *117*, 2479.
4. (a). Leahy, J. W., and Gardinier, K. M. *J. Org. Chem.* **1997**, *62*, 7089–7099; (b). de Muys, J. M., Rej, R., Go, B., Fortin, S., and Lavallee, J. F. *Bioorg. Med. Chem. Lett.* **1996**, *6*, 1111–1116; (c). Rej, R., Nguyen, D., Go, B., Fortin, S., and Lavallee, J. F. *J. Org. Chem.* **1996**, *61*, 6289–6295; (d). Salamoczyk, G. M., Han, K., Guo, Z., and Sih, C. J. *J. Org. Chem.* **1996**, *61*, 6893–6900.
5. Georg, G. I., and Ali, S. M. *Tetrahedron Lett.* **1997**, *38*(10), 1703–1706.
6. CPR&D, Eli Lilly & Co., unpublished work.
7. CPR&D, Eli Lilly and Co., unpublished work.
8. CPR&D, Eli Lilly and Co., unpublished work.
9. Prepared from DIBAL reduction of SAE route compound **15** to the aldehyde, followed by reduction to the alcohol with $NaBH_4$.
10. Pilot Plant Engineer, Eli Lilly & Co., Chemical Product Research and Development, **1997**.
11. This impurity can only be removed by chromatography.
12. Based on proton NMR analysis.
13. Scaled-up in Experimental Materials Laboratory, Eli Lilly & Co., CPR&D, 1997.
14. Melting point: 45°C.
15. Lawrence, N. J. *The Wittig Reaction and Related Methods: Preparation of Alkenes*, 1996, pp. 19–58.
16. (a) Vedejs, E., and Peterson, M. J. *Adv. Carbanion Chem.* **1996**, 2, 1–85, (b). Walker, B. J. *Organophosphorous Chem.* **1996**, *27*, 264–307; (c). Clayden, J., and Warren, S. *Angew. Chem., Int. Ed. Engl.* **1996**, *35*(3), 241–270; (d). Schlosser, Manfred *Top. Stereochem.* **1970**, *5*, 1–30.
17. CPR&D, Eli Lilly & Co., unpublished work.
18. DIBAL reduction of TBS protected 418217.
19. Brown, Karen, M., Lawrence, N. J., Liddle, J., and Faiz, M. *Tetrahedron Lett.*, **1994**, *35*(36), 6733–6736.
20. Boland, W., Ney, P., and Jaenicke, L. *Synthetic Commun.*, **1980**, 1015–1017.
21. Horner, L., et al. Ber. Dtsch. Chem. Ges., **1961**, *94*, 1987.
22. Scaled up in EML, Eli Lilly & Co., CPR&D, 1997.
23. This is a known silylating reagent.

24. The bis-silylated product has a higher R_f value, which makes it easier to monitor.
25. Maryanoff, B. E., and Reitz, A. B. *Chem. Rev.* **1989**, *89*, 863–927.
26. Boland, W., and Jaenicke, L. *J. Org. Chem.*, **1979**, *44*(26), 4189.
27. Einhorn, J., Einhorn, C., Ratajczak, F., and Pierre, J.-L. *J. Org. Chem.* **1996**, *61*, 7452–7454.

4

EFFORTS TOWARD A COMMERCIALLY VIABLE ROUTE AND PROCESS TO THE SYNTHESIS OF HIV PI GW640385X

John C. Roberts, Roman Davis, Thomas C. Lovelace, Daniel E. Patterson, Barry Sickles, Jennifer F. Toczko, and Ju Yang
GlaxoSmithKline, Synthetic Chemistry, Chemical Department Research Triangle Park, NC 27709–3398

Brian D. Doan
GlaxoSmithKline, King of Prussia, PA

Frank Roschanger
Boehringer Ingelheim, Richmond, VA

INTRODUCTION

Human immunodeficiency virus (HIV) protease has proven to be a challenging target for drug therapy largely due to its rapid and successful mutations. As resistant strains emerge to a given therapeutic, its effectiveness in the patient population is reduced. Thus, there is a continued effort to identify compounds of increased potency and activity against resistant strains.[1]

An HIV protease inhibitor (PI) candidate, GW640385X (**1**) (Figure 4.1), has proved promising when assayed against known drug-resistant strains. This compound entered development in mid-2001 as part of GlaxoSmithKline's ongoing antiviral program.[2–10] Compound **1** contains a 2-amino-3-phenylpropan-1-ol substructure found in other active pharmaceutical ingredients (APIs) of currently

Fundamentals of Early Clinical Drug Development: From Synthesis Design to Formulation, edited by Ahmed F. Abdel-Magid and Stéphane Caron

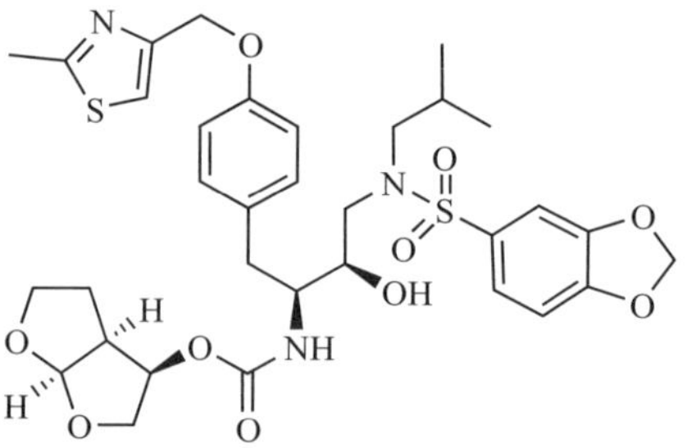

1

Figure 4.1. Structure of GW640385X (**1**).

marketed HIV PI products (Table 4.1). Indinavir and nelfinavir have related substructures. The series of compounds from which compound **1** arose was tyrosine-derived rather than phenylalanine-derived, and the *p*-hydroxyl moiety provided an opportunity for extensive derivatization, an opportunity not present for predecessor compounds of those in marketed API's. In the case of compound **1** itself, a thiazole-containing alkyl group is installed on the phenolic oxygen and the isobutyl amino moiety is part of a sulfonamide functionality (common to amprenavir and fosamprenavir). The most elaborate fragment contained within the structure of **1** is the (3*R*,3*aS*,6*aR*)-hexahydrofuro[2,3-*b*]furan-3-ol fragment (hereafter referred to as bisfuran). This bisfuran fragment also appears in other compounds reported to be under development. Seemingly small and innocuous, the identification of a cost-effective synthesis to this part of the molecule proved to be a challenge, and one that has earned a great deal of attention in recent years.[11–21]

Typically, when a compound is in preclinical development, the primary objective in Chemical Development is the delivery of API in the appropriate quantity and quality for each stage of the project. During the early phase of development, the identification of a route that might be implemented in long-term manufacture is a secondary objective, albeit with increasing importance over time. Nevertheless, it is important from the inception of a development project to address this secondary objective and to assess how difficult this secondary objective will be to achieve. The perspective taken in the following discussion comes from balancing these requisite deliveries and development objectives.

The first part of the chapter will focus on the management of impurities through one portion of the early supply route. The second part of the chapter will be devoted to further route-definition applied to the bisfuran fragment. Finally, a brief description will be given on how final modifications would be expected to produce a commercially viable manufacture route to compound **1**.

MANAGEMENT OF IMPURITY PROFILE GENERATED IN ROUTE A

Route A, the initial supply route, is outlined in Figure 4.2. *O*-Benzyl-protected phenylalanine-derived epoxide **2** is modified in six steps to compound **1** via epoxide opening, sulfonylation, benzyl deprotection, phenol alkylation, *t*-butoxycarbonyl

TABLE 4.1. Active Pharmaceutical Ingredients in Currently Marketed HIV Protease Inhibitors

Chemical Structure	Generic Name (Approval Date) Trademark (Company)	Chemical Structure	Generic Name (Approval Date) Trademark (Company)
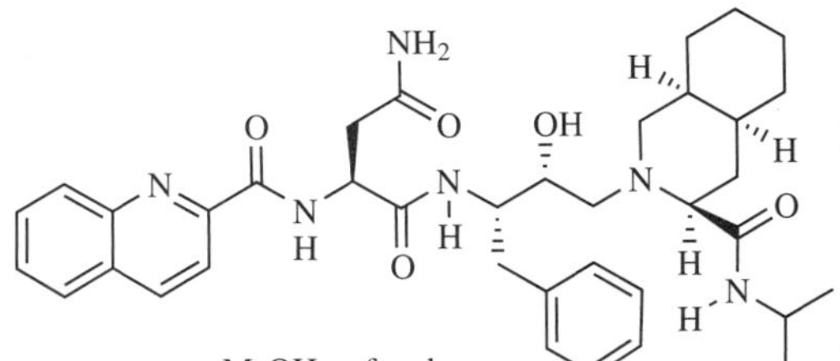 MsOH or free base	Saquinavir mesylate (1995) Invirase®(Roche) Saquinavir (1997) Fortovase®(Roche)	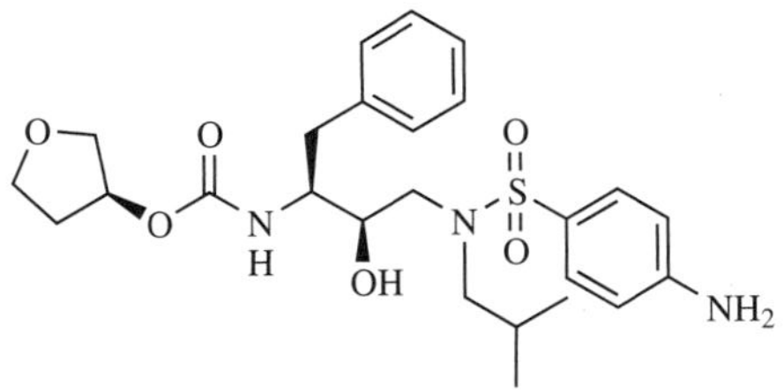	Amprenavir (1999) Agenerase® (GlaxoSmithKline)
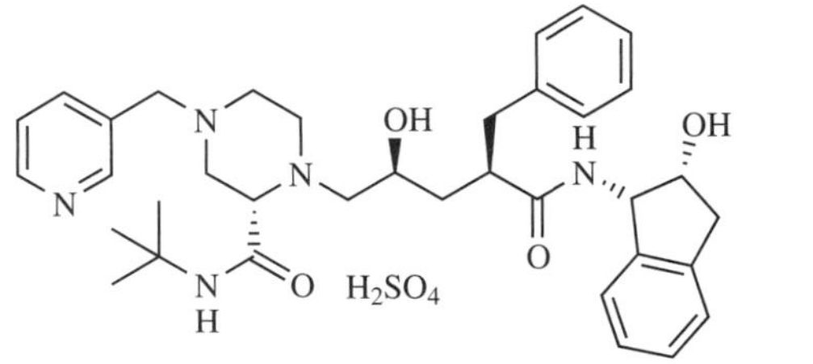	Indinavir sulfate (1996) Crixivan®(Merck)	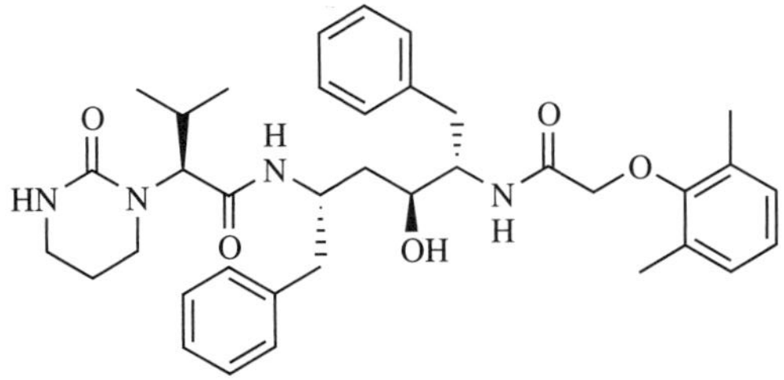	Lopinavir (2000) Aluviran®(Abbott) (with ritonavir) Kaletra®(Abbott)

(continued)

TABLE 4.1. (*Continued*)

Chemical Structure	Generic Name (Approval Date) Trademark (Company)	Chemical Structure	Generic Name (Approval Date) Trademark (Company)
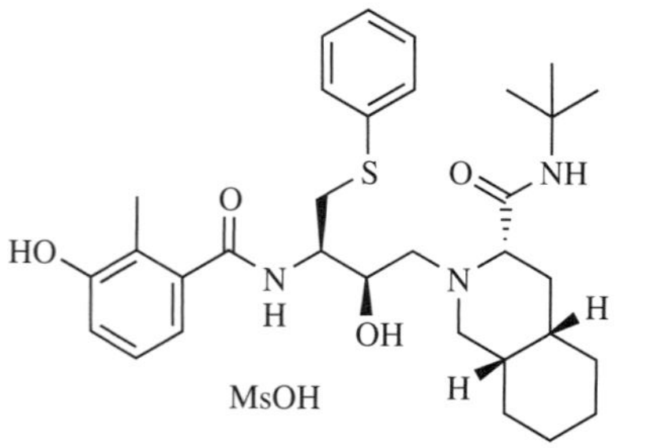	Nelfinavir (1997) Viracept® (Pfizer)	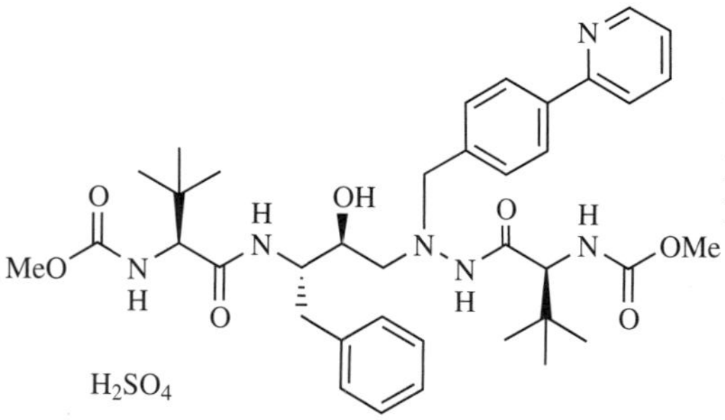	Atazanavir Sulfate (2003) Reyataz® (Bristol Myers Squibb)
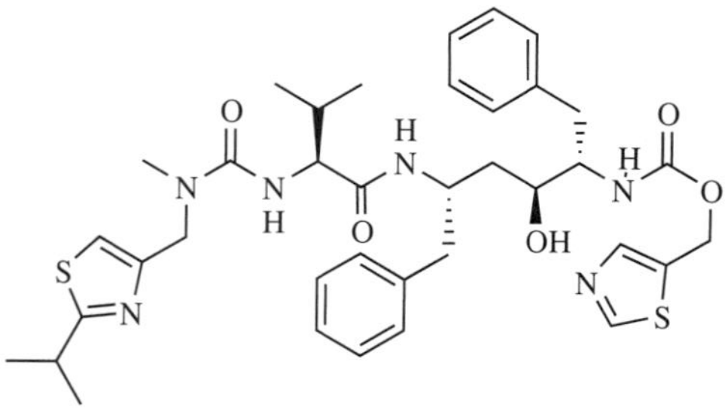	Ritonavir (1996); Norvir® (Abbott)	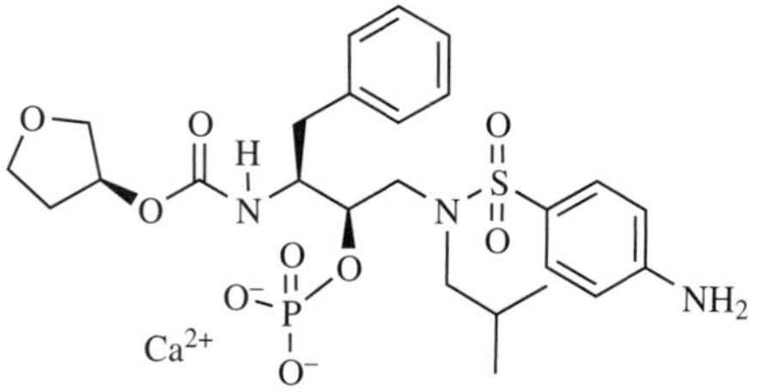	Fosamprenavir Calcium (2003) Lexiva® (Glaxo SmithKline)

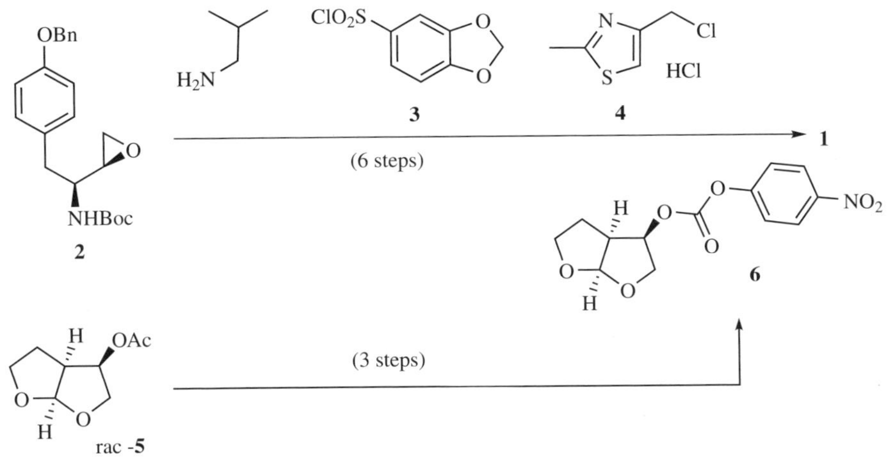

Figure 4.2. Outline of Route A (initial supply route).

deprotection, and final carbamate formation. In the Medicinal Chemistry route, to facilitate diversification efforts at the phenol substituent, the bisfuran unit was installed at an early step in the synthetic pathway. It was reasoned, however, that in order to reduce the cost associated with this unit, it would be advantageous to have its installation occur toward the end of the pathway. In both iterations of this early route, the carbamate-forming step was accomplished using *p*-nitrophenyl carbonate **6**, a highly crystalline and appropriately reactive intermediate.

STARTING MATERIALS FOR ROUTE A

The overall synthesis, from low-cost materials available in bulk, was greater than 20 steps, with the longest linear sequence being 14 steps. As such, there was an early assessment of where Route A stood with respect to the source of starting materials, a loose term that in this case includes intermediates synthesized at contract or even advanced intermediates that will eventually be synthesized at contract. The advanced intermediate, epoxide **2**, had been prepared at considerable cost from tyrosine (in six to seven steps). Experience with the sourcing of the corresponding phenylalanine-derived epoxide strongly indicated that cost for this material would eventually come down to acceptable levels through competitive bidding and without the application of internal resource. For the bisfuran starting material, however, the opposite was true: internal resource would indeed be needed both to support the current synthesis (seven steps) and the development of a less expensive one. Isobutylamine, while used in large excess, was known not to present a cost issue. Sulfonyl chloride **3** and thiazole **4** were not available in bulk, but they were easily prepared in one or two steps from commodity chemicals. Thus, as is usually part of the synthetic strategy in development with relatively long syntheses, internal resource focused primarily on the last steps. In this case, an elaborate supply strategy for the bisfuran starting material was required, which evolved to target acetate rac-**5** (i.e., the racemate of **5** where the ultimately desired stereochemistry is drawn).

DELIVERY STRATEGY WITH A TRANSITIONING ROUTE AND PROCESS

When compound **1** was being evaluated (along with another compound as a potential development candidate), at-risk development work had been carried out establishing the potential of the modifications already mentioned. In May 2001, candidate **1** formally entered into development and there was a need for 1 kg of drug substance approximately three months later. Subsequently, there would have to be an approximately 5-kg delivery in early 2002, 50 kg by the end of that year, and deliveries in excess of 100 kg thereafter. With an eye on the primary objective (delivery of appropriate quantity and quality to project timelines), an outlined strategy to address the secondary objective (a viable manufacture route) began to form. Nonetheless, the first delivery would have to be accomplished with only moderate modifications to the partially demonstrated Route A and an in-house synthesis of bisfuran rac-**5**.

In an ideal world, all route changes and significant process changes subsequent to the first delivery would be made all at once as part of the second or third delivery. It appeared, however, that the timing of these changes would likely not coincide. Eventually, a replacement for Route A that did not include a protection–deprotection approach (i.e., involving the benzyl ether) would have to be introduced. Similarly, there would have to be the inauguration of a new synthesis of the bisfuran fragment, one that would preferably use equipment standard to chemical manufacture, a synthesis that had not yet been conceived in May 2001 and that was not demonstrated until the end of 2002.

An alternative to the ideal solution was to gain a full understanding of impurities generated and derivatized throughout the synthesis. This understanding would allow for the introduction of changes at appropriate times with known and—by design—acceptable effects on these impurities. Thus, the impurity profile itself would be managed carefully during the prolonged use of a transitioning synthesis.

As mentioned earlier, the incorporation of the bisfuran unit in the last stage of the synthesis was an immediate deviation from the Medicinal Chemistry route. Since incorporation of this expensive fragment would eventually have to be performed at the end of the synthesis, making the change at such an early stage in development allowed for the possibility that the final manufacturing step to compound **1** would be defined and utilized in the first delivery. In turn, the impurities associated with this last step would be shared—at least to some extent—by all routes used to generate **1** through the course of development.

MANAGING IMPURITIES IN ROUTE A

In each of Steps 1 through 6 of Route A (Figure 4.3), impurities are formed. Some are removed to levels of less than 0.15% area under the curve (AUC) as detected by

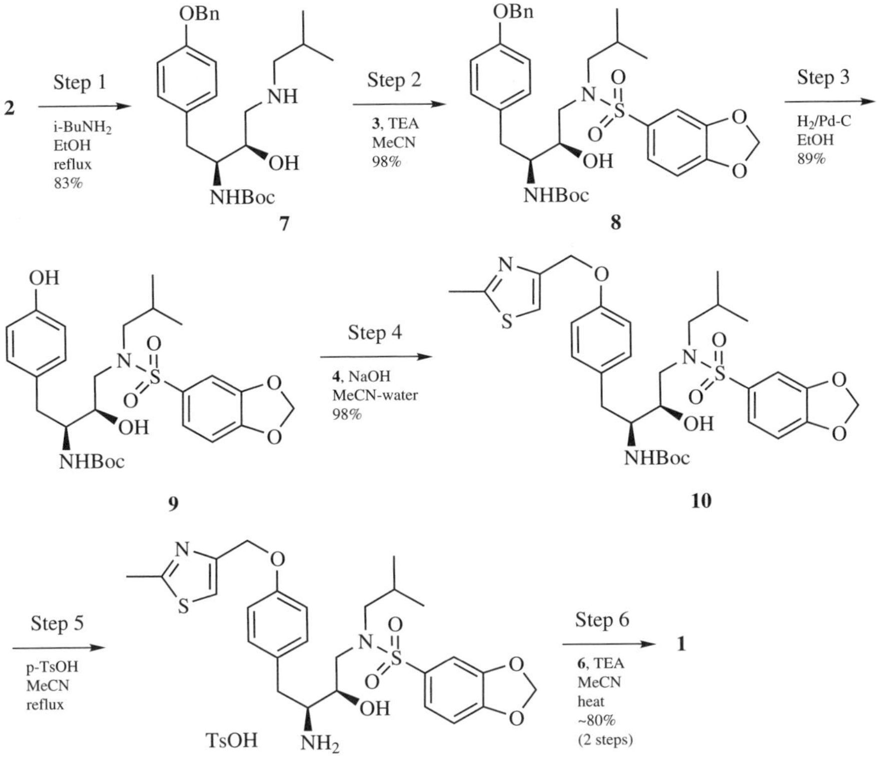

Figure 4.3. The Route A sequence from epoxide **2** to compound **1**.

high-performance liquid chromatography (HPLC), generally making them insignificant to early development considerations, while others are carried through at higher levels. It should be noted that % AUC is a unit that, while related to overall purity, does not include residual solvent and inorganics and is tied to the specific wavelength used for detection.

In any event, HPLC may be used to monitor apparent levels of impurities, which can transform through synthetic pathways. While some impurities remain as they were originally generated, others undergo chemical conversions in subsequent steps. As part of the strategy being pursued, the generation and fate of these impurities would have to be understood. As is routine in development, this type of investigative work began with the consideration of what the impurities were, what their precursors were in the last intermediate, what the precursors to those were in the penultimate intermediate, and so on. It included conclusions that certain impurities generated during the synthesis were easily removed and that others—ones that would have to be carefully monitored—were not (or at least not reproducibly). Early in development, this type of work can be somewhat speculative in nature, dependent largely on molecular weight information derived from HPLC mass spectrometry (LCMS) and the derivation of an overall consistent picture. In the case of the impurities in Steps 1 through 6, the LCMS method was further complicated by the poor response of compounds containing the *tert*-butoxycarbonyl (Boc) protecting group. As the project progressed, however, impurities that remained strategically important were synthesized as markers and identifications were made unambiguously. Nevertheless, the information collected and analyzed in this fashion guided the timing of process and route changes as the manufacturing route emerged.

Impurities Identified after Step 1

Let us consider Step 1 (Figure 4.3). As the process was first developed in Chemical Development, epoxide **2** was treated with excess isobutylamine in ethanol at reflux temperature. Secondary amine **12** and epoxide **2** were the main impurities in the crude reaction mixture (Table 4.2). The reaction mixture was treated with water to precipitate the amino alcohol product **7**. This proved to be a straightforward process that produced a relatively high yield of product, but there was negligible purification achieved. The 2–3% AUC levels of byproduct **12** was presumably generated from the desired product reacting with a second equivalent of epoxide **2**. It was reasoned that, if necessary in the short term, a much larger excess of isobutylamine may be used to reduce the levels of byproduct **13** and an extended reaction time could reduce the level of epoxide **2**. If, however, these impurities (or downstream derivatives thereof) were subsequently removed (which turned out to be the case), then the process was acceptable for the first delivery objective.

Impurities Identified after Step 2

A similar isolation strategy was initially employed in Step 2. Amine **7** was converted to sulfonamide **8**, a facile reaction carried out in acetonitrile with triethylamine at room temperature, and subsequently the reaction mixture was diluted with water effecting precipitation. All the previous impurities were carried through this isolation, and the level of a new one generated in the reaction, byproduct **13**, was not changed from the crude reaction mixture. The overall purity of sulfonamide **8** at this stage was approximately 95% AUC. The process itself was straightforward and the yield was very high, but, as in the first isolation, negligible purification was accomplished. Furthermore, it appeared that there was a balance between consumption of amine **7** and generation of byproduct **13** and it was noted that if one (or its downstream derivative) could be removed even if present at high levels, then a viable strategy would be to minimize the level of the other. In a later Route A process, conditions were utilized that reduced the levels of byproduct **13**.

Impurities Identified after Step 3

Step 3, hydrogenolysis of benzyl ether **8**, resulted in a clean conversion to phenol **9**, reaction conditions that also converted all of the earlier byproducts to the corresponding

TABLE 4.2. Impurities Generated in or Carried Through Steps 1–3

Step				Impurities
Step 1		2		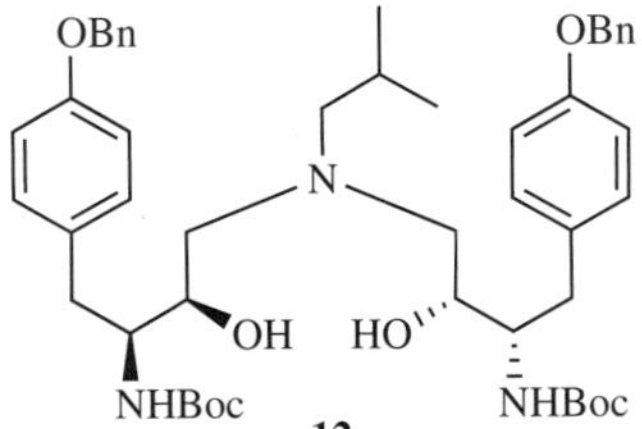 **12**
Step 2	2	7	12	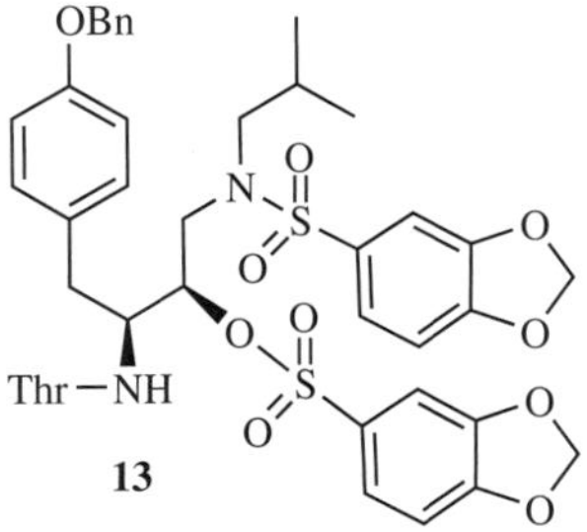 **13**
Step 3		8		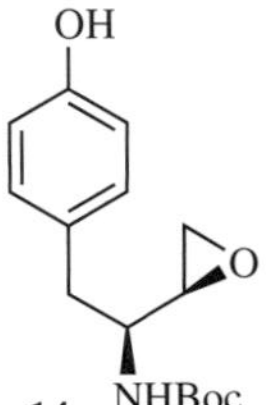 **14** 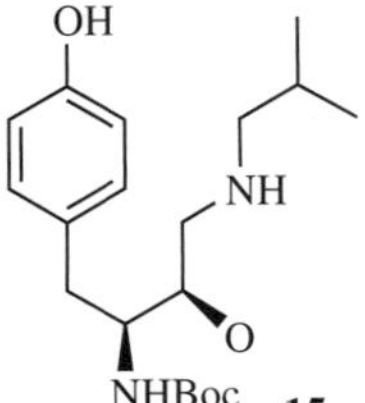 **15** 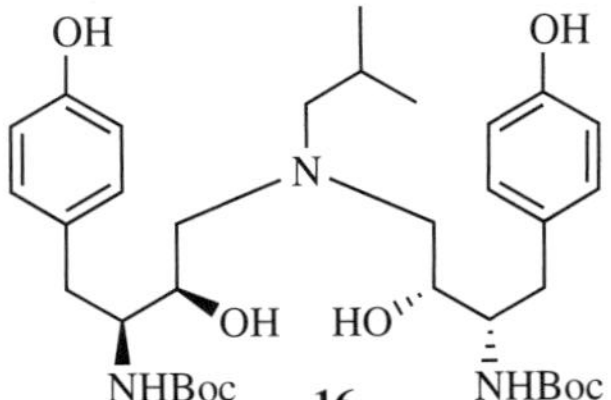 **16** 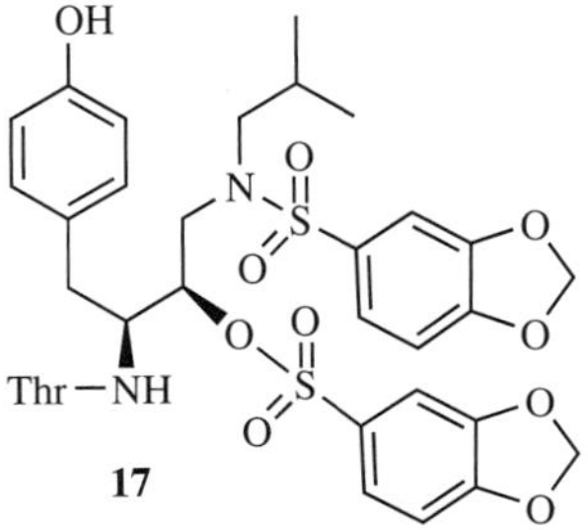 **17**

phenols (**14**, **15**, **16**, and **17**). The workup consisted of catalyst filtration, partial solvent exchange to ethanol, and addition of heptane to evoke crystallization. The filtration was very fast, and the isolated product proved to be greater than 99% AUC with a high overall yield. This result was obtained in a wide range of solvent ratios (i.e. residual tetrahydrofuran to ethanol to heptane) and heptane addition rates and therefore held promise as a robust isolation and purification. Furthermore, a consistent quality of phenol **9** could be isolated (i.e., near 100% AUC) regardless of the quality of benzyl ether **8** (over a range of a 90–95% AUC).

Phenol **9** was identified as a key intermediate because its high propensity to crystallize in nearly pure form would provide a great deal of flexibility with regard to how the earlier part of the synthesis could be modified. Expressed in general terms, with increased purity of a given intermediate, the less relevant is its history. Indeed, in a later iteration of the Route A process, Steps 1, 2, and 3 were combined into a single stage (i.e., three chemical conversions with one isolation step) and approximately 99% AUC product was still obtained.

Nitrophenyl Carbonate 6: Another Key Intermediate

At this point in the discussion, it is worth a brief mention of how *p*-nitrophenylcarbonate **6** (Figure 4.2) was also identified as a key intermediate. Bisfuran rac-**5**, a free-flowing oil, was generated in a very wide range of chemical purity through the course of the early development (typically 60–80% purity). There were also the issues of diastereomeric purity (rac-**5** was typically a 20:1 mixture of hydroxyl epimers) and enantiomeric purity of resolved intermediates (typically a 70:1 mixture of enantiomers). In short, conversion to the crystalline carbonate **6** followed by recrystallization improved the chemical purity to approximately 98% and enhanced both the diastereomeric and enantiomeric ratios of this fragment.

Impurities Identified after Step 4

Step 4 is the alkylation of phenol **9** using 4-chloromethyl-2-methylthiazole hydrochloride (**4**) in the presence of a base, a seemingly straightforward conversion. Indeed the desired conversion to alkylated product **10** was predominant in most of the conditions examined, and the product tends to crystallize out of the reaction mixture. The byproducts (Table 4.3) were the di-alkylated product **18** and cyclic carbonate **19**. The limited stability of the thiazole free base was of some concern. Furthermore, the dramatic differences in solubility between the free base and hydrochloride salt **4** initially made it difficult to identify conditions under which substrate, reagent, and base could be maintained in solution. As a result, the quality of product, particularly with respect to the levels of cyclic carbonate **19**, was variable. The use of concentrated sodium hydroxide in acetonitrile proved to be a balance that largely resolved these problems.

After some optimization, a reduced variability of the crude impurity levels was achieved, but the overall range continued at 92–95% AUC, troublesome for a step so late in the synthesis. After isolation the purity was somewhat improved and the

TABLE 4.3. Impurities generated in or Carried Through Steps 4–6

Step	Impurities
Step 4	9 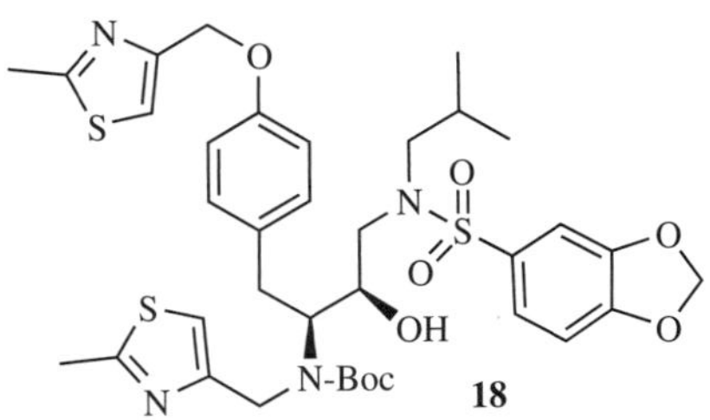 **18** 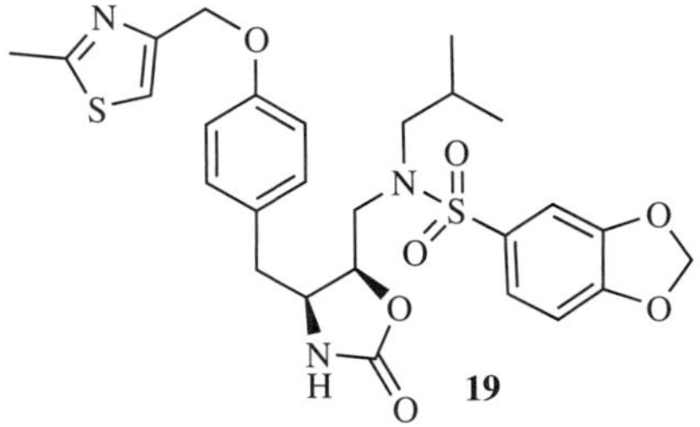 **19**
Step 5	10 19 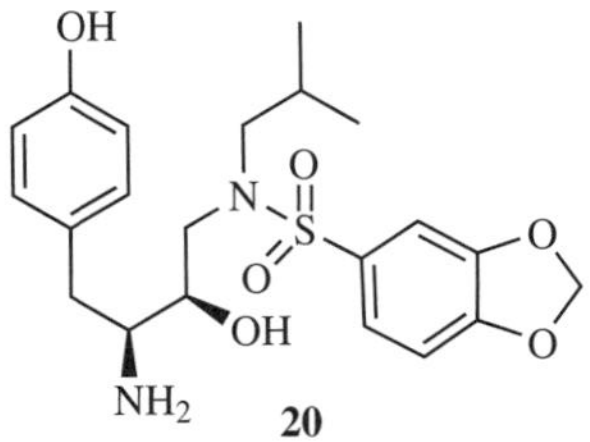 **20** 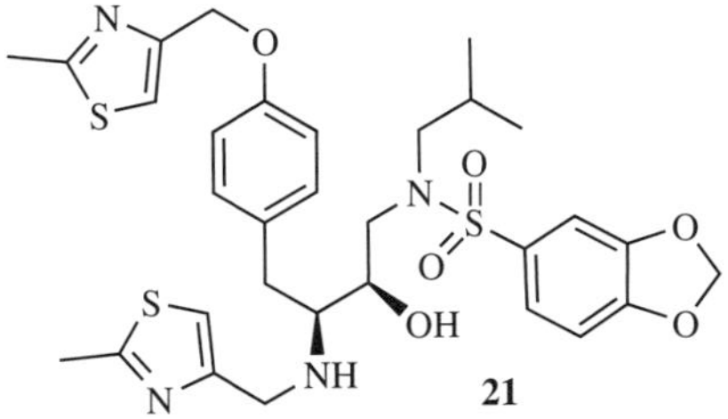 **21**
Step 6	11 19 21 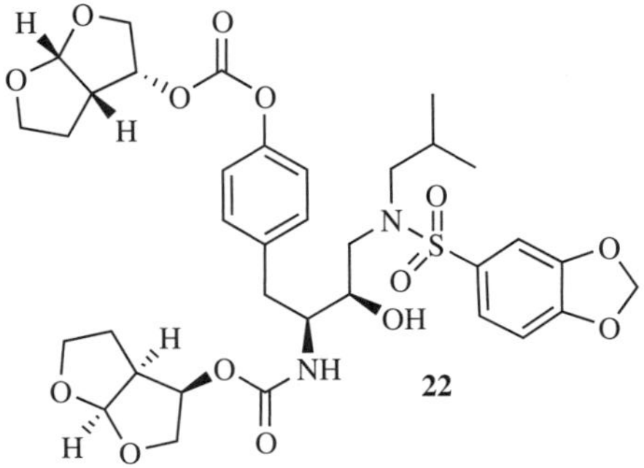 **22**

variability decreased to a 2% range (95–97% AUC), so various levels of phenol **9**, bis-thiazole **18**, and cyclic carbamate **19** would have to be carried through to the next step, a risk that continued to be taken through the life span of Route A. Incidentally, the carbamate moiety in the latter impurity is predominantly formed via cyclization when it is part of phenol **9** followed by *O*- and *N*-alkylation (versus cyclization on product **10** followed by *N*-alkylation), as evidenced by the facile cyclization of the starting material and the long-term stability of the product under the basic alkylation conditions.

Impurities Identified after Step 5

An excess of *p*-toluenesuflonic acid was needed in order to deprotect carbamate **10** to ammonium salt **11**. This excess was almost certainly due in part to the fact that 1 eq of acid was neutralized by the product amine as the reaction progressed. At 2 eq, however, the reaction was very sluggish, perhaps due to salt formation at the thiazole nitrogen. In any event, the preference at this stage would have been to isolate and purify either ammonium salt **11** or its corresponding free base rather than carry cyclic carbamate **19**, amino alcohol **20**, deprotected bis-thiazole **21**, and *p*-toluenesulfonic acid into the final step at appreciable (and variable) levels. However, isolation of the free base appreciably complicated the workup, and ammonium salt **11** tended to crystallize to very small particle size, making filtration (even on a 100-mg scale) extremely slow. In the meantime, an effort was made to use the crude reaction mixture from Step 5 directly in Step 6, an approach that turned out to be successful and that was later modified for use in the final manufacturing route and process.

Impurities Identified after Step 6

Carbonate **6** (Figure 4.2) was known to have limited stability in the presence of water under acidic conditions and also when dissolved in wet triethylamine. It was somewhat surprising, therefore, that the crude reaction mixture from Step 5, containing nearly 3 eq of water from the *p*-toluenesulfonic acid hydrate, could be successfully treated with **6** directly under conditions that involved excess triethylamine at elevated temperature used in Step 6 with a successful outcome. Nevertheless, application of the separately developed Step 6 method to the Step 5 mixture (both utilized acetonitrile as solvent) indicated that very little modification would be needed. With regard to the impurities, intermediate **11**, cyclic carbamate **19**, and bis-thiazole **21** did not react appreciably under the Step 6 conditions and were carried through from Step 5. One new impurity, bis-carbonylated **22** (derived from amine-phenol **20**), was generated. These impurities, together with residual carbonate **6**, the *p*-nitrophenol byproduct, and *p*-toluenesulfonic acid, afforded a relatively messy HPLC chromatogram. After transfer of the reaction mixture into water and isolation of the precipitated product, the levels of *p*-toluenesulfonic acid and *p*-nitrophenol were greatly reduced. Finally, recrystallization from isopropanol led to the final API in greater than 98% AUC. In early development batches all single impurities were maintained at less than 0.5% AUC. Close to this threshold was cyclic carbamate **19** (generated in variable levels in Step 4), and thus controlling this impurity was identified as quality critical.

Conclusion on Managing Impurities in Route A

By keeping close tabs on impurities generated and derivatized throughout the synthesis, impurity profile specifications were consistently met even as process and route changes were implemented. Indeed, when these changes were being considered for introduction, the effect on impurity profile was projected and the risks associated with meeting specifications were clearly assessed. For example, when a supplemental delivery for the first human study was unexpectedly required, it was known that Steps 1 through 4 would need to be executed in their original form, but that it was acceptable to use the more recently developed process for the last step and the final isolation. Another example involved the alternate processes and (eventually) route for carbonate **6** which were introduced based on convenience (since it was known by that time that impurities present earlier in its synthesis did not survive its isolation).

INTERMEDIATE SUPPLY STRATEGY FOR BISFURAN RAC-5

In anticipation of an HIV PI candidate being selected that would contain the bisfuran fragment, development resource was applied early to ensure the availability of a route and process that could provide at least initial kilogram supplies of carbonate **6** (Figure 4.4). Based largely on a literature preparation,[18] the synthesis began with the treatment of dihydrofuran with *N*-bromosuccinimide (NBS) and propargylic alcohol to generate bromoalkyne rac-**23**, which was then subjected to photochemical conditions that induced radical cyclization to afford alkene rac-**24**.[22]

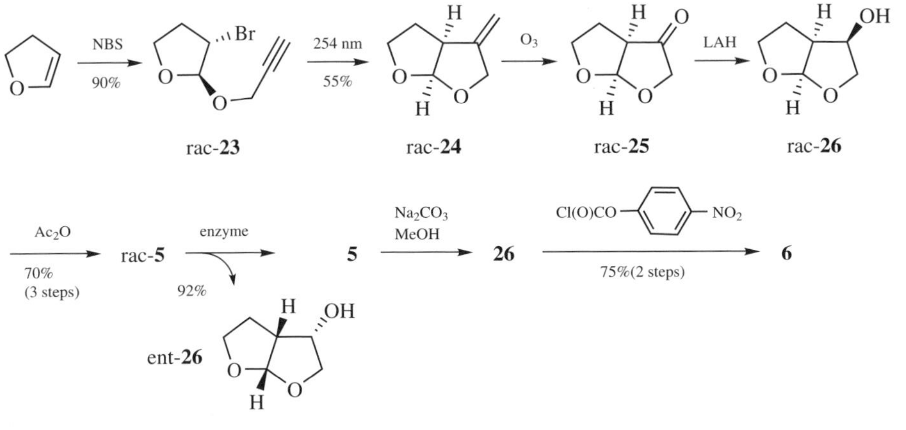

Figure 4.4. Early development syntheses of carbonate **6** via acetate rac-**5**.

Subsequently, ozonolysis was used to generate ketone rac-**25** and reduction from the less hindered face produced alcohol rac-**26**. Resolution of alcohol rac-**26** was performed indirectly starting with acylation to afford acetate rac-**5** and enzymatic hydrolysis of the undesired enantiomer to generate acetate **5** and alcohol ent-**26** (i.e., the enantiomer of the desired alcohol **26**). The deacetylated enantiomer was washed into the aqueous layer, leaving acetate **5** in the organic layer. The resolution was completed by hydrolysis of acetate **5** to alcohol **26**, which was then converted to carbonate **6**.

While other literature procedures (and most conceived routes based on literature precedence) presented problems with the cost of starting materials or involved chemistry that would be problematic at large scale,[11–21] the main concern with the in-house synthesis was the photochemistry. Due in part to the capital investment that would have been required had the photochemical process been taken to manufacturing, it was desired that a supply route be developed that utilized standard manufacture equipment. It should be noted, however, that before a replacement was available, the photochemical process had been used to generate more than 90 kg of alkene rac-**24**.

ALTERNATIVE APPROACHES TARGETING KETONE 25

Direct Acylation Approach

The conversion of ketone rac-**25** to carbonate **6** had been developed into quite an efficient process. Initially, alternative synthetic approaches to carbonate **6** were pursued that would intersect the old route at this intermediate. For example, as indicated in Figure 4.5, there appeared to be an opportunity to directly acylate dihydrofuran to generate 3-acetoxyacetyldihydrofuran (**27**), deacetylate to alcohol **28**,

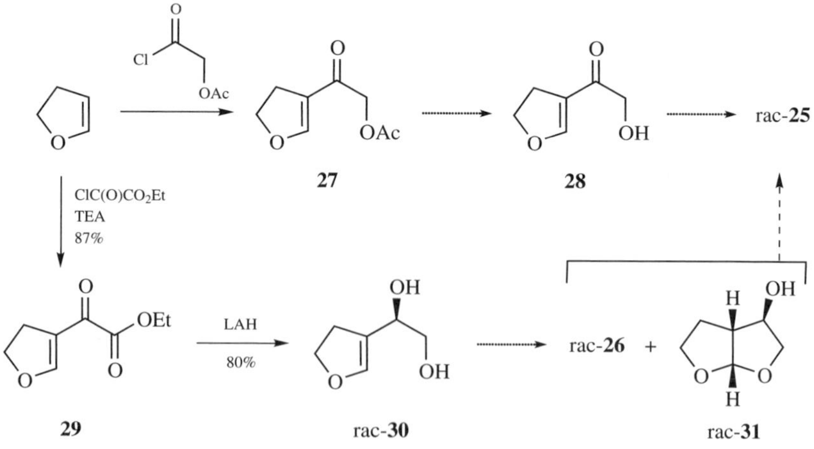

Figure 4.5. Unsuccessful acylation approach to ketone rac-**25**.

and effect ring closure. Indeed, the first step for which there is ample precedence involving highly reactive acid chlorides such as trichloroacetyl chloride and oxalyl chloride[23–26] proceeded smoothly at elevated temperature. It was found, however, that intermediate **28** was inadequately stable to the conditions required for its generation. Furthermore, under standard conditions for acetate hydrolysis or methanolysis, the acetate could not be removed without gross decomposition of the substrate. While it is very likely that a viable route based on this acetylation step could have been identified for small-scale use, it appeared less likely that one could be developed into a robust process suitable for industrial manufacture.

Indirect Acylation Approach

The facile formation of vinylogous ester **27** was encouraging and the use of other acid chlorides that might lead to less direct approaches to ketone rac-**25** were pursued. It was found that chloroacetyl chloride, bromoacetyl bromide, and oxalyl chloride all reacted to generate the desired acetylation product, albeit with similar (and much worse in the latter two cases) instability problems. While acid instability of such products suggested that a base should be used, their base instability or that of the associated acid chloride seemed to preclude such an approach. Ultimately it was found, however, that ethyl chloro(oxo)acetate was compatible with bases (e.g., pyridine or triethylamine) under the conditions necessary to effect the generation of dicarbonyl intermediate **29** (Figure 4.5). Reduction to diol rac-**30** turned out to be straightforward as long as the substrate was added to the reductant and not vice versa. It was reasoned that acid-catalyzed cyclic acetal formation would afford a mixture of rac-**26** and its diastereomeric pair rac-**31**, both of which could be oxidized to ketone rac-**25**. Unfortunately, diol rac-**30** was unstable to the conditions necessary to effect the cyclization step and rearrangement to the terminal aldehyde was a competing pathway.

ALTERNATIVE APPROACH NOT TARGETING KETONE RAC-25

The Key Reaction to the New Route

Efficient formation of the bisfuran ring system was accomplished by oxidative cyclization. The reaction of diol rac-**30** with NBS in wet tetrahydrofuran, afforded a mixture of desired rac-**32** and undesired rac-**33** in a three-to-one ratio (Figure 4.6).

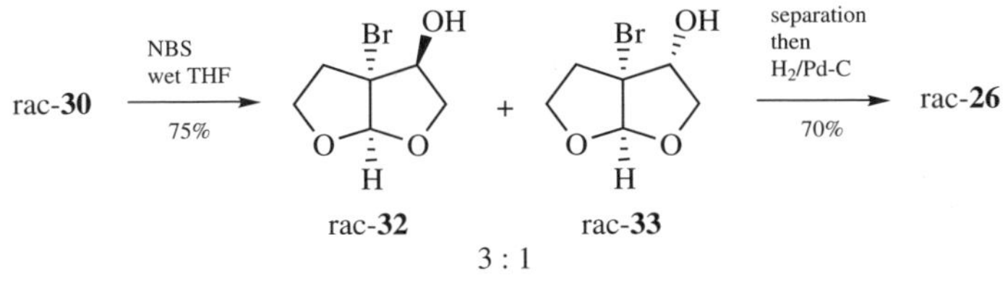

Figure 4.6. Successful oxidative cyclization approach to alcohol rac-**26**.

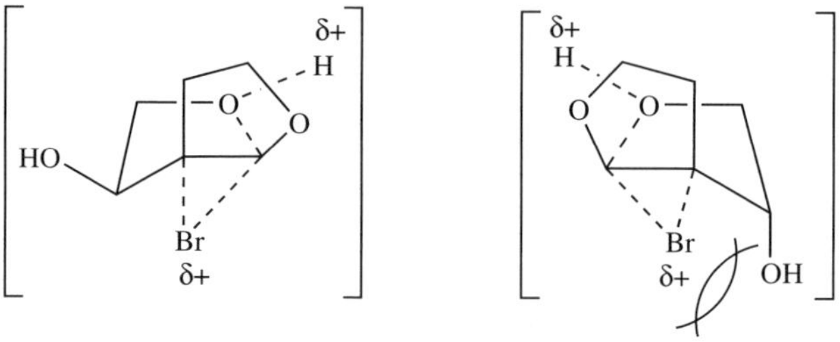

I II

Figure 4.7. Rationale for diastereoselection in the oxidative cyclization of diol **30**.

The diastereomeric purity of bromoalcohol rac-**32** could be easily improved by silica gel chromatography. Alternatively, simple extraction could be used to improve the diastereomeric purity, an approach that was more practical and very easily scaled. Furthermore, rac-**32** is crystalline and diastereomeric mixtures could indeed be enriched by recrystallization, but the recovery was limited by an approximately 1:1 ratio of diastereomers being left in the mother liquor.

The last step needed to demonstrate the viability of the new approach was the conversion of bromoalcohol rac-**32** to alcohol rac-**26**, a conversion that occurs under hydrogenation conditions. This hydrogenation is carried out in the presence of triethylamine to avoid the buildup of hydrobromic acid in the reaction mixture. The resulting triethylammonium bromide precipitates out of solution and coats the catalyst making the kinetics of the reaction variable from run to run. However, addition of water and optimization of the catalyst loading allowed for better control of this remaining issue.

The diastereoselection of the oxidative cyclization step deserves some comment. At first glance, the results may be rationalized by considering the predicted relative stability of transition states **I** and **II** (Figure 4.7).[27] In the former, the cyclic transition state that would lead to the observed major product, the pendant hydroxyl group on the forming five-membered ring is pseudo equatorial. In the alternative, transition state **II**, the hydroxyl group is pseudo axial. It should be noted, however, that when the reaction is run in water only a 1:1 mixture of rac-**32** and rac-**33** results. Furthermore, when wet dichloromethane is used as a solvent a 1:2 ratio is observed (i.e., a diastereomeric reversal). Perhaps the relative stability of transition states **I** and **II** is solvent dependent or an acyclic mechanism is also operative (or both). Alternatively, an S_N1-like process through the oxonium intermediates could be operative.

Four Pathways from the Bromoalcohols to Resolved Alcohol 26

After the viability of the three-step route to bromoalcohol rac-**32** had been demonstrated and hydrogenation was shown to generate debromonated bisfuran derivatives, four main pathways were conceived to generate resolved bisfuran alcohol **26** (Figure 4.8). Remarkably, all of them proved to be viable options.[28] In Pathway 1, reductive debromination of the bromoalcohol mixture and acetylation gave a

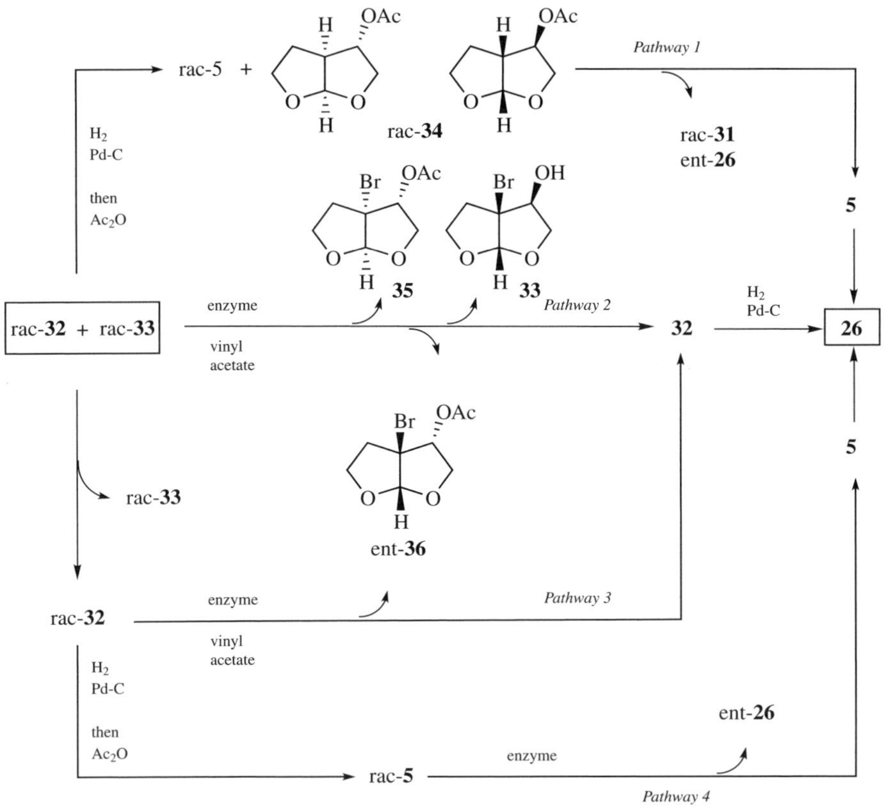

Figure 4.8. Four pathways pursued for converting bromoalcohol mixture to resolved alcohol **26**.

mixture of acetate rac-**5** and its diastereomeric pair rac-**34**. When this mixture is subjected to the enzymatic deacetylation conditions, the slowest acetate to hydrolyze is acetate **5** and the other (hydrolyzed) diastereomers may be washed into an aqueous layer. This constitutes a new route to resolved alcohol **26**.

In Pathway 2, the diastereomeric mixture is subjected directly to enzymatic acetylation conditions and bromoalcohol **32** remains unchanged while its enantiomer is acelyated (to acetate ent-**36**) and one of its diastereomers is acylated (to acetate **35**). The two bromoacetates and then diastereomeric bromoalcohol **33** can be removed by extraction, producing resolved **32** and another formal route to resolved alcohol **26**.

Pathway 3 reduces the complexity of the post-resolution extraction by including extractive purification of rac-**32** first. Thus, purified bromoalcohol rac-**32** is subjected to the enzyme acetylation conditions and extractive separation results in the unchanged **32** separate from its acylated enantiomer ent-**36** and a third route.

Finally, Pathway 4 intersects the earlier supply route through extractive purification of bromoalcohol rac-**32**, reduction of the bromide, and acetylation to generate

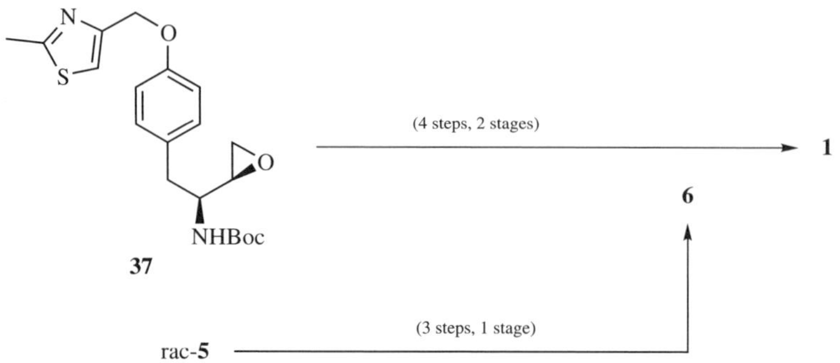

Figure 4.9. Outline of Route B (manufacture route).

acetate rac-**5**. As already mentioned, the conversion from this intermediate to carbonate **6** is quite efficient. Indeed, this conversion is executed as a single stage.

THE MANUFACTURE ROUTE

The cost associated with the protection-deprotection sequence in Route A was removed by starting with an already-derivatized epoxide (**37**, Figure 4.9). The added expense to the incorporation of this appendage (rather than the benzyl protecting group) was reduced to the cost difference between thiazole **4** and benzyl chloride. The solubility of intermediates was quite different through this modified sequence to compound **1**, and a variety of changes were required in order to define a manufacturing process.[29] Ultimately, the four chemical conversions required to convert epoxide **37** to GW640385X (**1**) were reduced to two stages (i.e., one intermediate isolation) in approximately 80% yield on 100-kg scale. The process was also developed without the need for an extra recrystallization of the final API. With regard to the bisfuran unit, it was acetate rac-**5** derived from the bromoalcohol route that was deemed to be the preferred manner in which to purchase this unit, and the extractive differentiation of the bromoalcohol diastereomers rac-**32** and rac-**33** was the preferred method (Pathway 4, Figure 4.9). As such, the overall yield from ethyl chloro(oxo)acetate was approximately 28% on 100-kg scale. The projected cost differential between the various possibilities was small, and selecting rac-**5** as a starting material had the added advantage that the resolution step—critical from a regulatory standpoint—would be performed in-house.

CONCLUSION

Ultimately, what started as a very long synthesis was developed to a simpler three-stage process executed internally with secured sources for strategically selected advanced intermediates as starting materials. While these advanced intermediates or their supply routes were modified from the first 1-kg delivery to the first 100-kg delivery, the quality of the final API was controlled by a detailed understanding of the factors that led to a given impurity profile.

ACKNOWLEDGMENTS

The authors would like to thank Bob Cooley, Roy Flanagan, Kerry Gombatz, Jiasheng Guo, Bill Hinkley, Mary Jackson, Purnima Narang, Tyler Sharkey, and Elizabeth Steere for their supporting contributions and useful discussions.

REFERENCES AND NOTES

1. Turner, D., Schapiro, J. M., Brenner, B. G., and Wainberg, M. A. *Antiviral Therapy* **2004**, *9*, 301–314.
2. Sherrill, R. G., Andrews, C. W., Bock, W. J., Davis-Ward, R. G., Furfine, E. S., Hazen, R. J., Rutkowske, R. D., Spaltenstein, A., and Wright, L. L. *Bioorg. Med. Chem. Lett.* **2005**, *15*, 81–84.
3. Hanlon, M. H., Porter, D. J. T., Furfine, E. S., Spaltenstein, A., Carter, H. L., Danger, D., Shu, A. Y. L., Kaldor, I. W., Miller, J. F., and Samano, V. A. *Biochemistry* **2004**, *43*, 14500–14507.
4. Kazmierski, W. M., Andrews, W., Furfine, E., Spaltenstein, A., and Wright, L. *Bioorg. Med. Chem. Lett.* **2004**, *14,* 5689–5692.
5. Miller, J. F., Furfine, E. S., Hanlon, M. H., Hazen, R. J., Ray, J. A., Robinson, L., Samano, V., and Spaltenstein, A. *Bioorg. Med. Chem. Lett.* **2004**, *14*, 959–963.
6. Kazmierski, W. M., Furfine, E., Spaltenstein, A., and Wright, L. L. *Bioorg. Med. Chem. Lett.* **2002**, *12*, 3431–3433.
7. Jain, R. G., and Lenhard, J. M. *J. Biol. Chem.* **2002**, *277*, 19247–19250.
8. Symonds, W., Cutrell, A., Edwards, M., Steel, H., Spreen, B., Powell, G., McGuirk, S., and Hetherington, S. *Clinical Therapeutics* **2002**, *24*, 565–573.
9. Sherrill, R. G., Hale, M. R., Spaltenstein, A., Furfine, E. S., Andrews, C. W., III, and Lowen, G. T. PCT Int. Appl. WO 9965870, 1999 (to Vertex Pharmaceuticals Inc., USA).
10. The subject of this chapter was presented in part at the New York City American Chemical Society national meeting in September, 2003: Roberts, J. Abstracts of Papers, 226th ACS National Meeting, New York, September 7–11, 2003.
11. Ghosh, A. K., Leshchenko, S., and Noetzel, M. *J. Org. Chem.* **2004**, *69*, 7822–7829.
12. Ghosh, A. K., Leshchenko, S., and Noetzel, M. W. PCT Int. Appl. WO 2004033462, 2004 (to The Board of Trustees of the University of Illinois).
13. Kesteleyn, B. R. R., Surleraux, D. L. N. G. PCT Int. Appl. WO 2003022853, 2003 (to Tibotec Pharmaceuticals Ltd., Ireland).
14. Ikemoto, T., and Piao, D. PCT Int. Appl. W O2004060895, 2004 (to Sumika Fine Chemicals Co., Ltd., Japan).
15. Doan, B. D., Davis, R. D., Lovelace, T. C. PCT Int. Appl. W O2003024974, 2003 (to SmithKline Beecham Corporation, USA).

16. Davis, R., and Lovelace, T. C. PCT Int. Appl. WO 2002060905, 2002 (to Glaxo Group Limited, UK).
17. Uchiyama, M., Hirai, M., Nagata, M., Katoh, R., Ogawa, R., Ohta, A. *Tetrahedron Lett.* **2001**, *42*, 4653–4656.
18. Ghosh, A. K., and Chen, Y. *Tetrahedron Lett.* **1995**, *36*, 505–508.
19. Ghosh, A. K., Thompson, W. J., Fitzgerald, P. M. D., Culberson, C., Axel, M. G., McKee, S. P., Huff, J. R., and Anderson, P. S. *J. Med. Chem.* **1994**, *37*, 2506–2508.
20. Ghosh, A. K., Thompson, W. J., and Mckee, S. P. PCT Int. Appl. WO 9426749, 1994 (to Merck and Co., Inc., USA).
21. Pezechk, M., Brunetiere, A. P., and Lallemand, J. Y. *Tetrahedron Lett.* **1986**, 27, 3715–3718.
22. Cossy, J., Ranaivosata, and J.-L., Bellosta, V. *Tetrahedron Lett.* **1994**, *35*, 8161–8162.
23. Moriguchi, T., Endo, T., and Takata, T. *J. Org. Chem.* **1995**, *60*, 3523–3528.
24. Effenberger, F., and Barthelmess, I. *J. Het. Chem.* **1995**, *32*, 599–602.
25. Hojo, M., Masuda, R., Sakaguchi, S., and Takagawa, M. *Synthesis* **1986**, *12*, 1016–1017.
26. Effenberger, F., Maier, R., Schoenwaelder, K. H., and Ziegler, T. *Chem. Ber.* **1982**, *115*, 2766–2782.
27. Dumeunier, R., Leclercq, C., and Marko, I. E. *Tetrahedron Lett.* **2002**, *43*, 2307.
28. Doan, B. D., Patterson, D. E., and Roberts, J. C. PCT Int. Appl. WO 2004002975, 2004 (to SmithKline Beecham Corporation, USA).
29. Martin, M. T., Roberts, J. C., and Toczko, J. F. PCT Int. Appl. WO 2005000249, 2005 (to SmithKline Beecham Corporation, USA).

5

DEVELOPMENT OF AN ASYMMETRIC SYNTHESIS OF ABT-100

ALBERT W. KRUGER, MICHAEL J. ROZEMA, BRIDGET D. ROHDE, BHADRA SHELAT, LAKSHMI BHAGAVATULA, JAMES J. TIEN, WEIJIANG ZHANG, AND RODGER F. HENRY

GPRD Process Research and Development, Department R450, Abbott Laboratories, North Chicago, IL 60064

INTRODUCTION

Mutation of the *ras*-oncogene regulating cell growth and proliferation is implicated in up to 25% of human cancers.[1,2] After transcription of the *ras*-protein, further activation via farnesylation is required before oncogenic activity can be exhibited. To exploit this requirement, inhibitors of farnesyl transferase (FT) have been targeted as potential chemotherapeutic agents. ABT-100, **1**,[3] has been identified as an FT inhibitor possessing excellent potency, bioavailability, and pharmacokinetics. Herein, we disclose the development of a process to deliver the first kilogram of ABT-100 to support further biological and clinical evaluation. Key to our success was the development of a method for the generation of the chiral tertiary alcohol bearing a heterocyclic substituent (Figure 5.1).

RESULTS AND DISCUSSION

Preclinical evaluation of ABT-100 was done almost exclusively on racemic material. By optimization of the Discovery route, a convenient three-step preparation of

Fundamentals of Early Clinical Drug Development: From Synthesis Design to Formulation, edited by Ahmed F. Abdel-Magid and Stéphane Caron

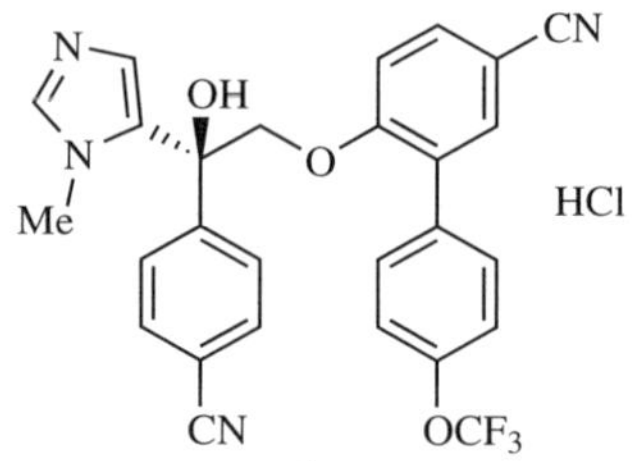

1

Figure 5.1. Structure of ABT-100.

racemic material was developed (Figure 5.2). Suzuki coupling of **2** and **3** allowed quick access to biaryl **4** in 87% yield. After S_N2 displacement to prepare **6**, the imidazoyl-Grignard reagent **7** added smoothly to give **rac-1** in 70% overall yield. This short sequence was rapidly amenable to scale-up. No chromatography or cryogenic conditions were used. In addition, kilogram quantities of the starting materials were available for use.

As ABT-100 proceeded to development, small amounts of the purified enantiomers became available via separation by HPLC on a chiral stationary phase. Not surprisingly, the biological activity resided exclusively in one enantiomer. Due to the limited solubility of **rac-1** in typical HPLC mobile phases, the scalability of a preparative HPLC separation was severely limited. So the goal of the program from the process chemistry perspective rapidly changed to development of an enantioselective preparation of **1**.

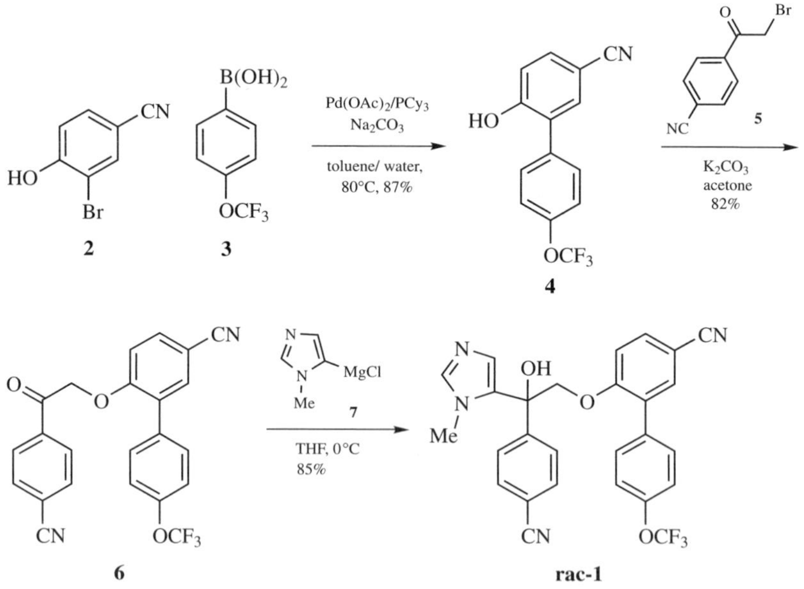

Figure 5.2. Enabling route to **rac-1**.

Given the easy and scalable enabling route to racemic material, a large effort was untaken to examine classical separation of **rac-1** using chiral acids. Despite the effort, no promising leads were found to warrant further investigation. Enzymatic methods to separate the stereoisomers also met with no success.

EVALUATION OF ASYMMETRIC ROUTES

Disconnection of the aryl ether linkage in **1** (via S_NAr)[4] reduces the synthetic problem to the preparation of **8** and **9**. Biaryl **8** can reasonably be assembled by a Suzuki protocol. Diol **9** could come from the diastereoselective addition of an imidazolyl organometallic to a ketone of the type **10** or **11** with an appropriate chiral auxiliary[5,6] (Figure 5.3).

The diastereoselective addition of organometallics to α-keto esters[5] and to a lesser extent α-keto ethers[6] bearing chiral auxiliaries constitutes an efficient method for the stereoselective synthesis of tertiary alcohols. In 1904, Mackenzie was the first to add Grignard reagents to chiral α-keto esters and recover optically active acids (after hydrolysis).[7] It would be several decades before Prelog used this and other observations to formulate what we now call Prelog's Rule[8] for the addition of nucleophiles to chiral α-keto esters. Several types of auxiliaries have been reported for this reaction, such as menthol,[5a,b] 8-phenyl-menthol,[5c] *trans*-2-substituted cyclohexanols,[5d,e] carbohydrate,[5f] and amino-indanol[5g]-based derivatives. Use of menthol as the chiral auxiliary provides many advantages: low cost, high

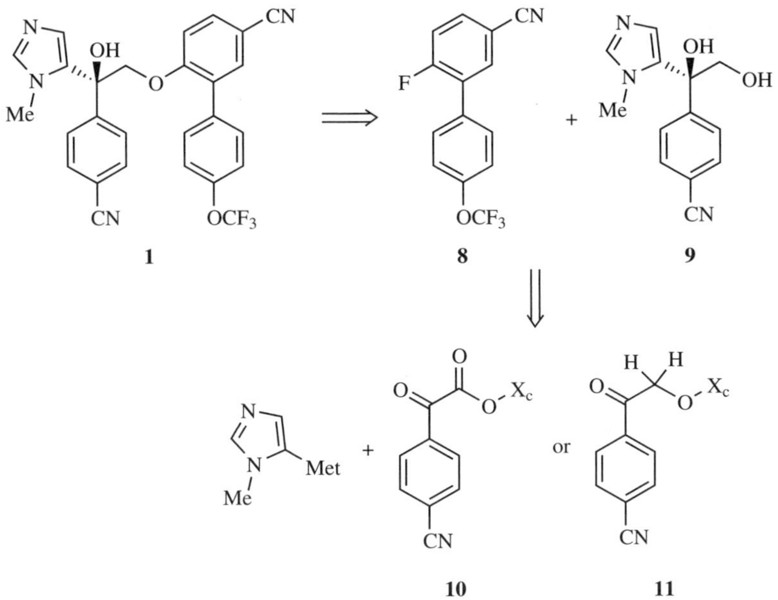

Figure 5.3. Retrosynthesis.

availability, good selectivity (*vide infra*), and a tendency for intermediates to be crystalline. Menthol is also readily available in both enantiomeric forms. The following discussion details the preparation of diol **9**.

PREPARATION OF DIOL 9 COUPLING PARTNER

Starting from commercially available ethyl (4-cyanophenyl)oxoacetate **12** (Figure 5.4), the (−)-menthyl ester was prepared using a modification[9] of the titanium (IV) alkoxide-catalyzed transesterification procedure developed by Seebach.[10] Thus, treatment of **12** with (−)-menthol (1.5 eq) in the presence of a catalytic amount of titanium (IV) ethoxide (15 mol%) in xylene at 80°C provided (−)-menthyl ester **13** in 79% yield. The reaction was pushed to completion by distillation of the ethanol generated by passing a stream of nitrogen through the reaction vessel. Unfortunately, using fewer than 1.5 eq of menthol led to incomplete conversion, presumably because the ethoxide ligands on the catalyst were exchanged for menthol during the reaction. Even though menthyl ester **13** is crystalline, it was difficult to separate the product from the excess (−)-menthol without resorting to a chromatographic purification. However, it was discovered that the remaining menthol could be rendered inert in the next step by *in situ* protection as an acetate.

Initially, the excess menthol was acetylated by the addition of DMAP and Ac_2O directly to the reaction mixture after the transesterification was complete. However, as the scale of the reaction was increased, a new product **14** was observed in up to 30% yield. A series of control experiments indicated the root cause for the appearance of **14** was the combination of DMAP, Lewis acid (Ti(IV)), and Ac_2O. This combination leads to ketene formation, which undergoes a [2 + 2] cycloaddition

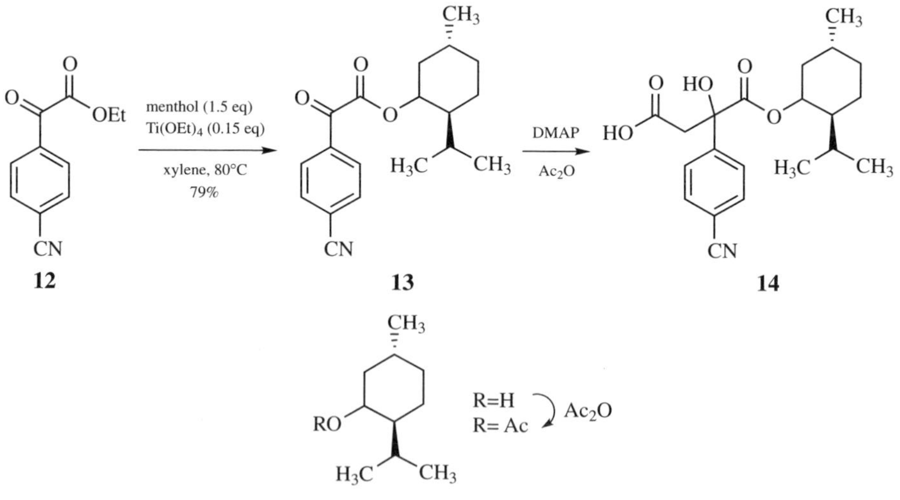

Figure 5.4. Preparation of (−)-menthyl ester **13**.

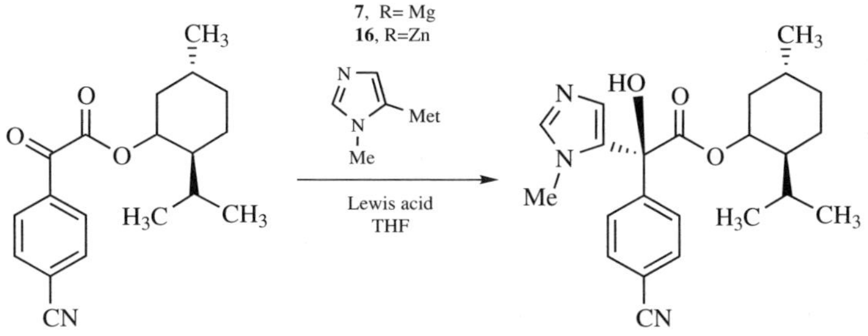

13 15

Figure 5.5. Diastereoselective addition to (−)-menthyl α-keto ester **13**.

with **13** followed by ring opening upon aqueous quench to form **14**. In the absence of DMAP, ketene formation is slow. Ultimately, it was discovered that simply adding Ac_2O to the reaction mixture would effectively acetylate the excess menthol and completely suppress the production of **14**. After an aqueous workup to remove the Ti residues, the mixture of menthyl ester **13** and menthyl acetate is carried into the next reaction as a solution in toluene.

Initial studies into the diastereoselective addition to (−)-menthyl α-keto ester **13** employed Grignard reagent **7**, prepared by magnesium-iodide exchange using ethylmagnesium chloride,[11] and resulted in **15** in a 2.3:1 diastereomeric ratio (dr) (Figure 5.5). Pretreatment of **13** with magnesium bromide etherate at low temperature (−40°C) afforded slightly better selectivity (dr 4:1). Boireau's precedent[5a,b] suggested that pretreatment of **13** with zinc chloride would give enhanced diastereoselection. Indeed these conditions gave a 10:1 mixture of diastereomers; however, low conversion was observed due to a large volume of gelatinous precipitates formed (Table 5.1).

The organozinc reagent **16** proved to be more selective and amenable to scale-up. In the absence of a Lewis acid, **16** does not react with the α-keto ester **13**. Of the handful of Lewis acids screened ($ZnCl_2$, $Zn(OTf)_2$, $BF_3 \cdot OEt_2$, $Ti(OEt)_4$, $MgBr_2 \cdot OEt_2$), only $MgBr_2 \cdot OEt_2$ effected the addition of the imidazolyl moiety to the ketone. Thus, in the presence of $MgBr_2 \cdot OEt_2$ in THF at 0°C, the tertiary alcohol **15** was generated with a diastereomeric ratio of 10–11:1 and in 85% yield.

Table 5.1.

Metal	Diastereomer Ratio (dr)	Lewis Acid	Conversion (%)	Temperature (°C)
Mg	2.3:1	—	>95	−40
Mg	4:1	$MgBr_2$	>95	−40
Zn	NA[a]	—	0	25
Zn	19:1	$MgBr_2$	<50	−40 (gel)
Zn	10:1	$MgBr_2$	>95	0
Zn	8:1	$MgBr_2$	>95	50

[a]NA, not available.

The selectivity of the addition under these conditions is not greatly affected by temperature. Even at elevated temperatures (50°C), the selectivity decreases only slightly to 7.5:1. At lower temperatures (−40°C), the reaction becomes gelatinous and difficult to stir and any increase in selectivity is negated by lower conversions.

Zinc reagent **16** was conveniently prepared using the procedure of Knochel[12] [Equation (5.1)]. On laboratory scale (<50 g), the reaction proved very reproducible; however, on larger scale, it was important to maintain adequate stirring and to control the addition rate of imidazolyl iodide **17** to the activated zinc to achieve good conversion to **16**. If the concentration of **17** becomes too high during the insertion reaction, a precipitate coats the surface of the zinc metal and conversion to the imidazolzinc reagent ceases. It was shown in laboratory experiments that in THF the addition of **17** to **16** forms an insoluble substance, the exact nature of which was not determined, but it is assumed to be a coordination complex between the iodoimidazole nitrogen and imidazolylzinc reagent.[13] Not surprisingly, it was also found that the stirring rate is important for this heterogeneous reaction. Slow stir rates in which the zinc metal settles to the bottom of the reactor increase the probability of the reaction stalling. Careful control of the addition rate of **17** and adequate stirring of the heterogeneous reaction mixture led to an efficient and reproducible reaction.

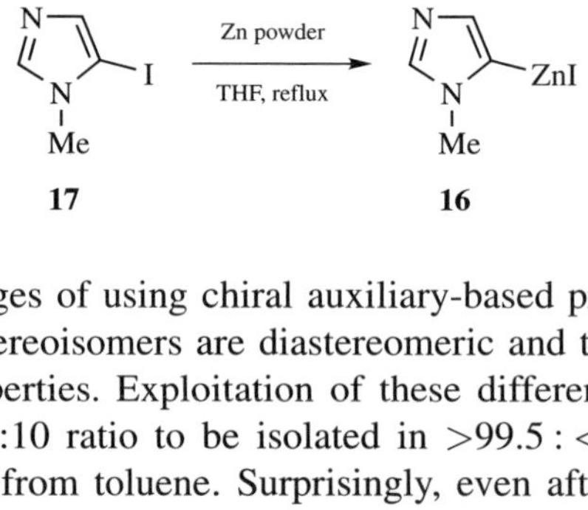

17 **16**

(5.1)

One of the advantages of using chiral auxiliary-based processes in asymmetric synthesis is that the stereoisomers are diastereomeric and therefore typically have different physical properties. Exploitation of these differences allowed the crude extract posessing a 90:10 ratio to be isolated in >99.5 : <0.5 diastereomer ratio (dr) by crystallization from toluene. Surprisingly, even after an aqueous workup, the product is isolated as a 2:1 complex **15-Zn** with zinc chloride. The exact nature of the zinc complex was established by single-crystal X-ray analysis of a sample crystallized from acetonitrile (Figure 5.6). The complex was crucial for effective purification of the desired diastereomer. In zinc free material, the purification of a 10:1 mix of diastereomers required several crystallizations to reach the 99:1 dr level and significant yield losses were unavoidable.

In order to proceed to the reduction of the ester, the product was liberated from the zinc complex. This was most effectively accomplished by treatment of the complex with the disodium salt of EDTA. Overall, tertiary alcohol **15** was obtained in 70–75% overall yield and with a diastereomeric ratio of >99.8:0.2.

Ester **15** was selectively reduced to diol **9** by reaction with $NaBH_4$ in methanolic THF at 50°C (Figure 5.7). Under these conditions, concomitant reduction of the

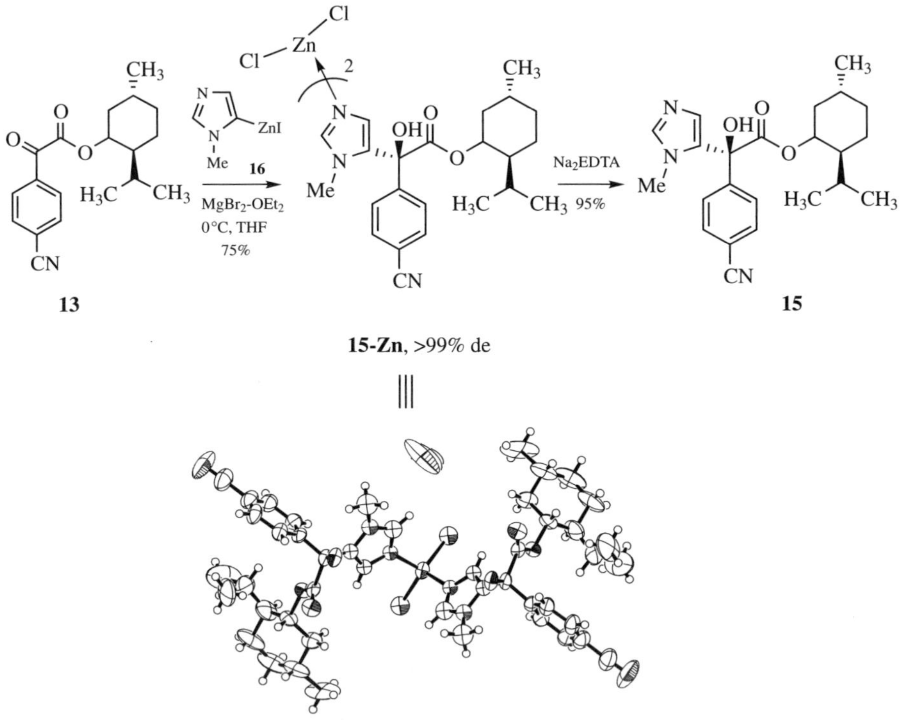

Figure 5.6.

nitrile was not observed. The only detected impurity was a trace amount (<0.5%) of the nitrile methanolysis product **18** which was not rejected on isolation. Nitrile reduction to the amine becomes more competitive when stronger reducing agents are employed ($NaBH_4$/HOAc or $LiBH_4$). Aluminum-based reagents gave predominantly aldehyde, from incomplete ester reduction, under a variety of

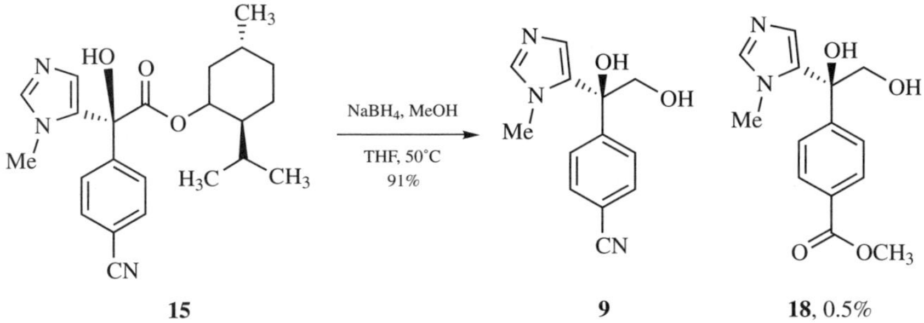

Figure 5.7. Selective reduction of ester **15** to diol **9**.

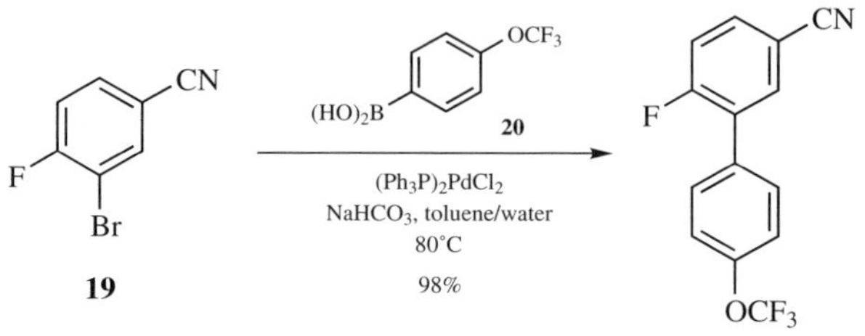

Figure 5.8. Preparation of biphenyl **8**.

conditions. Overall the $NaBH_4$-based procedure gave **9** in over 90% yield and excellent purity.

PREPARATION OF THE BIARYL-FLUORIDE COUPLING PARTNER

The biphenyl subunit of **1** was assembled from boronic acid **20** and bromofluorobenzonitrile **19** (Figure 5.8) using a Suzuki protocol.[14] While as little as 0.025 mol% $(Ph_3P)_2PdCl_2$ effects complete conversion in 15 hr, to obtain reasonable reaction times (<6 hr), a larger amount of catalyst (0.06 mol%) was typically used. Sodium bicarbonate was used as the base in toluene/water mixtures under an inert atmosphere. Biphenyl **8**[15] was consistently produced in 98% yield and excellent purity.

FINAL ASSEMBLY OF ABT-100

With the two coupling partners in hand, the aryl ether formation was examined. The S_NAr reaction could be accomplished using a variety of bases (LiHMDS, NaHMDS, KHMDS, KOtBu, KOH) in polar aprotic solvents (DMF, DMSO). The reaction is conveniently run with milled KOH in THF/DMSO at low temperature (<15°C) (Figure 5.9). The choice of base and stoichiometry influences not only the reaction rate but also the impurity profile. The alkylated diol **1** can react with the excess base to form epoxide **21** and phenol **22**. This degradation is more prevalent with the potassium bases. However, using lithium bases results in slow reactions and incomplete conversions. A balance of a reasonable conversion with acceptable levels of decomposition can be achieved by running the reaction at temperatures below 15°C. With KOH, trace amounts of nitrile hydrolysis products as well as dialkylated product **23** are also seen. These impurities are rejected in the precipitation of the freebase by the addition of methanol and by final salt formation.

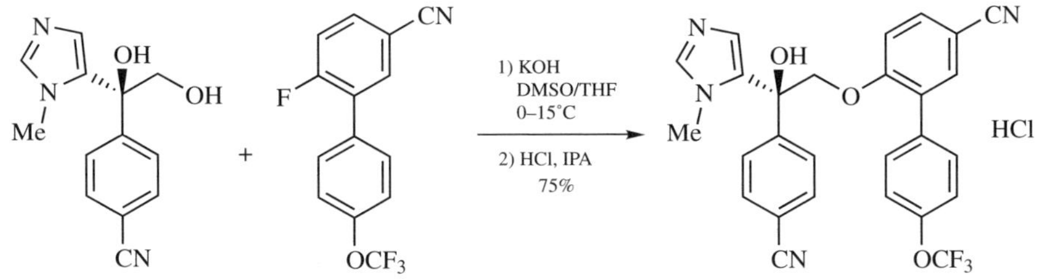

9 8 1

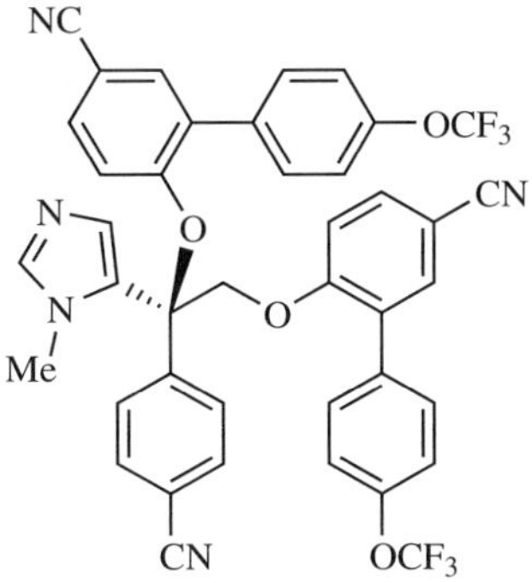

23

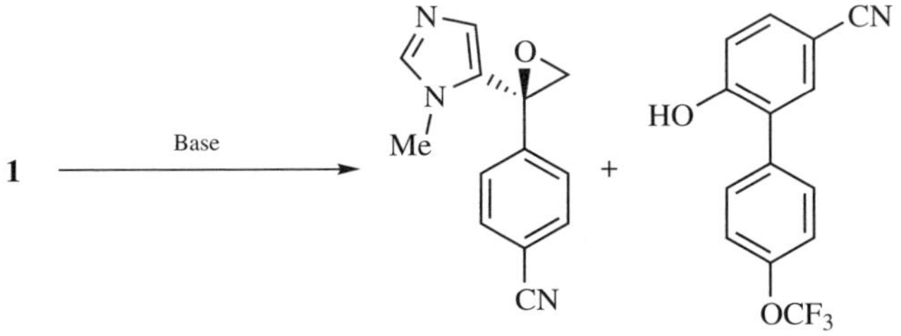

21 22

Figure 5.9. Preparation of epoxide **21** and phenol **22**.

The final isolation of ABT-100, **1** consists of dissolving the unpurified freebase in hot isopropanol, filtering, and converting to the HCl salt by the addition of aqueous HCl. The final product can be recrystallized from EtOH to produce larger particles with better handling properties.

In summary, we have developed a stereoselective and scalable synthesis of ABT-100, **1**, that produced material of >99% ee and in an overall yield of 37% in five steps on kilogram scale. The process is highlighted by a diastereoselective addition of an imidazolylzinc reagent to an α-ketoester to produce the stereogenic tertiary alcohol and the formation of the biaryl ether through an S_NAr reaction. In addition, an efficient Suzuki reaction to produce the biaryl moiety and a selective ester reduction to produce the requisite diol were also developed.

ACKNOWLEDGMENT

We thank Cathie Linton, Eric Kristensen, and Mike Moore for analytical support. We also thank Mike Fitzgerald for the preparative separation of the enantiomers of the ABT-100, and we acknowledge Howard Morton for his helpful discussions.

REFERENCES

1. Bos, J. L. In *Molecular Genetics in Cancer Diagnosis*, Cossman, J., Ed., Elsevier Scientific, New York, 1990, p. 273.
2. Prendergast, G. C., and Rane, N. *Expert Opin. Invest. Drugs* **2001**, *10*, 2105.
3. Claiborne, A. K., Gwaltney, S. L., II, Hasvold, L. A., Li, Q., Li, T., Lin, N-H., Mantei, R. A., Rockway, T. W., Sham, H. L., Sullivan, G. M., Tong, Y., Wang, G., Wang, L., Wang, X., and Wang, W.-B. Patent Appl. US2003/0087940 A1, May 8, **2003**.
4. For recent examples of aryl ether synthesis using the S_NAr reaction on electron deficient fluoro aromatics see: (a) Temal-Laib, T., Chastanet, J., and Zhu, J. *J. Am. Chem. Soc.* **2002**, *124*, 583. (b) Stark, H., Sadek, B., Krause, M., Hüls, A., Ligneau, X., Ganellin, C. R., Arrang, J.-M., Schwartz, J.-C., and Schunack, W. *J. Med. Chem.* **2000**, *43*, 3987–3994. (c) East, S. P., and Joullié, M. M. *Tetrahedron Lett.* **1998**, *39*, 7211. (d) Wells, K. M., Shi, Y.-J., Lynch, J. E., Humphrey, G. R., Volante, R. P., and Reider, P. J. *Tetrahedron Lett.* **1996**, *37*, 6439. For general reviews on the S_NAr reaction, see Zoltewicz, J. A. *Top. Curr. Chem.* **1975**, *59*, 33–64 and Bunnett, J. F., and Zahler, R. E. *Chem. Rev.* **1951**, *49*, 273.
5. (a) Boireau, G., Deberly, A., and Abenhaïm, D. *Tetrahedron Lett.* **1988**, *29*, 2175. (b) Boireau, G., Deberly, A., and Abenhaïm, D. *Tetrahedron* **1989**, *45*, 5837. (c) Whitesell, J. K., Deyo, D., and Bhattacharya, A. *J. Chem. Soc., Chem. Commun.* **1983**, 802. (d) Basavaiah, D., and Bharathi, T. K. *Tetrahedron Lett.* **1991**, *32*, 3417. (e) Basavaiah, D.; Krishna, P. R. *Tetrahedron* **1995**, *51*, 12169. (f) Loupy, A., and Monteux, D. A. *Tetrahedron* **2002**, *58*, 1541. (g) Senanayake, C. H., Fang, K., Grover, P., Bakale, R. P., Vandenbossche, C. P., and Wald, S. A. *Tetrahedron Lett.* **1999**, *40*, 819.
6. Charette, A. B., Benslimane, A. F., and Mellon, C. *Tetrahedron* **1995**, *36*, 8557.
7. McKenzie, A. *J. Chem. Soc.* **1904**, *85*, 1249.
8. Prelog, V. *Helv. Chim. Acta* **1953**, *36*, 308.
9. Krasik, P. *Tetrahedron Lett.* **1998**, *39*, 4223.
10. Seebach, D., Hungerbuehler, E., Naef, R., Shnurrenberger, P., Weidman, B., and Zueger, M. *Synthesis*, **1982**, 138.
11. (a) Holden, K. G., Mattson, M. N., Cha, K. H., and Rapoport, H. *J. Org. Chem.*, **2002**, *67*, 5913. (b) Panosyan, F. B., and Still, I. W. J. *Can. J. Chem.*, **2001**, *79*, 1110. (c) Burm, B. E. A., Blokker, P., Jongmans, E., van Kampen, E., Wanner, M. J., and Koomen, G-J. *Heterocycles*, **2001**, *55*, 495.

12. (a) Yeh, M. C. P., and Knochel, P. *Tetrahedron Lett.* **1988**, *29*, 2395. (b) Knochel, P., Yeh, M. C. P., Berk, S. C., and Talbert, J. *J. Org. Chem.* **1988**, *53*, 2390.
13. *Comprehensive Coordination Chemistry*, Vol. 5, Wilkinson, G., Ed., Pergamon Press, Oxford, **1987**, p. 925.
14. For reviews, see: (a) Miyaura, N., Suzuki, A. *Chem. Rev.* **1995**, *95*, 2457. (b) Suzuki, A. *J. Organomet. Chem.* **1999**, *576*, 147.
15. Biphenyl **4** is referenced in a Japanese patent with no procedure on its synthesis or characterization. Ikeda, A., Ozaki, M., Honami, R., Yumita, T., Yano, H., Nakano, Y, Kurihara, Y., and Hirano, T. Preparation of triazole derivatives as insecticides and acaricides PCT *Int. Appl.* (**1994**).

6

ASYMMETRIC HYDROGENATION: A NEW ROUTE TO PREGABALIN

Rex Jennings, William S. Kissel, Tung V. Le, Edward Lenoir, Thomas Mulhern, and Robert Wade

Chemical Research and Development, Pfizer Global Research and Development, Ann Arbor, MI 48105

INTRODUCTION

Pregabalin (**1**, Figure 6.1) is a new drug candidate being developed by Pfizer as an $\alpha_2\delta$ ligand with activity as a potent antiepileptic. Pregabalin is also being pursued against neuropathic pain for diabetic peripheral neuropathy and postherpetic neuralgia. Finally pregabalin is also being pursued against anxiety. Pfizer filed applications for pregabalin in both the United States and European Union in 2003. This chapter details work toward a second-generation synthetic process to pregabalin.

THE INITIAL MANUFACTURING ROUTE

The asymmetric hydrogenation route to pregabalin comes as a next-generation follow-up to the current manufacturing route, referred to as the malonate route (Figure 6.2). The malonate route has been used to produce drug substance through development for clinical studies and into manufacturing.[1] This route proceeds via a Knoevenagel condensation of two inexpensive commodity chemicals, isovaleraldehyde and diethylmalonate, using the combination of an amine (in most cases di-*n*-propyl amine) and acetic acid as catalyst. The resultant unsaturated diester **2** undergoes

Fundamentals of Early Clinical Drug Development: From Synthesis Design to Formulation, edited by Ahmed F. Abdel-Magid and Stéphane Caron

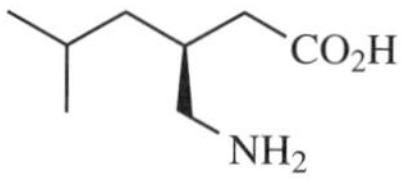

1, pregabalin

Figure 6.1. Structure of pregabalin.

conjugate addition of cyanide using either potassium cyanide or acetone cyanohydrin to give cyanodiester **3**. Subjecting cyanodiester **3** to a sequence of hydrolysis of one of the esters, decarboxylation, and hydrolysis of the second ester provides the intermediate cyanocarboylate **4**, which is hydrogenated over sponge nickel catalyst to give racemic **5**. This short sequence gives the pregabalin framework very rapidly. Unfortunately, pregabalin [the (*S*)-enantiomer] must now be separated from the undesired (*R*)-enantiomer via a classical resolution using mandelic acid as the resolving agent. Treatment of racemate **5** with (*S*)-mandelic acid provides the (*S,S*)-pregabalin mandelate salt **6** as a solid which is isolated, leaving the majority of the (*R*)-enantiomer in the filtrates. Simply heating the salt in a solvent mixture containing water and tetrahydrofuran splits the salt. Isolation of the resultant crude pregabalin solid separates it from the mandelic acid, which remains in the mother liquors where it can be recycled. Recrystallization of the crude pregabalin affords the desired crystal properties and purity.

The malonate route is a very effective process in that it starts with inexpensive, readily available starting materials and uses straightforward and robust chemistry to rapidly form pregabalin. It has proven quite reliable over a large number of lots and a variety of scales, and the process provided multi-metric-ton quantities of pregabalin to foster the candidate through clinical studies and into production. However,

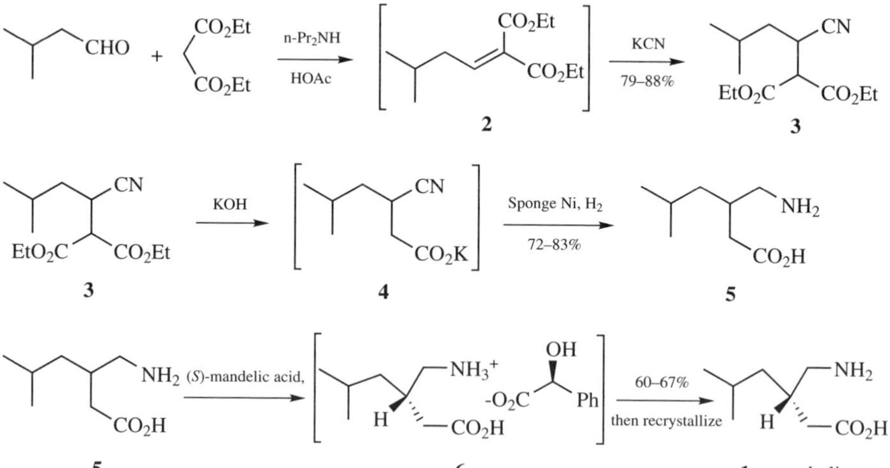

Figure 6.2. The malonate route to pregabalin.

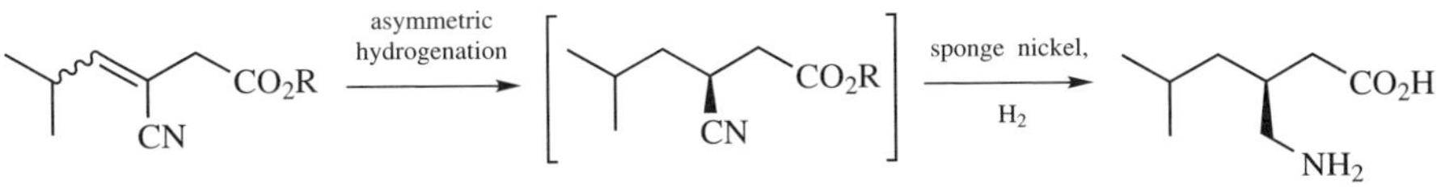

Figure 6.3. Proposed asymmetric hydrogenation to pregabalin.

this route does have some drawbacks. First and foremost it is a racemic process, which requires a classical resolution that adds two steps to the process and results in the loss of half of the material. To compound the problem, the route relies on the resolution as the final step of the process, which makes it necessary to carry the racemate all the way through to the end before separating out the desired enantiomer. Also, with four solids isolations, two of which result directly from the resolution, the process is solids-handling intensive. Solid isolation is a unit operation that can take a considerable amount of time and effort, therefore diminishing the throughput of a process.

With the inherent inefficiencies of a resolution process as mentioned above, we envisioned constructing pregabalin by way of a more efficient asymmetric route as shown in Figure 6.3.[2] Forming the stereocenter in the correct configuration in the first place would obviate the need for a resolution. For the sake of efficiency the stereocenter needed to be introduced using a catalytic method. Methods using stoichiometric auxiliaries already proved less efficient and much more expensive than the resolution route.[1] Instead, we chose to look at generating the stereocenter by way of an asymmetric hydrogenation. Starting with the properly positioned unsaturated nitrile **7,** asymmetric hydrogenation (and potentially hydrolysis if R is an alkyl group) would provide enatiomerically enriched analogue of cyanocarboxylate **8**, which is the single enantiomer form of intermediate **4** from the malonate route. Then, by analogy to the malonate route, sponge nickel hydrogenation would provide the amino acid, which in this case would be the enantiomerically enriched pregabalin. Recrystallization would provide the proper crystal properties and ensure the purity.

THE ASYMMETRIC HYDROGENATION PRECURSOR

Synthesis of the asymmetric hydrogenation precursor started with a published procedure for similar ester nitriles.[3] Baylis–Hillman condensation[4] of two readily available, inexpensive materials, isobutyraldehyde and acrylonitrile, catalyzed with triethylenediamine (DABCO) formed allylic alcohol **9** as shown in Figure 6.4. Like most Baylis–Hillman condensations, this reaction was painfully slow, taking up to 5 days to proceed to completion. Though the cycle time was extremely unfavorable for the reaction, throughput was one redeeming feature since the reaction could be run neat. Not satisfied by the time required, we looked into a method to speed up the reaction. We found that the use of polar protic co-solvents increased

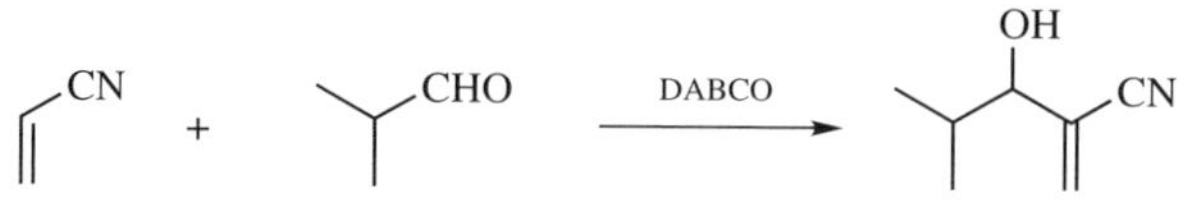

Figure 6.4. The Baylis–Hillman condensation.

the rate of the reaction while giving a similar purity profile, with water being the co-solvent of choice in terms of reaction rate, workup efficiency, and being a greener alternative.[5] Heating the condensation reaction to 50°C and using catalytic butylated hydroxytoluene (BHT) to stabilize the acrylonitrile at that temperature further accelerated the desired reaction to allow complete conversion to allylic alcohol **9** within a manageable 16 hr with good quality.[6]

As with most process development work, the impurity profile of Baylis–Hillman adduct **9** was an important diagnostic tool in determining the success of the condensation and leading to avenues to improve the synthesis. The three impurities in boxes (**10**, **11**, **12**) in Figure 6.5 were identified during the course of development and were found in increasing amounts during scale-up, especially during pilot runs. Aldehyde **10** results from the conjugate addition of the enolate of isobutyraldehyde into acrylonitrile. The impurity appears in the Baylis–Hillman product, and a portion carries through unchanged into the subsequent two product mixtures. Aldehyde **10** also undergoes further Baylis–Hillman condensation with acrylonitrile to give allylic alcohol **11**, which undergoes the same transformations as the desired Baylis-Hillman adduct for the subsequent three steps. In impurity **12**, the Baylis–Hillman alcohol (**9**) adds in conjugate fashion to acrylonitrile to give the ether. This product also carries through the subsequent two steps.

Upon scale-up to the pilot plant, we had noted that the impurities mentioned above were seen in increasing amounts. After a number of experiments including removal of the BHT antioxidant, we determined that the better oxygen exclusion and the continual use of fresh reagents at pilot scale might have related to the

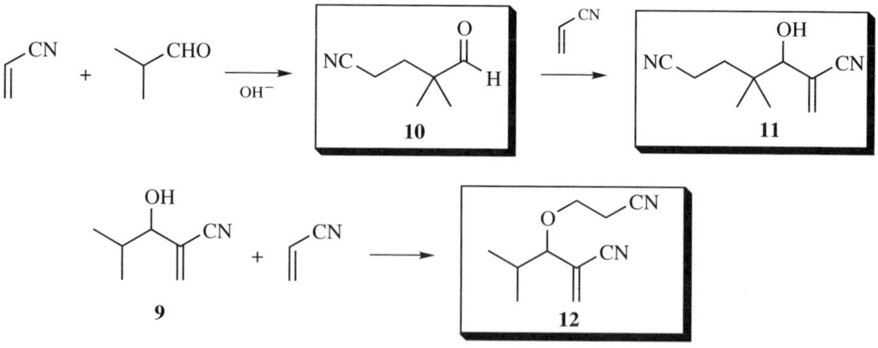

Figure 6.5. Typical impurities of the Baylis–Hillman condensation.

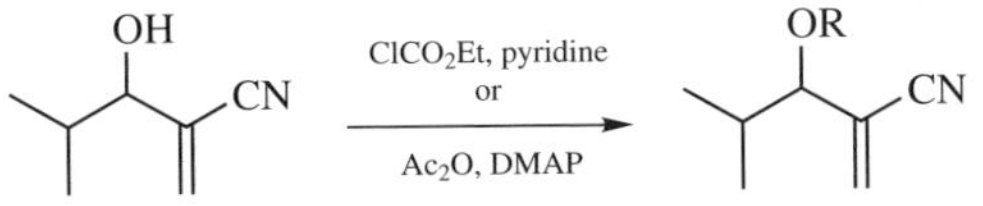

9

13, R = CO_2Et
14, R = Ac

Figure 6.6. Acylation of the Baylis–Hillman alcohol.

increased impurity levels. More specifically, we reasoned that the isbutyric acid that would likely form by oxidation of isobutyraldehyde either during the course of the reaction or in opened bottles sitting for long periods at lab scale might have helped the reactions. In that vein we examined the addition of acids to the reaction. Acetic acid did improve the purity of the product, unfortunately at the cost of reaction rate and completion. By contrast, the addition of a dilute solution of hydrochloric acid improved the purity while giving a comparable reaction rate to the original conditions, and the reaction proceeded to completion. Most importantly, the addition of hydrochloric acid made the reaction reproducible at larger scale. The product was typically carried directly on through the subsequent acylation step without isolation. However, analysis of the reaction mixture indicated that the material was usually greater than 95% pure (GC area).

Acylation of Baylis–Hillman adduct **12** proceeded in one of two different manifolds as shown in Figure 6.6. With our initial conditions, the π-allyl palladium carbonylation chemistry in the subsequent step required the use of an allylic carbonate. However, improvements in the downstream chemistry later allowed the use of an allylic acetate for the same transformation, which was greatly beneficial for the acylation as well as for the carbonylation.

Originally, treatment of the allylic alcohol with ethyl chloroformate in the presence of pyridine provided allylic carbonate **13**. Acylation with ethyl chloroformate to form the allylic carbonate frequently failed to proceed to completion and required large excesses of both ethyl chloroformate and pyridine. Also, the reaction only proceeded well in large volumes of environmentally unfriendly methylene chloride. The carbonate product also presented purity problems as diethyl carbonate formed by degradation of the chloroformate often contaminated the product and was difficult to remove.

By contrast, formation of allylic acetate **14** proved very easy. Treatment of allylic alcohol **9** with a small excess of acetic anhydride in the presence of a catalytic amount of 4-dimethylaminopyridine (DMAP), and a small amount of tetrahydrofuran as solvent cleanly formed allylic acetate **14** within a few hours. A basic workup provided the allylic acetate in very high purity. In addition to the excellent conversion, this reaction could be run almost neat, thus improving the throughput tremendously over the allylic carbonate route, which needed to be run in a large amount of solvent due to the pyridine hydrochloride byproduct. Additionally, the reduced solvent load, decreased reagent amounts, and use of hydroxide (workup) in place of an amine base greatly reduced the waste burden. At scales of up to

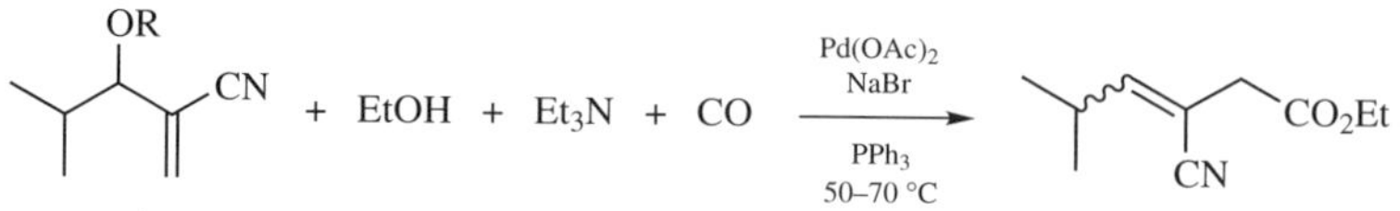

13, R = CO_2Et
14, R = Ac
15

Figure 6.7. The π-allyl palladium carbonylation.

800 kg, the Baylis–Hillman condensation and acetate formation typically proceeded in 81% yield (for the two steps) at purities of greater than 93% GC area.

In the following step either allylic carbonate **13** or allylic acetate **14** was subjected to a π-allyl palladium carbonylation to form nitrile ester **15** in approximately 3.5:1 *Z*:*E* mixture of olefin isomers as shown in Figure 6.7. At this stage the olefin ratio did not concern us because literature indicated that both would be viable substrates for the desired hydrogenation.[7] Use of the simple palladium acetate and triphenylphosphine catalyst system detailed in the initial literature conditions for the carbonylation required the use of a carbonate, because acetates were known to react poorly.[8] In fact, in our case the allylic acetate substrate gave no reaction under the original conditions. These conditions also required the use of very high palladium loadings (>3%) and moderately high carbon monoxide pressures (300 psig). Though the conditions were effective for preliminary studies, they clearly would not be desirable for later studies and commercialization for two reasons. First, pressures above 50 psig were not available in our pilot plant due to safety concerns. Thus, this hurdle would have required developing the chemistry and then transferring to a contract manufacturer capable of using carbon monoxide at high pressure, which in turn would have greatly slowed the development of the route. Second, the high loading of palladium in conjunction with the high cost and volatile market for palladium began to make this route cost-prohibitive.

Lowering the pressure required for the carbonylation became the first priority, with lowering the catalyst loading as a long-term goal. Modifications to the catalyst system and reaction stoichiometry made both possible. Literature revealed a modified catalyst system that allowed the carbonylation to run using alternate substrates to allylic carbonates.[9] Murahashi et al.[9] reported the carbonylation of allylic acetates and phosphates using a catalyst system consisting of tris(dibenzylideneacetone)dipalladium [$Pd_2(dba)_3$], triphenylphosphine, diisopropylethylamine DIPEA and sodium bromide. The DIPEA is proposed simply to neutralize the acetic or phosphoric acid formed in the reaction. Though Maurahashi found that less bulky amines led to decomposition through quaternary ammonium salts formation, we found that triethylamine worked as well as DIPEA. The sodium bromide is proposed to displace the acetate from the palladium in the π-allyl palladium species, thus preventing reversal back to the allylic acetate. The most important finding we discovered using the new catalyst system on our allylic carbonate substrate was that this catalytic system could effect the carbonylation at much lower pressures, 30–50 psig, compared to the previously required 300 psig. In our case, palladium acetate was as effective as the $Pd_2(dba)_3$ and much more cost effective and readily available at scale. Carbonylation on the allylic carbonate was not expected to form any acidic byproducts, so we tried the reaction in the absence of amine base. However, we found the amine necessary for any reaction to take place. This is potentially because triethylamine activates the palladium at the lower pressures, whereas carbon monoxide can activate the catalyst only at higher pressures.[10]

With the above catalyst modifications, the carbonylation could be run at pilot scale in our facility limited to a maximum pressure of 50 psig. However, a drawback to the use of the carbonate substrate became immediately apparent at 50 psig. With limited headspace in the reactor, during the course of the reaction the carbon dioxide by-product of the reaction began to build up and prevent additional carbon monoxide from entering the vessel. Thus, we needed to purge the reactor of the built up carbon dioxide and then recharge the reactor with carbon monoxide. This sequence needed to be repeated a number of times during the course of the reaction, leading to increased amounts of impurities such as ether **16**, which is formed in the absence of carbon monoxide (Figure 6.8). Fortunately, the modified catalyst system (discussed above) affected the carbonylation equally well with the allylic acetate **14** as with the carbonate **13**. This benefit was twofold. First, with use of the allylic acetate the by-product is acetic acid instead of carbon dioxide gas and ethanol, which alleviated the need to purge the reaction of gaseous byproducts. Second, the allylic acetate could be produced much more efficiently, cheaply, and in greater purity, making allylic acetate **14**a more appealing intermediate. Additionally, we reduced the excess of ethanol to between one to three equivalents, which allowed the reaction to be run essentially neat, thus improving the throughput and decreasing the waste burden for another step.

With the improvements from the modified catalyst system and the use of the more available allylic acetate substrate making the reaction workable at pilot scale, we turned our focus to lowering the catalyst loading. At the initial catalyst loading, the process was economically unfeasible due to the large expense of the palladium catalyst and volatility of the market price. The modified catalyst system provided some improvement. However, palladium loadings remained high, and at lower loadings the reaction became unreliable on larger scale and in different reaction vessel geometries. Analysis of the rate of carbon monoxide uptake during the carbonylation revealed that the reaction proceeded at the same initial rate regardless of the catalyst loading. However, with lower loadings the reaction simply died earlier, not allowing the reaction to go to completion. This finding led us to believe that improving the stability or lifetime of the catalyst would allow lower loadings. To this end, we studied the ratio of phosphine ligand to palladium with the idea that additional ligand would improve the catalyst lifetime in solution. We started by targeting a catalyst loading at which the reaction would not proceed to completion under the current standard conditions. At that loading, we increased the amount of phosphine ligand with the assumption that a portion of the phosphine was oxidizing; and at the lower catalyst loadings, that amount of depleted phosphine became significant. This study showed us that the reaction was significantly more robust using two to three equivalents of phosphine per palladium than it had been at the standard loading with one equivalent per palladium. With this new ligand ratio the reaction became robust at all scales and reactor geometries, and the catalyst load could be dropped by a factor of more than 25 from the initial literature conditions. By dropping the catalyst load to this level, the process became economically attractive and independent from the variable price of palladium. Also, the use of lower palladium loading minimizes the possibility of palladium contamination downstream and makes the process greener because less metal waste is produced.

We also studied the use of alternative ligands extensively with the goal to improve both the rate and catalyst load for the carbonylation. Due to issues with the (*E*)-isomer which will be discussed below, we also used the ligand study to explore the possibility of altering the (*Z*)- to (*E*)-olefin isomer ratio. The study found no ligands to be more effective than triphenylphosphine for reduction of catalyst loading or increasing the rate of reaction. Unfortunately, we also found no ligands that significantly altered the (*Z*)- to (*E*)-olefin isomer ratio.

As we scaled the reaction up into 2000- to 4000-liter reactors with newly improved reaction conditions and the times for additions and transfers naturally increased, we found that the carbonylation again performed more erratically and generally poorer. This behavior was reminiscent of the behavior with catalyst loads that were too low. To probe the cause, we looked on lab scale at extending the time between the addition of reagents and heating the reaction under carbon monoxide. The results clearly showed that extended hold periods between mixing all of the reagents and charging the carbon monoxide and heating led to poor carbonylation. More specifically, long hold periods after addition of the triethylamine to the remaining catalyst components caused deterioration of the catalyst. Presumably, this problem occurs because triethylamine begins to activate the palladium catalyst upon mixing, and the activated catalyst begins to decompose rapidly without the proper reaction conditions.[10] As a simple remedy, we held addition of the triethylamine to the reaction mixture until immediately before heating the reaction and charging the carbon monoxide. With this change the reaction proceeded well at up to 4000-liter scale with no significant catalyst degradation issues. Pilot runs using the updated catalyst system and charging procedures gave some of the purest and highest yielding (81%) reactions to date at scales of up to 765 kg.

With the great deal of work that went into optimizing the catalyst loading and reaction robustness for the carbonylation, the impurity profile became an integral part of assessing the reaction quality. Figure 6.8 shows the major impurities from the carbonylation. Cycloadduct **18** was one of the most important diagnostic tools for the reaction. The appearance of larger amounts of **18** in conjunction with residual starting allylic acetate **9** indicated that the catalyst had died, and elimination of the acetate began to preferentially occur. Interestingly, diene **17**, which is the initial product formed by the elimination of acetic acid, was only seen in small amounts. The diene undergoes facile Diels–Alder dimerization to form cycloadduct **18** even upon storage at room temperature.[11] As mentioned above, in the absence of sufficient carbon monoxide, the intermediate π-allyl palladium species undergoes addition of ethanol to form allylic ether **16**. One series of impurities specific to the allylic acetate substrate are the allylic inverted acetate **19** and alcohol **20** olefin isomers. Though the addition of sodium bromide helps to remedy the reversible

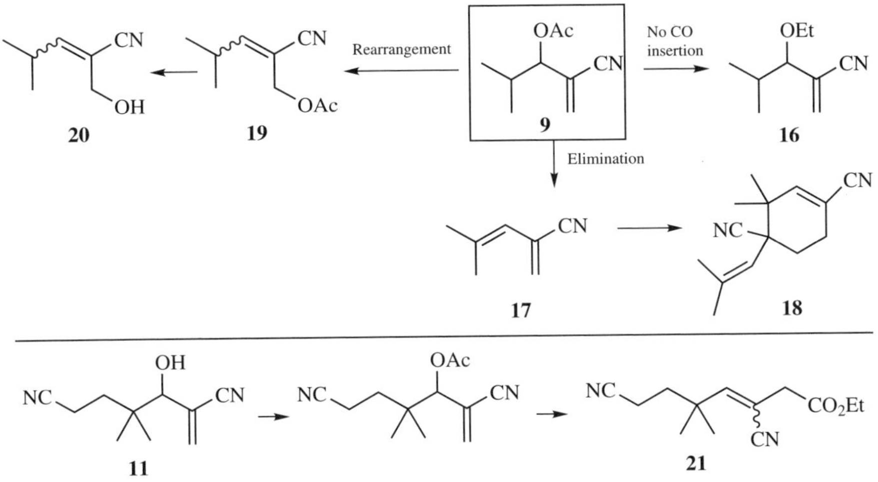

Figure 6.8. Typical impurities from the carbonylation.

addition of palladium into the acetate bond, the π-allyl palladium species can still undergo some allylic transposition and reductive elimination to give **19** as a mixture of olefin isomers. Isomer **19** is a potential substrate for the π-allyl palladium chemistry; however, it seems less prone to participate in the carbonylation and builds up to some degree. Hydrolysis of the acetates leads to allylic alcohols **20**. Levels of these impurities are seen to grow to some extent as the catalyst dies, but also when the reaction is starved of carbon monoxide. Finally, impurity **21** comes from the chain extended allylic alcohol **11** carrying through the analogous acetylation and carbonylation sequence.

ASYMMETRIC HYDROGENATION OF THE ESTER NITRILE

With the unsaturated nitrile system in hand, for simplicity sake we considered performing the asymmetric hydrogenation at the ester stage and, in course, performed some studies in conjunction with ChiroTech examining asymmetric hydrogenation on ester **15** (Figure 6.9), which was purified by distillation.[2] Using a wide range of chiral rhodium complexes under typical hydrogenation conditions, ester **15** proved to be a less-than-ideal substrate for asymmetric hydrogenation with low conversions and only 8–42% enantiomeric excess. Though the reaction could be pushed to completion with heating to 55°C, the selectivity issue could not be easily resolved.

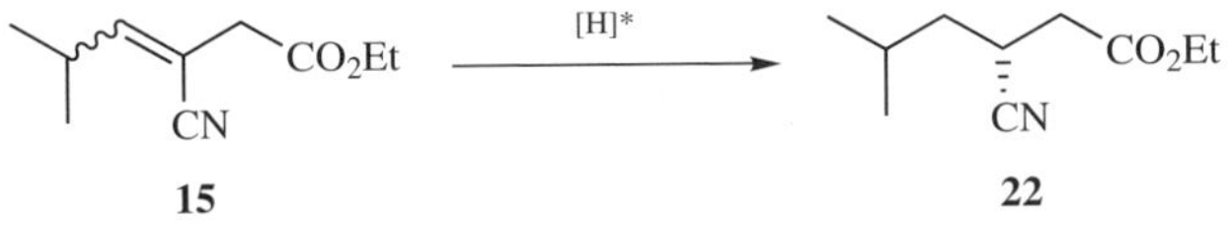

Figure 6.9. Potential hydrogenation of the ethyl ester.

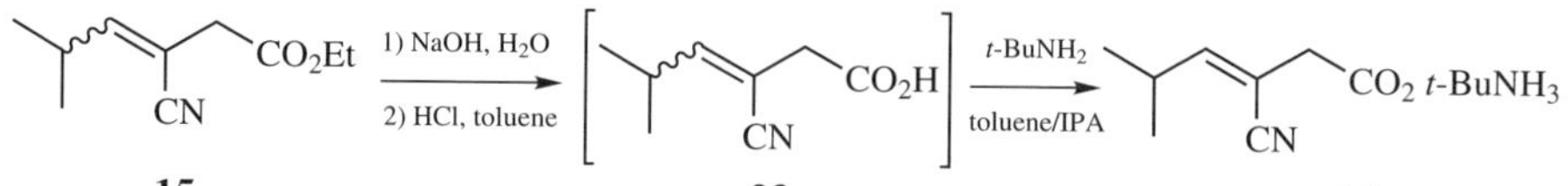

Figure 6.10. Formation of the *tert*-butylammonium salt.

CARBOXYLATE SALT FORMATION

Following Burk's experience with the hydrogenation of itaconic acids, we chose to look at carboxylate salt derivatives of the ester.[7] We examined both the potassium salt and the *tert*-butylammonium salt, but we went forward with the *tert*-butylammonium salt due to ease of synthesis and isolation. Saponification of the ester with sodium or potassium hydroxide, acidification with hydrochloric acid to form the free acid, and subsequent treatment with *tert*-butylamine provided *tert*-butylammonium salt **24** as shown in Figure 6.10. We found isolation of a pure salt at this stage to be fairly important for two reasons. First, the elimination of impurities at this stage excludes contaminants that have the potential to foul the expensive asymmetric hydrogenation catalyst and thus increase the loading. Second, removal of impurities at this stage lowers the impurity burden on the subsequent crystallization of pregabalin ensuring the isolation of pure drug substance. As the first isolable solid in the process, this purification was the first opportunity to remove the number of by-products that accumulated in the preceding steps. Fortunately, isolation of the salt proved extremely good at removing impurities. During development of the route, crystallization of the salt continually provided material of excellent quality, generally greater than 99% pure, from starting ester that had been 60–80% pure. Though typically not required, the salt could be recrystallized or reslurried to improve the purity further.

Though the crystallization and recrystallizations afforded excellent purification, this cleanup comes the expense of the loss of the majority of the (*E*)-isomer, because the (*E*)-isomer does not crystallize well out of solution under any conditions examined. The loss is unintentional, since the (*E*)-isomer is a viable substrate for the asymmetric hydrogenation.[12] Samples of the *tert*-butylammonium salt that were enriched up to 60% in the (*E*)-isomer provided rapid hydrogenation with acceptable induction. However, the desire to ensure the purity of the *tert*-butylammonium salt overrode the quest for a higher yield, and we determined that the loss of the (*E*)-isomer was most equitable.

As noted above, the purity of the *tert*-butylammonium salt was typically very high, but we were interested in identifying some of the impurities that were seen in the 0.1–0.3% range. Displayed in Figure 6.11 are nine impurities seen in our lots through development. These impurities are shown as the free acids that were isolated, though they most likely exist as *tert*-butylammonium salts. Notably, four of the impurities **25–28** come from degradation of the parent through hydrolysis of the nitrile. Though nitrile hydrolysis is unusual under such mild conditions, we have frequently seen hydrolysis of similar nitriles most likely due to an anchimeric

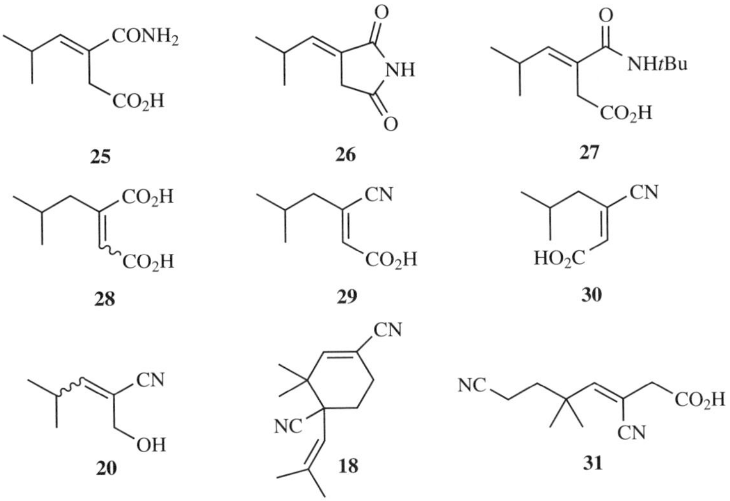

Figure 6.11. Typical impurities in the *tert*-butylammonium salt.

effect from the adjacent carboxylate. Cyanoesters **29** and **30** are simply position isomers of the desired product and could come from the carbonylation or isomerization of **15**. Alcohol **20** and cycloadduct **18** are impurities that carry through from cyanoester **15,** while **31** derives from **21**. With careful control of the crystallization, all impurities could be minimized and recrystallization can nearly remove all impurities again at the expense of the (*E*)-isomer.

An interesting note on the positional stability of the olefin in cyanoester **15** comes from work aimed at synthesizing the positional isomers **34**, which were examined as part of our asymmetric hydrogenation studies. Horner–Emmons condensation of phosphonate **32** with ethyl glyoxylate shown in Figure 6.12 formed the

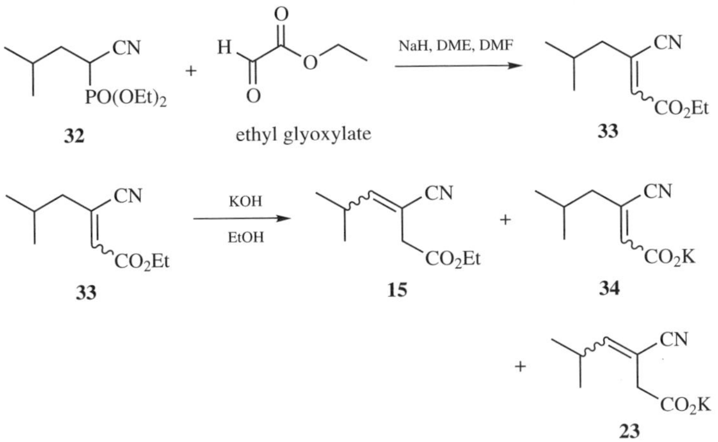

Figure 6.12. Horner–Emmons approach to the olefin positional isomer.

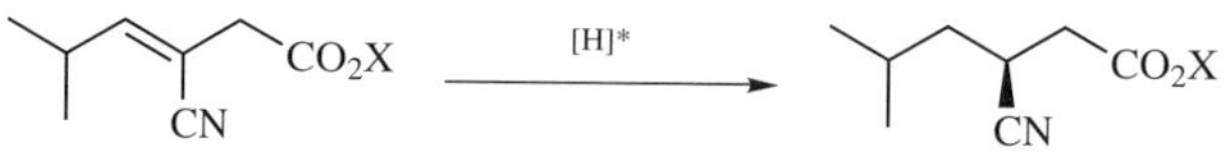

Figure 6.13. Asymmetric hydrogenation of the carboxylate salts.

cross-conjugated cyanoesters **33** mixture of (*E*)- and (*Z*)-isomers. Upon subjecting **33** to the same saponification conditions that were initially used for cyanoester **15** (potassium hydroxide in ethanol), a substantial amount of olefin migrated to the position out of conjugation with the ester, giving a mixture of the esters and acids shown in Figure 6.12. Though we eventually found conditions that did not effect the olefin migration, this experiment indicates the olefin outside of conjugation with the ester (or acid) is favored over the cross-conjugated olefin, because cyanoester **15** showed no evidence of migration during hydrolysis using the same conditions.

ASYMMETRIC HYDROGENATION

With a suitable route to unsaturated cyano carboxylate salt **24** in hand, we could study the asymmetric hydrogenation sequence (Figure 6.13). Hydrogenation of both *tert*-butylammonium salt **24** and potassium salt **35** was successful with a number of catalyst systems with high selectivity and turnover as shown in Table 6.1. As mentioned previously, we chose to move forward with the *tert*-butylammonium salt. We selected the catalyst [(*R,R*)-(Me-DuPHOS)Rh(COD)]BF_4 for further development based on a number of criteria, including its excellent stereoselectivity and its commercial availability on large scale due its use in other processes. Though the

TABLE 6.1. Asymmetric Hydrogenation of *tert*-Butylammonium 3-cyano-5-methyl-hex-3-enoate 24

Entry[a]	Precatalyst	Reaction Time	Conversion (%)[b]	% *S* or *R*[b]
1	[(*R,R*)-(Me-DuPHOS)Rh(COD)]BF_4	15 min[c]	100	97.5 (*S*)
2	[(*R,R*)-(Et-DuPHOS)Rh(COD)]BF_4	15 min[c]	100	98.7 (*S*)
3	[(*R,R*)-(iPr-DuPHOS)Rh(COD)]BF_4	6 hr	72	62.0 (*R*)
4	[(*R,R*)-(Me-BPE)Rh(COD)]OTf	45 min[c]	100	91.7 (*S*)
5	[(*R,R*)-(Et-BPE)Rh(COD)]BF_4	45 min[c]	100	90.5 (*S*)
6	[(*S,S*)-(iPr-BPE)Rh(COD)]OTf	6 hr	59	54.0 (*S*)
7	[(*R,R*)-(Me-FerroTANE)Rh(COD)]BF_4	20 min[c]	100	97.7 (*S*)
8	[(*R,R*)-(Et-FerroTANE)Rh(COD)]BF_4	20 min[c]	100	92.3 (*S*)

[a]One millimole of substrate in 5 ml methanol was hydrogenated with 10 μmol of precatalyst in a glass-lined stainless steel pressure vessel with hydrogen at 90 psi at room temperature.
[b]Conversion and enantiomeric excess were determined by GC.
[c]Time within which hydrogen uptake had ceased.

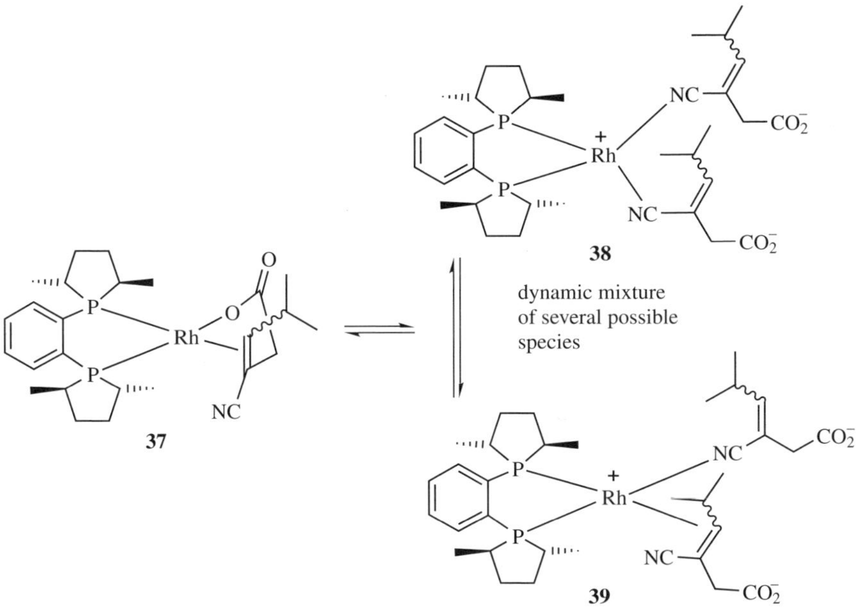

Figure 6.14. Dynamic mixture of species indicated by NMR studies.

molar loading of 2700:1 seen in these studies for Rh(DuPHOS) is lower than that seen for other substrates, this loading is impressive considering the nitrile-containing substrate and product could hinder the reduction (see below).

An important mechanistic detail for the hydrogenation came from the comparison of the asymmetric hydrogenations of the carboxylate salts and of the cyanoester. Hydrogenation of the carboxylate salts with [(*R*,*R*)-(Me-DuPHOS)Rh(COD)]BF_4 gave the (*S*)-configuration at the new stereocenter while reduction of cyanoester **15** with the same catalyst gave the opposite stereochemistry. This result implies that the nature of the carboxylate residue plays an important role in the stereoselectivity of the hydrogenation, which is consistent with reports that describe bidentate binding to the cationic rhodium by both the carboxylate and olefin.[13] NMR studies indeed indicated the presence of bidentate chelate **37** (Figure 6.14) within a complex mixture. The major amount of catalyst appeared to be nitrile-bound, which is consistent with species **38** and **39**. Thus the minor amount of chelate **37** (~10%) acts as a conduit for the hydrogenation, which is consistent with results reported for other asymmetric hydrogenations.[14] The relatively low amount of the reactive intermediate **37** is likely why higher amounts of catalyst are required as compared to other substrates, since both the starting material and product contain nitriles, which are good ligands for rhodium and probably bind the majority of the catalyst.

With asymmetric hydrogenation conditions worked out, we moved on to the end-game of carrying **24** on to pregabalin. For optimal processing, *tert*-butylammonium salt **24** was taken through the asymmetric hydrogenation as an ~10% solution in methanol with an effective catalyst loading of greater than 2000:1 molar ratio, at

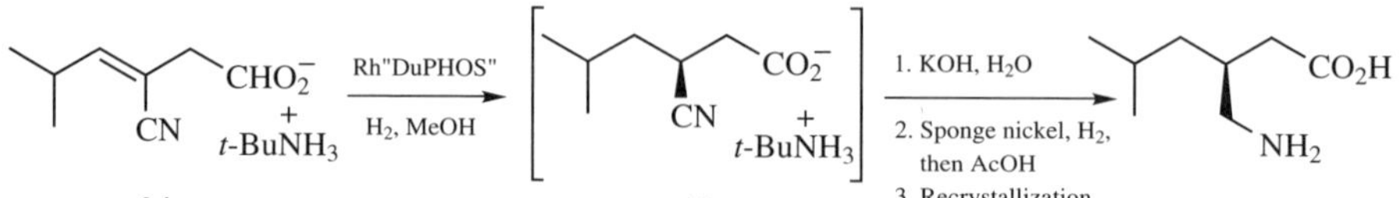

24 **40** **1, pregabalin**

Figure 6.15.

about 50 psig of hydrogen and at ambient to slightly greater than ambient temperature (Figure 6.15). Throughout development, the asymmetric hydrogenation performed very well under these conditions, and stereoselectivity was generally greater than 97.5% of the desired (*S*)-enantiomer. Driving the reaction to completion also turned out to be an important factor for the asymmetric hydrogenation, more so than simply due to the yield loss associated with an incomplete reaction. Since the subsequent transformation is a sponge nickel hydrogenation of the nitrile, that step would also reduce any residual unsaturated nitrile **24** in a racemic fashion, thus effectively eroding the stereoselectivity of the asymmetric hydrogenation. Fortunately, under our optimized conditions the asymmetric hydrogenation typically proceeded to greater then 99% conversion, very cleanly.

For the nitrile reduction following the asymmetric hydrogenation, potassium hydroxide needed to be added to the reaction to prevent secondary amine formation (dimerization) during the reduction. This addition presented the opportunity to remove the *tert*-butylamine (which had been freed up by the base addition) by distillation and perform a solvent switch simultaneously, giving us very similar nitrile reduction conditions to those used in the malonate route. After the nitrile hydrogenation, adjustment of the solvent ratio and addition of acetic acid to neutralize the potassium salt crystallization gave pregabalin in good yield and high purity. Recrystallization from an isopropyl alcohol and water mixture afforded the proper crystal properties of pregabalin in the desired purity. Typically, none of the (*R*)-enantiomer could be detected at this point.

The true test of the asymmetric hydrogenation sequence came from testing the process at pilot and manufacturing scale, especially considering the air sensitivity of the catalyst. The asymmetric hydrogenation process has now been successfully run a number of times at tens to hundreds of kilogram scale and has given pregabalin of comparable quality to the malonate route and the yields have remained consistently around 65–75% for the three-step end-game.

CONCLUSION

With this work, we demonstrated viability of the asymmetric route to pregabalin. To make the route workable, we developed an efficient, cost-effective route to the unsaturated nitrile substrate precursor to the asymmetric hydrogenation. An efficient route to the precursor was required for the route to be viable, to offset the unavoidable expensive of the asymmetric hydrogenation catalyst. With a route to the hydrogenation substrate in hand, we demonstrated the asymmetric hydrogenation sequence to be effective for the formation of pregabalin whose purity is similar to that derived from the malonate route.

The new route has shown a number of benefits over the malonate route, including lower cost of goods, higher throughput, and lower waste burden. Our projections show that the asymmetric hydrogenation route would have a 20% improvement in cost of goods. Additionally, we expect that the asymmetric hydrogenation route has some room for further improvement as a young process, whereas the malonate route, as a more mature process, is less likely to see drastic reductions in cost of goods. The new process is also a green process with a projected drastic reduction 30% in waste compared to the malonate route, which is due to both (a) the large number of reactions that are essentially run neat and (b) not needing to carry through 50% of the wrong enantiomer. Also, this process obviates the need for cyanide use and disposal and reduces quantity of the of sponge nickel required. Finally, and perhaps most importantly, the process shows great improvement in throughput. Depending on the model used to calculate the throughput at production scale, the asymmetric hydrogenation process is projected to give greater than a 50% improvement in throughput. This number can be as important as an improvement in cost of goods because it allows plant space to be cleared for uses on other products or more importantly it helps avoid the capital expense of building a new plant to produce a product. All of the above factors have made to the asymmetric hydrogenation route an important option for the manufacture of pregabalin and have led to its further development.

REFERENCES AND NOTES

1. Hoekstra, M. S., Sobieray, D. M., Schwindt, M. A., Mulhern, T. A., Grote, T. M., Huckabee, B. K., Hendrickson, V. S., Franklin, L. C., Granger, E. J., and Karrick, G.L. *Org. Proc. Res. Dev.* **1997**, *1*, 26–38.
2. For details on initial studies on the asymmetric hydrogenation see Burk, M. J., de Koning, P.,D., Grote, T. M., Hoekstra, M. S., Hoge, G., Jennings, R. A., Kissel, W. S., Le, T. V., Lennon, I. C., Mulhern, T. A., Ramsden, J. A., and Wade, R. A. *J. Org. Chem.* **2003**, 68, 5731–5734.
3. Yammoto, K, Deguchi, R, and Tsuji, J. *Bull. Chem. Soc. Jpn.* **1985**, 3397–3398.
4. (a) Baylis, A. D., Hillman, M. E. D German Patent 2155113 (1972). (b) For a review see Ciganek E. The Catalyzed α-Hydroxyalkylation and α-Aminoalkylation of Activated Olefins (The Morita–Baylis–Hillman Reaction), in Paquette, L. A., Ed., *Organic Reactions*, John Wiley & Sons, New York, 1997, pp. 201–343.
5. Aggarwal, V. K., Dean, D. K., Mereu, A., and Williams, R. *J. Org. Chem.* **2002**, *67*, 510–514. Auge, J., Lubin, N., and Lubineau, A. *Tetrahedron Lett.* **1994**, *35*, 7947–7948.
6. For examples of Baylis–Hillman reaction at elevated temperatures see ref. 4 b, p. 212.
7. Burk, M. J., Bienewald, F., Harris, M., and Zanotti-Gerosa, A. *Angew. Chem. Int. Ed.* **1998**, *37*, 1931–1933.
8. Tsuji, J, Sato, K, and Okumoto, H. *J. Org. Chem.* **1984**, *49*, 1341–1344.

9. Murahashi, S.-I, Imada, Y., Taniguchi, Y, and Higashiura, S,-Y. *Tetrahedron Lett.* **1988**, *29*, 4945–4948. Murahashi, S.-I, Imada, Y, Taniguchi, Y, and Higashiura, S. *J. Org. Chem.* **1993**, *58*, 1538–1545.

10. Hegedus, L. S., Weider, P. R., Mulhern, T. A., Asada, H., and D'Andrea, S. *Gazz. Chim. Ital.* **1986**, *116*, 213–220.

11. For a report on the homo-Diels–Alder cyclization of diene **17** during distillation see Brannock, K. C., Bell, A., Burpitt, R. D., and Kelly, C. *J. Org. Chem.* **1964**, *29*, 801–812.

12. Reference 7 shows that the Rh(DuPHOS) reduces the (*E*)- and (*Z*)-isomers equally well.

13. Burk, M. J., Bienwald, F, Challenger, S., Derrick, A, and Ramsden, J. A. *J. Org. Chem.* **1999**, *64*, 3290.

14. Landis, C. R., and Halpern, J. *J. Am. Chem. Soc.* **1987**, *109*, 1746–1754.

7

RESPONSIBILITIES OF THE PROCESS CHEMIST: BEYOND SYNTHETIC ORGANIC CHEMISTRY

Stéphane Caron

Chemical Research & Development, Pfizer Global R&D, Groton, CT 06340

INTRODUCTION

The process chemist plays several critical roles throughout the early-, middle-, and late-stage development of new drug candidates in the pharmaceutical industry. The primary reason for this is that the manufacture of the active pharmaceutical ingredient (API or drug substance) fuels a plethora of activities ranging from development of analytical methods and appropriate formulations, safety and toxicology evaluation, and, ultimately, administration of the drug candidate in a clinical study. In early development, the process chemist is often confronted with a decision between (a) committing to a synthetic route that is believed to be less than optimal in order to meet a bulk delivery timeline or (b) developing a new, more efficient, cost effective and environmentally friendly synthesis with the risk of delaying the program, which would be of major impact. What the world sees of process chemistry are the identification of beautiful solutions to difficult synthetic challenges and the demonstration of the chemistry at increasingly larger scale. As exemplified in several chapters of this book, the process chemist often starts with complex syntheses designed for efficient analog preparation used by the medicinal chemists in their effort to identify a candidate. At the outset of the development process, such a synthesis may be modified to support the initial preparation of the API and then later develop the process to a synthetic route that could implemented in a commercial facility.[1,2] However, the role

Fundamentals of Early Clinical Drug Development: From Synthesis Design to Formulation, edited by Ahmed F. Abdel-Magid and Stéphane Caron

and responsibilities of process chemists extend far beyond synthetic organic chemistry. Process chemists must excel not only as synthetic chemists but must also understand the basic principles of physical organic chemistry, chemical engineering, analytical methodology, material science, methods of isolation, purification and crystallization of molecules, and many other areas. Process chemists also apply the principles of green chemistry, since the search for the most efficient and cost effective process will normally result in a synthetic route where the amount of solvents, excess reagents, and waste generation are minimized.

Process chemists should also be capable of understanding a multidisciplinary environment through personal interactions while performing within one of the most rigorously regulated environments. This chapter will detail some of the key responsibilities required of the process chemist to be successful in pharmaceutical research and development. While no academic training fully prepares a student to be a process chemist, one's passion for chemistry will often be the most important factor in initiating a successful career in the discipline.

GENERAL EXPECTATIONS

Above everything else, a process chemist is expected to be the synthetic expert leading the efforts to identify the synthetic route that will be selected for the preparation of the API. Multiple other responsibilities stem from the fact that the knowledge around the chemistry of a given molecule often resides with the process chemist (Figure 7.1).

Multidisciplinary Interactions

The process chemist is often a member of project teams charged with the task of supporting the development of drug candidates. At the interface between discovery and early development, an attempt is frequently made during the drug candidate selection phase to identify potential drug compounds where process chemistry

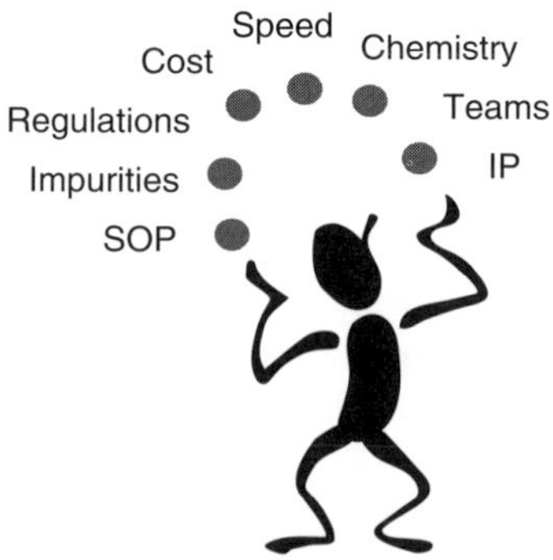

Figure 7.1.

improvements would be beneficial to the program and expedite the progression of the selected compound. As a candidate progresses through development, the chemist has the primary responsibility of communicating the status and issues related to the technology of the synthetic route and preparation of the API. Additionally, it is expected that strategic direction, perspective, and necessary input in discussions on topics or issues at the interface of their discipline will be provided by process chemists. Examples of such interactions include working with chemical engineers on the transfer and implementation of a procedure on a large scale as well as issues regarding technical aspects of scale-up such as mixing and reactor configuration for a critical step, working with an analytical group on the structural identification of an impurity or a degradation product or when setting specifications for future control of the purity of an API, working with formulators when selecting a final form or the desired particle size, working with quality assurance (QA) representatives on the resolution of an investigation which resulted of a process deviation, or working with regulatory personnel in the preparation of chemistry manufacturing and controls (CMC) documentation and the establishment of a regulatory strategy. Because of the complexity of drug development, the process chemist needs to interact efficiently with a number of partners and be able to assess the impact of a change in the manufacturing process on the API. This requires skills as a communicator and negotiator as well as a sound understanding of the drug development process. For example, a change in the manufacturing process which might appear minor in terms of chemistry could be significant to another discipline and might require an amendment in regulatory documentation.

In this case, the chemist is often in a situation where he or she needs to decide between what would be easiest and less resource intensive for his or her department in the short term versus what appears to be optimal for the long-term development of the compound. There is never a simple answer and many factors must be considered. The ultimate decision might arise from the confidence the organization has in the clinical efficacy, safety, and ultimately commercial success of the candidate.

An additional challenge resides in the establishment of timelines for a specific candidate, since the bulk prepared fuels the progression of the compound. In early development, it is often difficult for the chemist to accurately estimate either a delivery date or a certain quantity of API, because so little is known about the process and the manufacturing experience is limited. Still, promises of a "weight by date" are often made. The commitment to deliver on this promise is crucial because of the necessary planning of all activities that will occur once the first precious few kilograms of API are synthesized. The successful chemist relies on confidence in the chemistry, experience on how well a certain type of chemistry generally performs on scale, and, more often than not, their own natural instinct.

Intellectual Property

Intellectual property is at the cornerstone of the pharmaceutical industry. It is a principal mechanism for protecting the value of marketed products, and it provides the justification for the investment that a corporation makes in research and development of future products. Because of their responsibility in the identification of the synthetic route to a drug candidate, process chemists play an important role in the generation of intellectual property. Opportunities to patent a given process, novel synthetic intermediates, salt forms, polymorphic forms, and so on, will present themselves in the course of drug development and require input and documentation from the chemist. In the majority of cases, the process chemist has received no formal training in patent law yet will be asked to collaborate with patent attorneys in the preparation of a patent application. It is expected that the process chemist will become familiar with the basic structure and components of a patent, provide the necessary information for its preparation, and review its content prior to filing with a patent office. This can result in challenging interactions when the chemist is not familiar with legal jargon and the attorney is not fully versed in the art of synthetic chemistry.

While a corporation ensures that its patents are enforced in the jurisdictions where they have been granted, it also has the obligation to ensure that the patents of others are not infringed. More often than not, it is the responsibility of the chemist to work with attorneys and conduct the appropriate searches in the patent literature to confirm freedom to operate for a given process. This is especially challenging in a crowded competitive area where new subject matter constantly appears. One major change for process chemistry in recent years has been the growing trend to patent synthetic methodology in academia, especially in the area of catalysis.[3] When a desired transformation can be accomplished only by a patented reaction for which a license to operate is not available, the process chemist may be discouraged from evaluating this chemistry because the reaction may not be used for the manufacture of desired product. If a license is granted for the patented invention, it is the responsibility of the process chemist to remain within the terms and context of the license to operate. This scenario raises the question; How can one assess the usefulness of the chemistry and justify the expenditure of obtaining a license when not necessarily having the right to fully evaluate the technology? Furthermore, the patented technology may have minimal value for a candidate in early development with unproven safety and efficacy. However, the regulatory environment is such that commitment to a synthetic route must be made early in order to deliver an appropriate and consistent purity profile, show control points at key intermediates, and so on. In order for the patented technology to be useful, mutually acceptable licensing terms must be reached as early as possible in development. Otherwise, unacceptable licensing terms and/or restriction to fee-for-service may result in avoidance of these important reactions for the preparation of the API. If a license cannot be obtained or if the licensed technology does not provide a cost effective solution, the process chemist must attempt to design around the patented technology. Such examples include the preparation of novel phosphines shown in Figure 7.2[4–6], development of new processes to known compounds (Figure 7.3)[7], or alternative reagents for asymmetric reductions as depicted in Figure 7.4.[8,9]

Intellectual property is an important part of the job of process chemists who have the responsibility of ensuring freedom to operate a manufacturing process

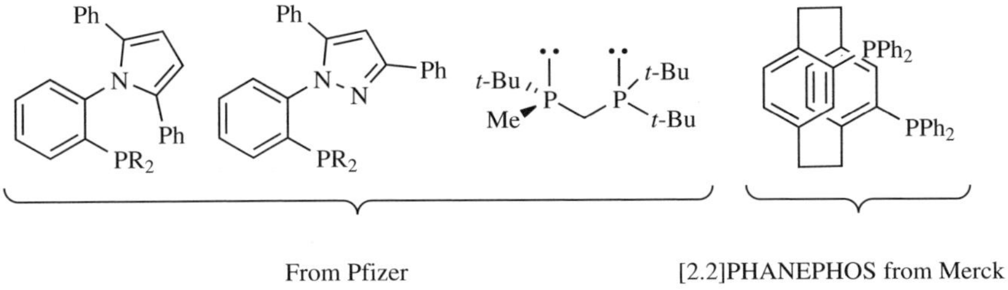

Figure 7.2. Novel phosphines from Pfizer and Merck.

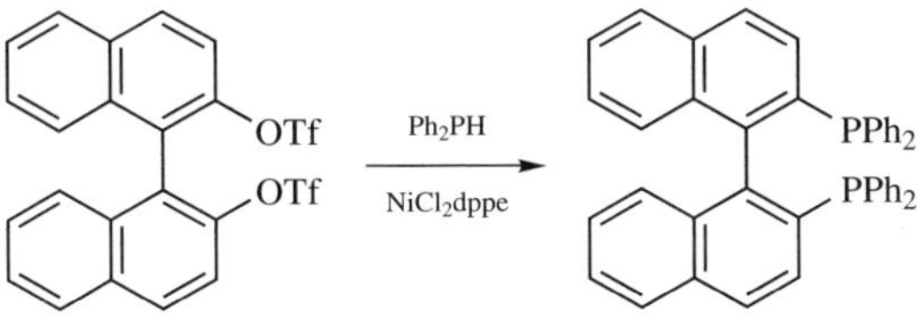

Figure 7.3. Merck's synthesis of BINAP.

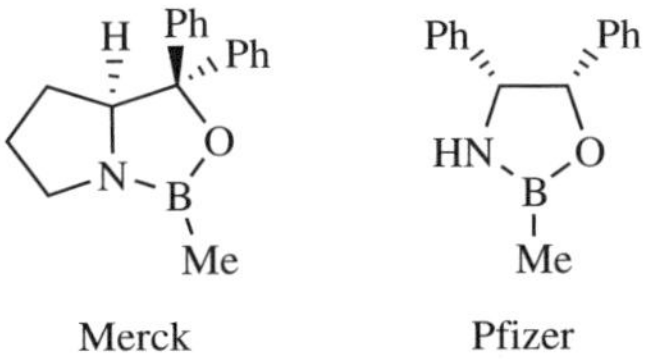

Figure 7.4. Oxazaborolidines reagents for chiral reductions.

and identifying opportunities to patent novel chemistry that could provide additional patent protection for a drug candidate.

LIVING IN A REGULATED ENVIRONMENT

While the day-to-day job of a process chemist working toward the identification of a synthetic route in the laboratory might be very similar to a organic chemist working in academia, the most drastic difference between the two resides once the route selected is used to prepare the molecule for potential use in humans because this aspect of our work is highly regulated.

Good Manufacturing Practices (GMP)

Current Good Manufacturing Practices (cGMP) are regulations that have been put in place to ensure that a product is produced in a consistent and controlled environment according to specific quality standards. Good Manufacturing Practices impact all aspects of pharmaceutical manufacturing, from selection of the starting material, facilities, and equipment used in manufacturing, to the training of the staff involved in the preparation of the API. GMP should provide control and minimize risk throughout manufacture of the API rather then rely solely on testing of the final product. In the United States, cGMPs are regulations covered in the Code of Federal Regulations, 21 CFR parts 210 (Current Good Manufacturing Practice in Manufacturing, Processing, Packing, or Holding of Drugs: General) and 211 (Current Good Manufacturing Practice for Finished Pharmaceuticals). For regulatory toxicology studies, it is not necessary that the API used be prepared under strict GMP, although the study is conducted under Good Laboratory Practice (GLP).[10] However, compliance with cGMP regulations is required if the material produced is intended for clinical trials in humans, or animals in the case of veterinary medicines. Therefore, if a single lot of material is prepared and intended for both toxicology and clinical trials, it must be under cGMP. While most agree that GMP control increases throughout the phases of development as knowledge of the process increases, in reality, this concept is very difficult to achieve when the same facility is used for manufacturing multiple drug candidates at different stages of development. The default is often to adhere to the strictest procedure.

Training and documentation are two of the most important aspects of cGMP. The regulations state that each person engaged in the manufacture of a drug product must have the education, training, and experience, or any combination thereof, to be able to perform the assigned function. While education and experience is easy to confirm, adequate training is a more challenging task, especially in light of constantly changing regulatory environment. Therefore, the chemists engaged in manufacturing need to receive proper training on cGMP and understand and follow all relevant standard operating procedures. It is also crucial to ensure that adequate training records are generated and maintained.

Documentation is another cornerstone of cGMP. If the proper documentation does not accompany the API manufactured, this API might as well not exist! Records must clearly show how material was prepared and in which facilities and equipment. As stated previously, documentation must also show that the personnel involved in the manufacture were qualified to do so. As such, manufacturing groups must work closely with quality assurance to ensure that the integrity of the API prepared is not compromised because of inadequate documentation.

Overall, GMPs are in place for the preparation of a drug substance to ensure the integrity and quality of the material manufactured. For the experienced process chemist, much of this becomes second nature, however, one must stay alert to any new regulatory obligations and not assume that what has been done in the past is still appropriate.

Standard Operating Procedures

Standard operating procedures (SOPs) provide one of the mechanisms by which an organization ensures that it will adhere to cGMP requirements. An SOP is a document, clear in scope and purpose, describing how a certain task should be conducted regardless of who performs it. As a result of SOPs, an organization can operate with consistency. SOPs can be specific to a unique facility, process, or piece of equipment. They can be broad in scope and cover an entire operation and applies to several locations. All SOPs, however, are relevant to every compound at any phase of clinical development.

Anyone working in a GMP environment must not only be aware of SOPs but must also be able to demonstrate that they have been appropriately trained and understand their scope and content.[11] Furthermore, they must have access to and always follow the most current version of any SOP. This can become a daunting task. As new GMP facilities are built, new equipment is purchased, new unit operations are identified, and soon the number of SOPs can rapidly increase and makes it extremely difficult for the process chemist operating in the GMP environment to fully remember the context of all of them.

The process chemists will often be asked to draft or review the content of a new or updated SOP. As such, it is necessary for the chemist to understand the importance of the SOP in the context of the regulations, ensure that it does not conflict with another approved procedure, and ensure that it can be followed consistently. Additionally, the process chemist, being the subject matter expert, can be asked to train other colleagues in the deployment of a newly created or updated procedure. While SOPs are seldom the process chemist's favorite topic, it is an important aspect of our business and their importance cannot be taken lightly.

Regulatory Interactions. The chemist also plays an important role in the generation and submission of information to regulatory agencies around the world. Along with meeting the technology challenges associated with designing a commercial process, the chemist must also be mindful that this commercial process must meet standards, in terms of process design and product purity, set by regulatory agencies across the world. In early development, any questions regarding the synthesis of the drug candidate, its purity profile, proposed acceptance criteria, and so on, will require the input of the process chemist, usually in collaboration with the analytical chemist. The process chemist will also often participate in discussions through regulatory meetings or teleconferences with regulatory agencies to either (a) answer any regulatory queries regarding the drug substance or (b) seek regulatory guidance for the strategy proposed in the development of a drug candidate. In the later stage of development, the chemist has several important responsibilities in the assembly of the Drug Substance section of the module 3 (quality section) of common technical document (CTD, section S3.2). Important information such as the description of the manufacturing process and process controls, together with characterization and justification of acceptance criteria, relies on the expertise of the process and analytical chemist. Once again, in a collaborative environment, the scientific knowledge of the process chemist is primordial for successful regulatory interactions and the advancement of new therapeutic agents.

ADDITIONAL CHALLENGES

The fact that we live in a regulated environment also brings along additional challenges, especially when factors from the external environment force the pharmaceutical industry to change its normal business practice.

Impurities

The process chemist operates in a constantly changing and sometimes inconsistent regulatory environment. When preparing the API, the final drug substance is expected to be pure, but impurities will always be present to a certain level. The API provided for the toxicology assessment allows a clinical trial sponsor to qualify the impurity profile of the given lot of material. *Qualification* is the process of acquiring and evaluating data that establish the biological safety of an impurity or a given impurity profile at the level specified. If a single impurity reaches a certain level, it is imperative for it to be *identified*, meaning that the structure of the impurity should be elucidated and characterized and its impact assessed.

One recent example of changing regulation is "Q3A Impurities in New Drug Substances," which was revised in February 2003 by the International Conference on Harmonization of Technical Requirements for Registration of Pharmaceuticals for Human Use (ICH). Prior to February 2003, the 1996 guidance stated that the qualification threshold was 0.1% as long as the maximum daily dose did not exceed 2 g per day. Attachment 2 from this guidance is shown in Table 7.1.[12] The revised guidance clearly showed that any new impurity exceeding 0.10% now requires identification and that "rounding down" is no longer acceptable.

This exemplifies the fact that as synthetic chemistry has evolved as a field, so has analytical chemistry. For example, the evolution of HPLC and mass spectroscopy techniques has made it such that it is now common for an analytical group to have

TABLE 7.1. Attachment 2 from Q3A Impurities in New Drug Substance Guidance

Attachment 2: Reporting Impurity Results in an Application

The following table is an illustration of reporting impurity results for identification and qualification in an application.

"Raw" Result (%)	Reported Result (%)	Action	
		Identification (Threshold 0.10%)	Qualification (Threshold 0.15%)
0.066	0.07	None	None
0.0963	0.10	None	None
0.12	0.12[a]	Yes	None[a]
0.1649	0.16[a]	Yes	Yes[a]

[a]After Identification, if the response factor is determined to differ significantly from the original assumptions. it may be appropriate to remeasure the actual amount of the impurity present and reevaluate against the qualification threshold.

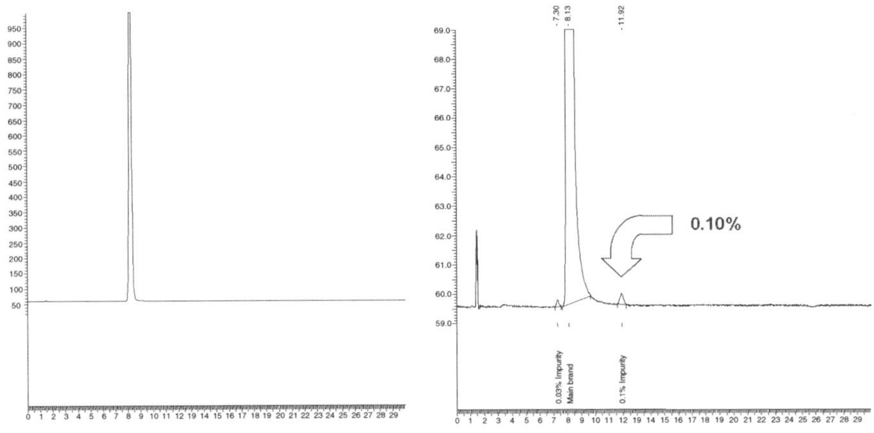

A B

Figure 7.5. Chromatograms showing a 0.10% impurity.

the ability to detect part per million (ppm) quantities of an impurity. Figure 7.5 shows two chromatograms. Chromatogram A represents a material that appears to possess an excellent purity profile. Chromatogram B is a depiction of what a process chemist would be more accustomed to. It is the same chromatographic trace as the one on the left where the main band has been expanded and attention focused on the presence of impurities. The peak to the right of the main band represents the maximum allowable level of an impurity where no action would be required (0.10%). Any higher level of this impurity would require identification of its structure, and if it were to exceed 0.15% in a subsequent batch, the new lot of material would not be of acceptable quality because it would not have been qualified.

Potentially Genotoxic Impurities. Another recently emerging regulatory change has been with regards to potentially genotoxic impurities.[13] While there is no formal guidance on the topic, the European Medicines Agency (EMEA) and Food and Drug Administration (FDA) have independently taken action toward minimizing the risk associated with the presence of potential mutagens and carcinogens. These are potential residues from typical organic reactions such as alkylating agents or intermediates with functionality that is considered to present a potential mutagenic and/or carcinogenic hazard. The EMEA issued a draft guideline on the limits of genotoxic impurities in June 2004. The EMEA states in its guideline that "it is extremely difficult to experimentally prove the existence of a threshold for the genotoxicity of a given mutagen. Thus, in the absence of appropriate evidence supporting the existence of a threshold for a genotoxic compound, it may be prudent to assume that no safe exposure level exists." With that being the case, the EMEA proposed that for genotoxic compounds with sufficient evidence for a threshold-related mechanism, a permitted daily exposure (PDE) be based on the no-observed effect level (NOEL) or the lowest-observed effect level (LOEL). In the case of compounds without sufficient evidence for a threshold-related mechanism, it is proposed that the levels be controlled to "as low as reasonably practicable" (ALARP) where avoidance is not possible. Since in the majority of cases the toxicological assessment of a potentially genotoxic impurity is very limited, the EMEA acknowledged that a "pragmatic approach is therefore needed which recognizes that the presence of very low levels of genotoxic impurities is not associated with an unacceptable risk." The approach proposed is based on the concept of a threshold of toxicological concern (TTC)[14–16] which had been developed for the food industry. This approach comes down to the risk probability of getting cancer if administered a given dose of a compound chronically. The EMEA suggests a TTC of 1.5 μg/day, which represents a high probability that a 10^{-5} lifetime cancer risk will not be exceeded for the chronic daily intake of a genotoxic impurity with unknown carcinogenic potential. The TTC value would not apply for highly potent carcinogens, such as an aflatoxin, but could be exceeded under certain conditions, such as short-term exposure, treatment of a life-threatening conditions without a safer alternative, or when life expectancy is less than 5 years or when the impurity is a known substance and human exposure will be much greater from other sources.

On the other hand, the FDA has not issued any guidance yet, but industry experience indicates a reluctance to apply the TTC approach. Limited interactions suggest that FDA may take a very conservative position such that a compound with a structural alert must be considered as a potential mutagen and purged from the API, unless proven otherwise. This position places a substantial challenge for the synthetic and analytical chemists in early development since robust control of impurities has not yet been demonstrated because of the low level of understanding of the synthetic process and the limited experience with the analytical methodology. Manufacturing and laboratory development experience is limited in early development; and identification, characterization, and control of potentially genotoxic impurities to the ppm levels are at best difficult and often unachievable without deployment of a considerable amount of resources and time. However, in the absence of clear guidance, these impurities are treated as "guilty until proven innocent." Impurities represent additional challenge to process chemists in the design of a chemical process and further exemplify the importance of process robustness and control of purity profile late in the synthesis.[17]

Changes in Required Documentation

Another example of the impact of regulation comes from the risks associated with "mad cow disease."[18] In 1986, the first case of bovine spongiform encephalopathy (BSE) was recognized in the United Kingdom. In 2001, the European Agency for the Evaluation of Medicinal Products (EMEA) issued a note for guidance on minimizing the risk of transmitting animal spongiform encephalopathy agents via human and veterinary medicinal products (EMEA/410/01). For the process chemist, this guidance applies to all raw materials and reagents and necessitated confirmation that raw materials and reagents such as lactose, amino acids, fatty acids, and so on, were not from animal sources. For example, this would be especially problematic if a reaction used an enzyme such as pig liver esterase (PLE) to effect a transformation. This guidance required active collaboration with raw material suppliers for confirmation that a raw material was not from an animal source. A change in the selection of starting material(s) or reagent(s) is required if this confirmation could not be provided.

CONCLUSION

The role of the process chemist extends far beyond the daily development of chemistry for the synthesis of pharmacologically active molecules. As exemplified in the selected topics of this chapter, our responsibilities cover a number of topics ranging from team participation to following regulations from diverse regulatory agencies around the world. While synthetic organic chemistry remains at the heart of the process groups throughout the pharmaceutical industry, process chemists have the opportunity to learn and contribute in a number of different ways and will continue to do so for many years to come as our industry changes and evolves in the face of many new challenges.

ACKNOWLEDGMENTS

The author acknowledges the comments and suggestions from Jim G. Baxter, Tamim F. Braish, David M. De Antonis, Keith M. DeVries, Kieran Dignam, Krista L. Dobo, Sally Gut, Lynne A. Handanyan, Keith E. McCarthy, and John A. Ragan in the preparation of this chapter.

REFERENCES

1. Nafissi, M., Ragan, J. A., and DeVries, K. M. *From Bench to Pilot Plant; Process Research in the Pharmaceutical Industry*, 1st edition, ACS Symposium series 817, Washington DC, 2002.
2. Abdel-Magid, A. F., and Ragan, J. A. *Chemical Process Research; The Art of Practical Organic Synthesis*, 1st edition, ACS Symposium series 870, Washington DC, 2004.
3. Hawkins, J. M., and Watson, T. J. N. *Angew. Chemi. Int. Ed.* **2004**, *43*, 3224–3228.
4. Hoge, G., Wu, H.-P., Kissel, W. S., Pflum, D. A., Greene, D. J., and Bao, J. *J. Am. Chem. Soc.* **2004**, *126*, 5966–5967.
5. Singer, R. A., Caron, S., McDermott, R. E., Arpin, P., and Do, N. M. *Synthesis* **2003**, 1727–1731.
6. Pye, P. J., Rossen, K., Reamer, R. A., Tsou, N. N., Volante, R. P., and Reider, P. J. *J. Am. Chem. Soc.* **1997**, *119*, 6207–6208.
7. Cai, D., Payack, J. F., Bender, D. R., Hughes, D. L., Verhoeven, T. R., and Reider, P. J. *J. Org. Chem.* **1994**, *59*, 7180–7181.
8. Quallich, G. J., and Woodall, T. M. *Tetrahedron Lett.* **1993**, *34*, 4145–4148.

9. Mathre, D. J., Jones, T. K., Xavier, L. C., Blacklock, T. J., Reamer, R. A., Mohan, J. J., Jones, E. T. T., Hoogsteen, K., Baum, M. W., and Grabowski, E. J. J. *J. Org. Chem.* **1991**, *56*, 751–762.
10. GLP usually refers to Good Laboratory Practice for non-clinical laboratory studies and is covered in the Code of Federal Regulations, 21 CFR part 58.
11. Vesper, J. L. *BioPharm* **2000,** 28–32.
12. Several guidance documents can be found at http://www.fda.gov/cder/guidance/. Q3A is listed in the International Conference on Harmonisation section under Quality.
13. In the ICH safety guideline (ICH S2B, 1997; A Standard Battery for Genotoxicity Testing in Pharmaceuticals) it is agreed that the genotoxic potential of drug substance consists of a test for gene mutation in bacteria, an *in vitro* test with cytogenetic evaluation of chromosomal damage in mammalian cells or an *in vitro* mouse lymphoma tk assay and an *in vivo* test for chromosomal damage in rodent hematopoietic cells.
14. Kroes, R., and Kozianowski, G. *Toxicology Letters* **2002,** *127,* 43–36.
15. Kroes, R., Renwick, A. G., Cheeseman, M., Kleiner, J., Mangelsdorf, I., Piersma, A., Schilter, B., Schlatter, J., van Schothorst, F., Vos, J. G., and Wurtzen, G. *Food Chem. Toxicol.* **2004,** *42,* 65–83.
16. Munro, I. C., Kennepohl, E., and Kroes, R. *Food Chem. Toxicol.* **1999,** *37,* 207–232.
17. Muller, L., Mauthe, Robert, J., Riley, Christopher, M., Andino, Marta, M., Antonis, David, D., Beels, C., Degeorge, J., De Knaep, Alfons, G. M., Ellison, D., Fagerland, Jane, A., Frank, R., Fritschel, B., Galloway, S., Harpur, E., Humfrey, Charles, D. N., Jacks, Alexander, S., Jagota, N., Mackinnon, J., Mohan, G., Ness, Daniel, K., O'Donovan, Michael, R., Smith, Mark, D., Vudathala, G., Yotti, L. *Regulatory toxicology and pharmacology: RTP* **2006**, 44, 198–211.
18. Hileman, B. *Chem. Eng. News* **2004,** *82*, 21–25.

8

OUTSOURCING—THE CHALLENGE OF SCIENCE, SPEED, AND QUALITY

SIMONE ANDLER-BURZLAFF

CarboGen AG, Neulandweg 5, CH-5502 Hunzenschwil, Switzerland

JASON BERTOLA

CarboGen and AMCIS, Whitehouse Station, NJ 08888

ROGER E. MARTI

Zurich University of Applied Sciences, Winterthur, CH-8401 Winterthur, Switzerland

During recent years, the competition has significantly increased in the pharmaceutical industry, with more intense pressure on pricing and scrutiny on drug safety. Although the demand for new medicines remains high, the pharmaceutical companies are finding it much more difficult to bring new molecular entities (NMEs) to market (Figure 8.1). R&D spending has been steadily increasing, now accounting for approximately 15–20% of sales.

Starting in the early 1990s the outsourcing industry experienced significant growth in Europe and the United States as pharmaceutical companies grew rapidly and relied on outsourcing companies to complement internal resource and new emerging companies, with limited or no internal capabilities, were established. Since then, diminishing pipelines and mega mergers have created an overcapacity situation. Nevertheless, a lot of pharmaceutical companies still see outsourcing as a viable business model. This trend can be seen in the establishment of a supply network not only for bulk products under manufacturing agreements but also with growing interest in the early-phase development and production service. Although there is significant price pressure on the

Fundamentals of Early Clinical Drug Development: From Synthesis Design to Formulation, edited by Ahmed F. Abdel-Magid and Stéphane Caron

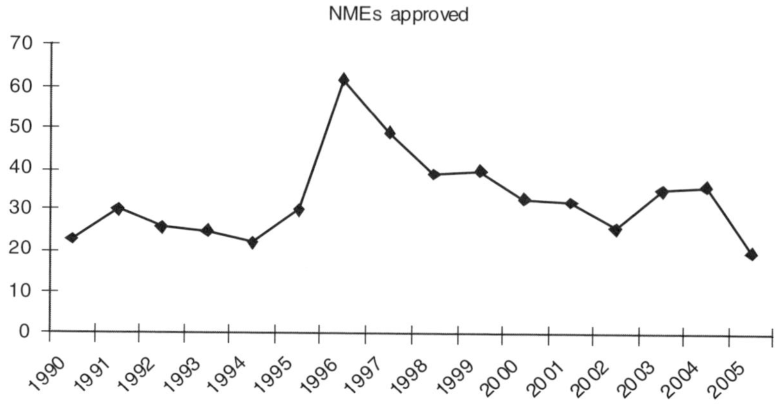

Figure 8.1. Approved new molecular entities (NME) per year.

service industry, specialization in niche markets—such as, for example, highly active pharmaceutical ingredients (HAPIs)—yield good returns.

Since the pharmaceutical companies tend to rely more and more on a system of preferred suppliers, the initial phase of identifying the right supplier for their scope of work is the foundation of this process—client outsourcing collaboration—a process that will be described more in detail below.

The pharmaceutical industry is a highly regulated environment, and the rules change frequently. Medicines regulations (FDA, ICH), environment (EPA), transportation (DOT), employee safety (OSHA), patient rights, and animal welfare (NIH, USDA) all impact the pharmaceutical and health care industry. The ability to understand the regulatory process and implement the frequent changes is important to pharmaceutical companies and their vendors.

Selecting, qualifying, and managing vendors are essential components of the outsourcing process. The rules and regulations[1] clearly state that the contracting firm has the responsibility to see that the outside vendor used is qualified to do the work, and that the work performed is satisfactory. Withhold recommendations resulting from pre-approval inspections (PAIs) are not uncommon. When you consider the cost of delaying the approval of an NDA, the importance of selecting the right service provider cannot be overstated. In addition to time, selecting an inadequate service provider can damage relationships with regulatory agencies, negatively impact precious funding, and product market dynamics.

SUPPLIER SELECTION PROCESS

Quality Audit

From the quality aspect, there are four parts to the auditing process. First, the process starts with developing an understanding of one's needs and wants in order to set the standards necessary to judge vendors audited; the second part

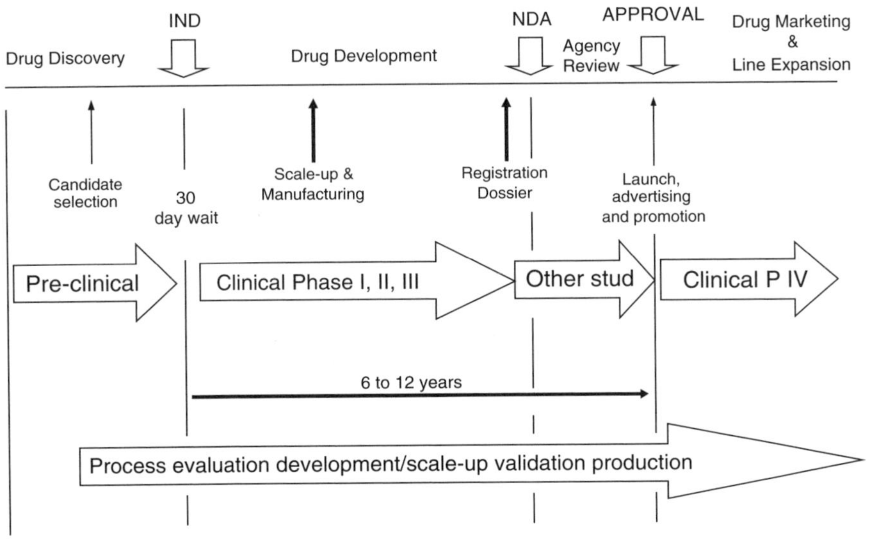

Figure 8.2. Milestones for pharmaceutical companies.

deals with the vendor's capabilities, the third is the implementation of a quality agreement, and the fourth is a program of regular inspections to ensure continued compliance.

Pharmaceutical Company's Needs and Wants. This is an extremely important part of the process where the quality requirements for the project to be outsourced are defined and grouped into needs, which are necessary for the success of the project and compromise is not possible, and wants that are desirable but would not preclude you from working with a vendor. However, if available, it could favorably impact your decision to work with a vendor. This function is project-specific and based on project milestones (Figure 8.2) and timelines.

The Assessment of a Company's Capabilities

Initial Assessment. Typically, the evaluation process starts with a questionnaire (example—Appendix 1). The purpose of the questionnaire is to ensure that the company being evaluated meets some of the basic requirements necessary before more time and effort is allocated to the evaluation process. It is also a good way of prioritizing vendors based on some general characteristics such as location, experience working with the FDA, resources, and technical capabilities.

Site Audit. Prior to the site audit, a detailed agenda or themes for discussion (example—Appendix 2) must be prepared and supplied to the vendor to ensure that the appropriate

Table 8.1. Requirements for the Site Audit

Scientific/Technical Personnel:	This group evaluates the vendor staff experience, technical competence, ability to understand the project, and its challenges. Project staffing flexibility and continuity of staff are other important factors.
Quality Control:	QC procedures, sampling procedures, and sample tracking. Ability to handle QC requirements in-house versus using a third party.
Quality Assurance:	SOPs; vendor knowledge of laws, regulations, and recent case studies that provide the historical knowledge specific to the pharmaceutical company's program. Data relevant to regulatory inspections (483s, warning letters). Employee training program.
Procurement/Alliance Management:	Assess responsiveness to client demands, joint service mentality, ownership of IP requirements, security of information, crisis management, and pricing structure.

staff is available, present at the time of the audit, and prepared to address all issues. Table 8.1 indicates what the auditing team should include.

During the audit, observing a vendor's working culture and environment provides a lot of information on working habits and their compatibility with the pharmaceutical company.

Implementation of a Quality Agreement. The quality agreement should be discussed and negotiated with the contract to ensure that there are no conflicts between the two documents. A good agreement will define terms (GMP, release testing), outline vendor and customer quality and regulatory obligations, and specify any pre-shipment (if manufacturing) and/or documentation required, failure investigation and change control procedures, criteria and frequency of audits, and ultimately a responsibility matrix for raw materials and change control (specifications, manufacturing procedures). The formation of a Quality Working Group (QWG), with staff from both the vendor and pharmaceutical company, is a useful way of quickly addressing issues or concerns that may arise during the course of a project. The QWG should plan to meet regularly, outside the scope of inspections, to discuss continuous improvement processes.

Continued Compliance Through Regular Inspections. Vendors should be audited regularly, at a minimum every year, to ensure continued compliance. Additional audits are scheduled as needed to support or prepare for regulatory milestones such as an NDA filing. The regular follow-up audits do not need to be as thorough as the initial inspection, unless there are major deficiencies in a vendor's performance. Additional audits, such as mock FDA inspections, to prepare a vendor for a pre-approval inspection (PAI) are desirable.

The Causes of Compliance and Quality Failures in a Vendor

a. *Mixed Regulatory Signals:* Typical when vendor receives feedback on their quality systems by different sources and is unable to process and implement the recommendations.
b. *Lack of Understanding of the Regulatory Process:* Inability to interpret and implement regulations caused limited or no understanding of the industry's requirements.
c. *Using the Customer Audits for Training Purposes:* Effective training should be part of every company's quality system. Audits provide a way of testing and benchmarking one's training program, but do not replace the program itself.
d. *Inability to Support Regulatory Positions Different from the Clients:* Vendors serve multiple clients and therefore must be comfortable and capable to defend their regulatory systems. A vendor that tries to amend the quality system for each customer will create confusion and ultimately noncompliance.

The first look at the outsourcing process focused on the supplier selection process. However, the requirement's between pharmaceutical companies can vary significantly. The easiest way to describe this is following the drug development process.

Overview of Drug Development Process. Looking at the overall process of drug development from lead identification to the filing of a New Drug Application (NDA), the process is long and expensive and most products fail to meet preclinical/clinical endpoints. Vendors play significant roles in all parts of the development process. Looking at the large pharmaceutical companies, they invest a significant amount of their revenue in funding research and development programs. Due to their in-house capabilities, the first grams of the new potential drugs are often synthesized in-house by their experienced chemists. Once a synthesis is defined, occasionally the product is outsourced primarily to free up internal capacity or to access a technology or expertise that is not available in-house. These companies typically provide defined synthetic procedures to achieve the scope of the project requiring scale-up modification and analytical work. Pharmaceutical companies may require further development work on the provided process to define a robust scalable procedure for future clinical manufacturing. The outsourcing demands are therefore, not driven by the inability to deliver a scalable process, but instead by the desire to speed up the process eliminating capacity constraints as well as to attain quick access to technology and expertise. In addition to the need for kilogram quantities of their potential NCEs, there is an increasing demand for cost-efficient supply of competitor compounds for reference testing and sometimes metabolites. On the opposite side of the spectrum, biotech and small emerging pharmaceutical companies, either virtual or research-based, rely exclusively on outsourcing partners to fulfill their requirements. With the exception of specialty pharmaceutical companies, the biotech and emerging pharmaceutical companies are very strong research-based organizations but normally have little or no capacity nor the capabilities to synthesize more than a few grams, if at all, of their new potential lead based on medicinal chemistry routes in-house. As a result of this, the supplier will be involved in the drug development process much earlier than with the big pharmaceutical companies. Consequently, the level of detail in the technical information provided is often limited, requiring a significant amount of assumptions and alternative development plans in the initial phase of preparing a proposal.

Most of the biotech companies have identified one compound that will be further developed ideally to the first proof of concept in men and possibly facilitate an out-licensing process to an interested big pharmaceutical companies. This, however, also means that the foundation of the company strongly relies on the success of the single compound, especially with respect to further funding by the financial community. Given the attrition in this business, the risk is high.

Biotech and emerging pharmaceutical companies look for assistance early in the drug discovery process supporting their chemistry efforts or identifying second-generation compounds in parallel. Identifying a good and reliable partner at the beginning is key for these companies, since efficiency, technical and regulatory expertise will help them move forward quickly in the drug development process, and support regulatory filings.

The ideal outsourcing partner should be able to adapt to the variable demands of the different clients as well as to the different stages of a process/project in the overall development circle (Figure 8.3). Whereas the big pharmaceuticals have the expertise and in-house know-how in addition to a very good understanding of what is expected from a vendor, the start-up and biotech company will rely more on the competencies of vendor and include them in the development process. Therefore, good suppliers need to have the regulatory expertise to support their

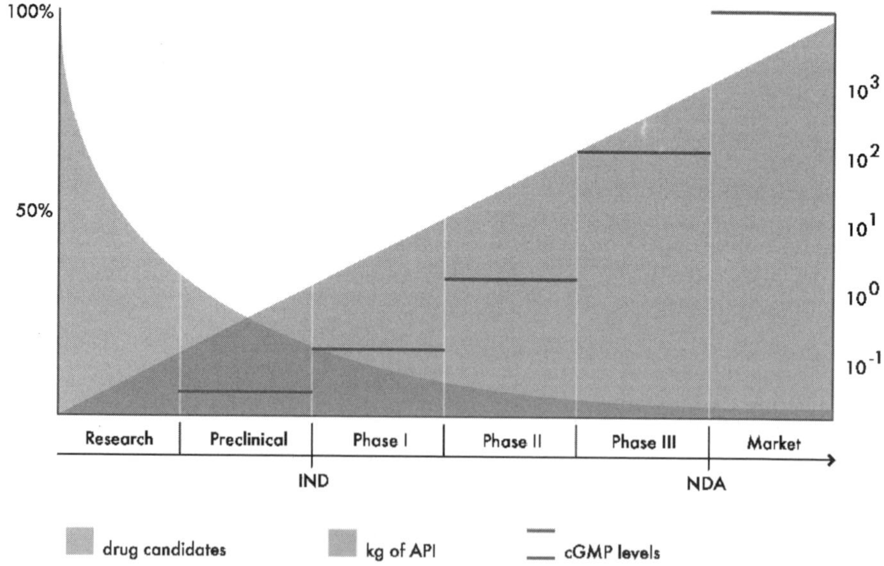

Figure 8.3. Different stages of a process/project in the overall development circle.

clients and satisfy the regulatory agencies as well as the expertise in-house to support the small and biotech industry in the development process. All of them appreciate good science combined with high speed but not at the expense of a suitable quality system. The combination of the three key aspects will be described in more detail in subsequent chapters of this book. What holds all three components together in the outsourcing process is an efficient project management system that not only allows for modification from the supplier side but also simplifies and expedites communication with clients.

SCIENCE

The topic of science in chemical process development may often be associated with highly sophisticated multistep syntheses of chemically and stereochemically complex molecules. But in reality the challenges are often as simple as inadequate analytical methods for in-process control, which can result in the wrong assumptions about conversion and selectivity or unexpected side reactions, which form difficult impurities. In all these cases, good science means strong synthetic knowledge combined with excellent analytical support. The talented chemist needs to understand structure and reactivity of organic molecules and apply this knowledge to any chemistry challenge. The equivalent analytical person needs to gather reliable analytical data including structural information and information about purity and strength. For an outsourcing company to be successful, qualified personnel and a mix of synthetic chemists and analytical chemists need to manage different types of requests such as straightforward chemistry projects compared to complex route finding projects. A strong organization will usually have one analytical chemist per two to three synthetic chemists, depending on the analytical requirements within the projects.

The broad mix of projects and the fast turn-around makes is attractive for chemists to join an outsourcing company. The direct hands-on approach to chemistry problems and their practical solutions in production is a stimulating and rewarding atmosphere. The most important aspect is the delivery of good science and the ability to analyze a problem carefully and develop a sound solution based on creative and innovative suggestions. But having the right people on board is only one part; investments in state-of-the-art technologies and equipment and also recognizing new trends are crucial for an outsourcing company to be successful. A company can only be competitive when it can handle projects in a cost-effective and timely fashion. That means it can be a competitive advantage to have invested in the right combination of equipment and technologies to bring value to the client's project. The level of analytical and synthetic efforts and equipment must match the scope of the project.

SPEED

The two most common questions that a proposal addresses are "How much will it cost?" and "When can I expect delivery?"

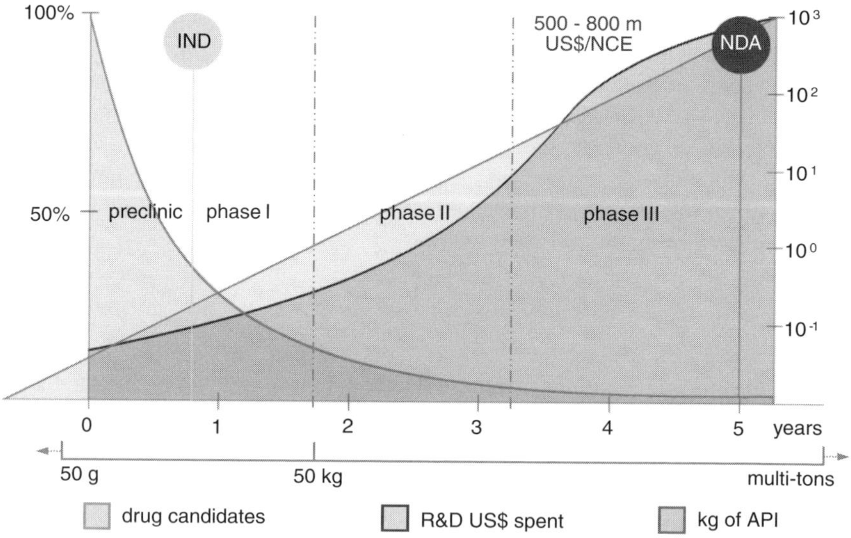

Figure 8.4. Exponential increase in the cost of development.

Everyone understands that speed is important in the drug development process: The faster the product can be delivered and developed, the faster the material can be further tested (e.g., in animals, on humans, physicochemical tests, etc.), and so on. A seamless development from lead optimization to market is the ultimate goal for all companies and is a big challenge for an outsourcing company. This is especially true since nowadays the timeframe from the first hit to getting a compound into the market has significantly increased from around 8 years to up to 15 years at the beginning of the twenty-first century. The increase in development times directly impacts patent life of a new drug. With the cost of development increasing exponentially (Figure 8.4) in the clinical phases, it is to the company's advantage to generate as much information as early as possible on any given project to enable a more educated decision about moving forward or termination of the project.

Estimating timelines for delivery is based on the quantity and quality of technical information available and the experience in the interpretation of this information. If during the development process chemical and technical problems arise, it is the outsourcing company's responsibility to communicate the finding to the client and propose solutions to overcome the non-forecasted challenge. All projects require strong collaboration between the client and service provider. The challenge of the outsourcing company is to give realistic expectations and not over promise. The speed has to be in competition with reliability of product delivery.

Because often the customer's requirements change in the course of the performance of a project, flexibility is an important aspect of the process. It is not unusual to be asked to expedite delivery. The solution, whenever possible, is to work in parallel and add resources to the project. This is often only possible in convergent syntheses. If the process is linear, the increasing speed is accomplished by increasing the risk—for example, shortening the lab evaluation to develop a practical and robust route before going into scale-up and/or do the scale-up after each lab phase and not wait for the results for the subsequent steps. The outsourcing company needs to balance these risks and ensure that the client is fully aware.

An outsourcing company can shorten timelines/development times by seamlessly transferring processes from the laboratory to the kilo plant. It is important to have a lead chemist directly involved in both the hands-on work in the lab and the subsequent scale-up and development phase in the kilo plant. This way, there is no loss of information from the laboratory to the first kilograms of production. This process can also be very helpful in troubleshooting the chemistry.

An example of a technical solution to minimize the interface is the combination of a facility for production of cytotoxic/cytostatic compounds with regular production equipment. The ability to handle a wide range of chemistry and technology tasks persuades the pharmaceutical company to place the entire project with one provider, thereby eliminating the need to manage different supplies and eliminating time-consuming technical transfers. There are other ways of finding speed in collaboration. For example, it may be beneficial to have a supplier geographically close in order to facilitate direct communication and allow close supervision. However, if too much supervision is required, some of the advantages of outsourcing, such as freeing up internal resources, will diminish.

On the other hand, working with suppliers in different time zone may be beneficial. This can result in work continuing at the supplier after normal business hours.

QUALITY

With respect to quality, there are several aspects that a service/outsourcing company must fulfill to satisfy the customer's needs. These aspects are the product quality (good science and the delivery of material within the desired specifications), the regulatory quality, and last, but not least, the service quality.

The basis for fulfilling all the quality aspects lies in the definition of internal processes in general by management handbook and in more detail by standard operation procedures (SOPs). Having clearly defined processes enables a service company to provide consistent expectations and good service. Maintaining these processes requires constant training and a senior management team that is supportive and not willing to compromise. This must be part of the culture of the organization.

Service providers with a good-quality culture are more likely to have a strong regulatory track record and a business based on repeat business.

Depending on the internal expertise of a customer, more or less guidance may be required. For example, a small biotech company does not always have qualified regulatory personnel in house and may require guidance right from the start. In such a case the service company with its quality system is the driving force and must develop a long-term partnership with the customer. The goal of the vendor is to understand the client's needs and develop a solution appropriate for each case. For example, for the synthesis of a simple building block or an intermediate under non-GMP (good manufacturing practice) conditions, documentation by ^{1}H NMR spectra and thin-layer chromatography (TLC) may be adequate. But this is not sufficient for first current GMP (cGMP) batch for the same material! Comprehensive analytical method development by HPLC or even coupled with MS/MS and documentation of all data within a LIMS (Laboratory Information Management System) is a "must have" for an outsourcing company to deliver to the customer.

Medium and large pharmaceutical companies with in-house regulatory expertise have enough know-how in-house to have a clear picture of what is required and expected. In such a case the outsourcing company needs to have the flexibility to be able to adapt certain internal process to the desires of customer without jeopardizing the system.

Ultimately, both service provider and customer must benefit from the relationship for the partnership to produce valuable results.

PROJECT MANAGEMENT

Once a supplier has been identified based on the quality assessment procedure described above, several other aspects become important to build a long-term strategic partnership between the two parties and maximize the outcome of the initial investment in the supplier selection process. In addition to the ability of the supplier to deliver material in the right quality, the real test of the strength of a relationship occurs if a significant challenge is encountered in the performance of a project. The ability to identify a problem is certainly important, but often the challenge is to communicate the problem quickly and effectively. If possible, having some ideas on a solution already in place differentiates a good from a bad partner. As in all close collaborations, communication plays an important role in that process. The role of a project manager starts at the beginning during the process evaluation and continues throughout the life of a project. It is very important in every relationship to have a project manager/alliance manager at both ends that owns the process and is responsible for its success. Project management deals not only with trouble-shooting in the event of a problem, but also with keeping internal and external budget controls and anticipating the customer's need for information. This will allow the customers to really focus on their competences, minimize their involvement with monitoring their projects at the contractor, and keep track of the overall project scope and the related budget.

CASE STUDIES

Science—Focus on Speed

A nice example of the good interaction of science, speed, and quality is the following synthesis of a "simple" molecule shown in Figure 8.5. Although this compound is commercially available, the customer quickly required larger quantities than what was currently available. First of all, the issue of quick delivery was solved by assigning enough qualified chemists and analytical resources to the project. The quality

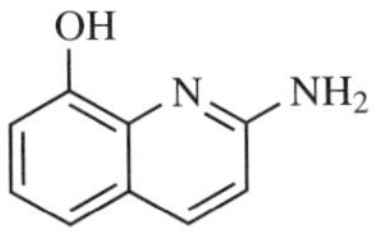

Figure 8.5. Synthesis of a "simple" molecule.

aspect was also easily solved because this compound would be used in synthesis research laboratory and no special regulatory efforts were required. Standard specifications (>95% pure by NMR assay) were applicable. What about science? No specific technical information was available from the customer, only the general information that the Tschichibabin reaction was a mess and involves a significant safety concern. As a result, several groups had to interact for the success of this project.

A detailed retro-synthetic analysis and extensive literature search revealed only a few entries for this compound. All this information and some alternative ideas were summarized, and the different synthetic approaches were discussed and prioritized based on the ease of synthesis. Within a few days, four different synthetic approaches were available for evaluation and then discussed with the customer. In this discussion we agreed on a parallel evaluation of different synthetic approaches, having first a short period of route finding, followed by a review on the best route, and then finally a quick move on to the synthesis of a 50-g sample. If needed, the route could then be further developed, resulting in a process for the production of kilogram quantities.

After two weeks from the start of the project, the most successful route was identified from the four different approaches and, after a discussion with the customer, a decision was made to proceed with the laboratory synthesis of the 50-g sample and to focus all efforts on delivery. By doing so, and after some hard work to solve some "simple" problems like filtrations or phase separation, two weeks later we were able to send the first 50-g sample to the customer. After meeting the first milestone of identifying a synthetic route and preparing the initial sample, a subsequent project was initiated to develop a reliable and scalable process. The outcome of that was a kilo-laboratory production of a first batch with a good and safe process as subsequently published.[2]

Second-Generation Compounds for Small Biotech Company Leads to Strong Partnership and Further Development Projects

The next two case studies are examples of understanding the customer's needs, developing innovative ideas to solve problems, and bringing value through process improvements. Both cases are about the synthesis of second-generation lead compounds.

The first case study is about the start of new and strong partnership. The project started with the drawing of an NCE on a piece of paper. Good retro-synthetic analysis capabilities allowed the identification of some good and practical strategies to deliver a set of different but related compounds on a small scale for initial screening. An important aspect of the project was the excellent analytical work performed on LC-MS, NMR, and IR. Without this support the project may not have been as successful. A good working relationship was established through regular teleconferences and face-to-face meetings. The synergy between the two technical teams produced great results. Project management's involvement facilitated the exchange of information in this fluid environment. During the course of the project, new ideas for new compound classes were discovered and incorporated into the project. As the testing of one of the prepared compounds showed a promising activity, the customer required more of that material. That was easily accomplished within the vendor's by adding resources to the team without losing any valuable technical information. The project team was expanded to include scale-up chemists, and the compound moved on into the kilo lab where it was synthesized on a 100-g scale with a process ready for cGMP production.

The second case study started with the preparation of a 100-g batch of a compound by a one-step synthesis. However, the yield was low and the process was not scalable. The goal was to deliver the requested amount and brainstorm on some alternative ideas to prepare this compound in a more efficient way and/or prepare alternative structures. Within this new project, we prepared a series of new second-generation lead compounds based on customer's ideas and our chemical input. This demonstration of our capabilities and the maximum amount of services in the customer favor was the starting point to initiate a second project, namely, the development of a phase-2 compound. Because the customer was a small start-up pharmaceutical company with no chemical development capabilities, the technical input was highly appreciated and the successful completion of this first project was the basis for future business.

These examples demonstrate that a flexible and customer-oriented approach to solving "simple" chemistry problems by applying the right tools from science (synthetic chemistry and analytical), speed and flexibility (working in parallel and focused on solving the problem), and the "right" quality at the right time brings a competitive advantage to our customers.

CONCLUSION

The use of vendors and contracting by pharmaceutical companies is increasing. This is true for large and small pharmaceutical companies, as well as biotech companies. There are economical and logistical benefits that can be gained from outsourcing, but liabilities can result from the misunderstanding of regulatory roles and responsibilities of vendors to their customer. The process is complex and time-consuming but, if approached systematically, can be very rewarding.

REFERENCES

1. Federal Register, Vol. 43, No. 190, p. 45034 (September 29, 1978).
2. Storz, T., Marti, R. E., Meier, R., Nury, P., Roeder, M., and Zhang, K. *Org. Process Res. Dev.* **2004**, *8*, 663–665.

APPENDIX 1: THE EVALUATION PROCESS

Table 8.2. Questionnaire for the Evaluation Process

Company			
Location			
Public or Private			
Organization Chart (attach)			
Personnel			
Key Contacts:			
	Quality/Regulatory		
	Business		
	Scientific		
Educational Background:			
		Ph.D	
		M.S.	
		B.S.	
		Other	
Facility			
Location			
Age			
FDA inspected?			
Last audited date			
	Any 483s?		
		How many findings?	
		Were they resolved?	
	Any warning letters?		
		How many were there?	
		Were they resolved?	
DMFs			
Contributions to IND			
Contributions to NDA			
DEA licensed?			
Equipment			
	list of equipment type and size		
Material Handling (Yes/No)			
	Sterile (Class)		
	Cytotoxics		
	Potent Material (Class)		
	Controlled Substance		
	Beta-lactams		
	Broth Handling-biomass handling (UF, NF)		
	Recombinant Organisms		
Safety			
	Number of days without loss time accident		
	OSHA Reportable Incidents(#)		

APPENDIX 2: AUDIT PROCESS THEMES

1. **Personnel**

    - Organization charts
    - CVs
    - Staffing paradigm (5 days/40 hr or 3 shifts/7 days, for example)
    - QA/QC structure in the organization

2. **Laboratories**

    - Tour of the laboratories
    - Adequately staffed and equipped with all necessary equipment

3. **Production facilities**

    - Tour of the facility
    - Cross-contamination issue
    - Warehouse and distribution systems
    - People/material/equipment flow
    - Technologies and products that can or cannot be handled
    - Standard housekeeping and pest control practices/procedures
    - Any cGMP, OSHA, or EPA issues

4. **Quality systems (SOPs to support the following areas)**

    - Batch Disposition
    - Batch records/process documentation
    - Deviations system
    - Change control system
    - Procedures system
    - Audit program—internal and external
    - Training system
    - Equipment qualification/validation/PM program
    - Cleaning—equipment and facility
    - Complaints
    - Environmental monitoring
    - cGMP, GLP

5. **Process engineering**

    - Scale-up and technology transfer procedures

6. **Project management**
    - Processes to ensure that a project will meet its objectives
    - What systems are used to monitor and manage a project?
    - How is the project status communicated?
    - Is risk management performed on the project and how is it communicated?
7. **Documentation and know-how**
    - How is documentation managed? What document management systems are employed?
    - What types of information are held in a document management system?
    - Knowledge management tools that are in use
8. **Communication**
    - Provision for communicating and retaining information
    - Frequency of communications
    - IP security
9. **General Financial**
    - Size of the company
    - Financial stability
    - Access to capital and investment programs
10. **Vendor strategy and vision**
    - 3- and 5-year plan
    - Customer mix (large pharma versus biotech)
11. **Contracting and metrics**
    - Discussion and agreement on metrics to measure performance
    - Number and type of other contractual agreements
12. **Demonstration of respect for diversity**
    - Diversity of workforce
    - Policy and programs to support diversity
13. **Health and safety management**
    - Processes for inspecting and handling facility hazards
    - Written safety program
    - History: fatalities or serious injuries

14. Environmental management

- Procedures for spill and environmental releases
- Local and governmental regulation for waste disposal
- Any history of environmental problems?
- On-site water treatment and incineration
- Does the facility have all necessary permits?

15. Inspection history

- Regulatory agencies' visits, reports, and current status on replies

9

AUTOMATION AND THE CHANGING FACE OF PROCESS RESEARCH IN THE PHARMACEUTICAL INDUSTRY

EDWARD J. DELANEY, MERRILL L. DAVIES, BRENT D. KARCHER, VICTOR W. ROSSO, A. ERIK RUBIN, AND JOHN J. VENIT

Process R&D Department, Bristol-Myers Squibb Co.

INTRODUCTION

Over the past 10 years, dramatic changes have taken place within the pharmaceutical industry as new strategies have been adopted to identify new drug candidates. To sharpen their competitive edge in rapidly creating intellectual property and identifying leads for product development, pharmaceutical companies have invested heavily in automated instrumentation and informatics tools to (1) increase the pool of compounds available from which to guide the generation of new leads and (2) speed the acquisition of biological activity data from which informed decisions can be made on drug candidacy and further development.

Process research and development functions have also seen changes over time. With the passage of the Waxman–Hatch Act in 1984, pharmaceutical companies began an era of shortened product life cycles and enhanced emphasis on new product innovation. These changes have in turn intensified the need for at-risk process development at an earlier stage and have created a shift from a "learn-while-doing" approach toward a "learn-before-doing" approach, as described by Pisano in *The Development Factory.*[1] Advances in analytical instrumentation and greater accessibility to analytical tools have also played a role in shifting the primary objective from

Fundamentals of Early Clinical Drug Development: From Synthesis Design to Formulation, edited by Ahmed F. Abdel-Magid and Stéphane Caron

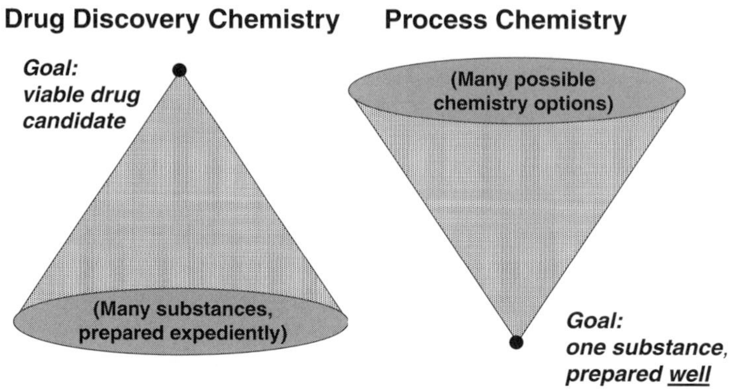

Figure 9.1. Goals of drug discovery and process R&D.

one of *making material* to one of *creating knowledge*. Given this scenario, it would seem appropriate to ask why Process R&D organizations have not embraced automation and informatics technology to the same extent as Drug Discovery organizations.

The answer lies to a large extent on the divergent nature of the goals of the two functions. Drug discovery units seek to expediently prepare and screen a large number of compounds from which a chosen few are selected for development. The informational deliverable is the structure of each chosen candidate, along with its associated biological and physicochemical data supporting the development decision (Figure 9.1, left). In this environment, automation and informatics technologies are primarily employed based on the repetitive nature of compound preparation and assay execution (e.g., parallel synthesis based on split-and-pool strategies, and the use of high-throughput assay technology). Process R&D units have the very different goal of designing appropriate process technology for each candidate molecule moving into and through clinical development. The informational deliverable in this case is the choice among an array of potential synthetic pathways and an understanding of each chosen transformation in sufficient depth to ensure that each candidate can be successfully prepared in a timely manner, at any scale desired (Figure 9.1, right).

The goals of Process R&D also vary according to the stage of development (Figure 9.2). When projects are first transitioned to development, relatively small quantities (1–5 kg) are needed, and the probability for overall project success is still low. Typically, Process R&D organizations employ some expedient variant of the synthetic route used during drug discovery to make initial development supplies. Thereafter, however, they seek to identify the future commercial route and develop manufacturing processes as effectively as possible, and this is the timeframe where the greatest opportunity exists for productivity enhancement.

The diagram in Figure 9.3 can be used to represent the challenge faced by process scientists at the onset of the search for a long-term process. The top surface of the overall cone represents the broad number of hypothetical chemical processing options that might be attempted (possible experimental approaches), whereas the point at the bottom represents a hypothetical end point, an "optimal process"

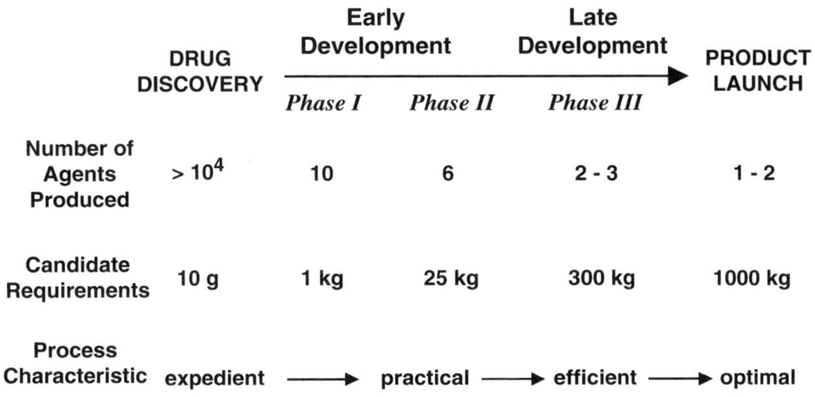

Figure 9.2. Stage-dependent goals of process R&D.

that meets all criteria of interest (including minimal cost of goods, fewest overall steps and operations, minimum environmental impact, and greatest safety provided in materials handling). In reality, such a process is rarely (if ever!) achieved since the goals are often conflicting and subject to trade-off. The top surface of the smallest cone ("past") can be used to represent the traditional approach to process research and development that relies heavily on the individual training and experiences of process chemists and engineers. Experimentation has historically been carried out more or less sequentially in three-necked round-bottomed flasks, limited by the experimentation space available (i.e., the typical laboratory hood). Although the experimentalist can usually conceive of many possible approaches to solving a problem, the above limitations naturally force scientists to choose the most likely to succeed based on their experience and intuition, often leading them down a sequential path of trial and error. Similarly, time limitations for development force

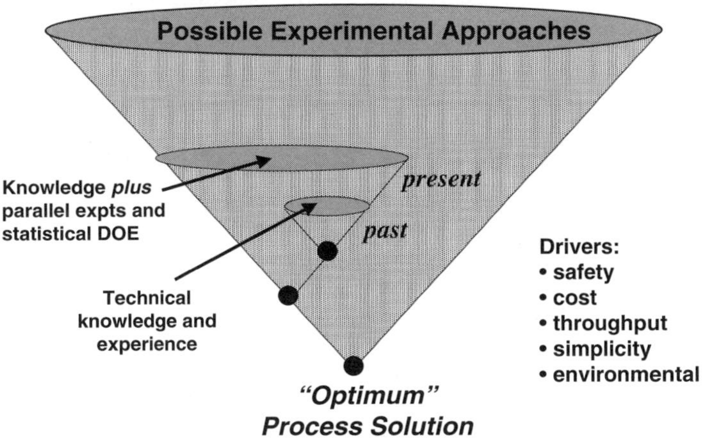

Figure 9.3. Process R&D productivity gain opportunity.

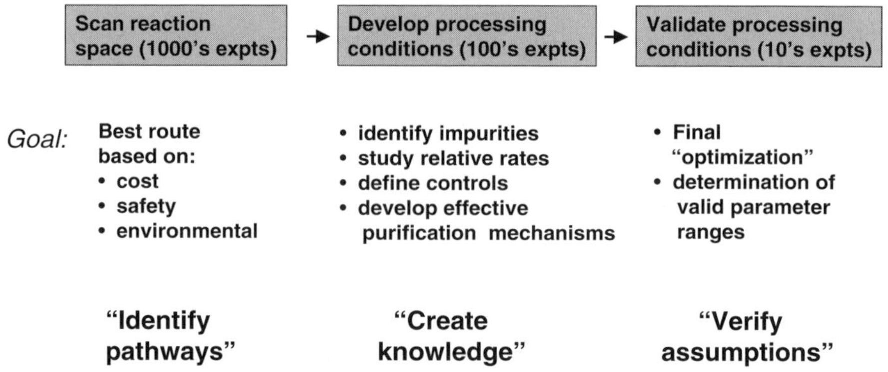

Figure 9.4. Future approach based on availability of tools that enhance experimentation capability.

process scientists toward empirical problem-solving over more extensive experimentation that could lead to greater depth of mechanistic understanding and control. Consequently, it should not be surprising to find that development of final manufacturing processes frequently requires two or more rounds of laboratory development (i.e., route changes or major process modifications) accompanied by scaled demonstration.

To break out from this paradigm process, scientists require vehicles enabling greater *parallelism* in experimentation, as well as mechanisms that enhance rates of reliable data gathering and analysis. For example, while it would be wasteful to attempt every conceivable synthetic approach to a target molecule (the top surface of the cone in Figure 9.3), high risk/high reward approaches are frequently overlooked or under-researched using the traditional model. If tools that enable systematic and rapid evaluation of a wider range of process possibilities are made accessible to process scientists, it would be logical to expect that critical (and frequently nonobvious) discoveries would be made earlier, allowing for long-term process technology to be developed sooner and with greater thoroughness (top surface of the larger cone of Figure 9.3, "present," leading to a process result closer to optimum). With tools that enable higher throughput of *experimentation*, a model for process research and development could easily look like that described in Figure 9.4.

PROCESS KNOWLEDGE AND THE "ART" OF PROCESS DESIGN

A poignant example follows on how such tools could be applied and leveraged. Very early in his career, one of the authors was assigned to develop process technology for an experimental steroidal anti-inflammatory agent. Route selection was not a priority in this instance, as a specific steroid intermediate was a clear choice as starting material (i.e., commercially available for a reasonable cost). The key transformation from that starting material, at its initial point of development, is described in Figure 9.5. The central problem in this step was achieving regioselective ketalization at the 17-ketone position in the face of competing thioketalization at the 3-ketone position.

Product mix (undeveloped):

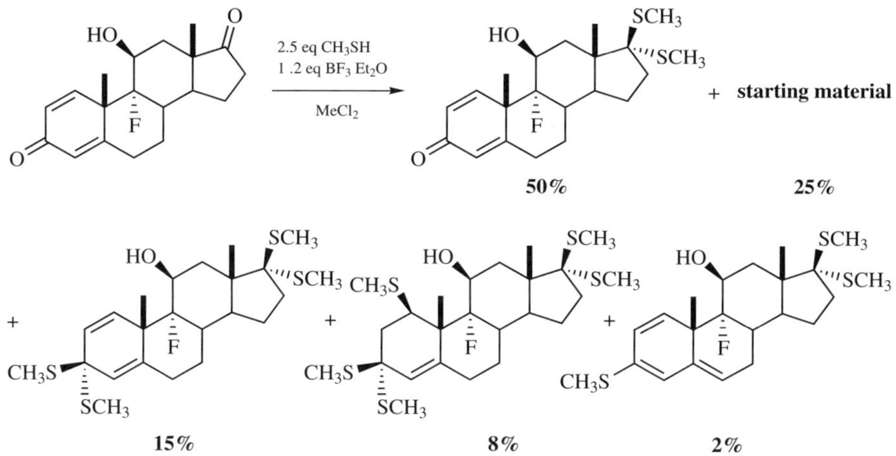

Figure 9.5. Example of early-stage process development.

Initial experimentation revealed the extent of the difficulty, because yields could not be raised above 50%. Shorter reaction times lowered the relative amount of "over-thioketalization," but left considerably more starting material, while longer reaction times consumed a greater proportion of starting material, but raised the impurity levels. In either case, chromatography was required to remove both residual starting material and the byproducts to allow further processing, because efforts to remove these materials by crystallization failed to show promise. In the face of pressure to produce supplies for clinical development, the procedure might have remained as described in Figure 9.5 much later into the development cycle, had it not been for a fortuitous error that revealed an important subtlety. In the course of carrying out one specific small-scale experiment, the researcher found that the desired product had precipitated from the reaction mixture, starting material had been completely consumed, and impurity levels had dropped dramatically. Upon investigation, he realized he had made a careless miscalculation resulting in a 10-fold overcharge of both boron trifluoride etherate catalyst and methanethiol!

In hindsight, the author pieced together what had been at play (Figure 9.6). Empirically, a sufficient level of catalyst had been chosen to produce the desired reaction within a reasonable length of time (30–60 minutes), and a limited excess of methanethiol was employed since experiments to date suggested that regioselective reaction would not be achievable. Chemical intuition would thus suggest that additional methanethiol would result only in increased production of the "over-thioketalization" impurities. In fact, however, the two ketones are vastly different. The carbonyl oxygen at the 3-position is more electron-rich, and it competes more effectively for available boron-trifluoride catalyst. However, once complexed by boron trifluoride, it is less reactive than the complexed form of the carbonyl oxygen at the 17-position, since the complexed 3-carbonyl is highly stabilized by the

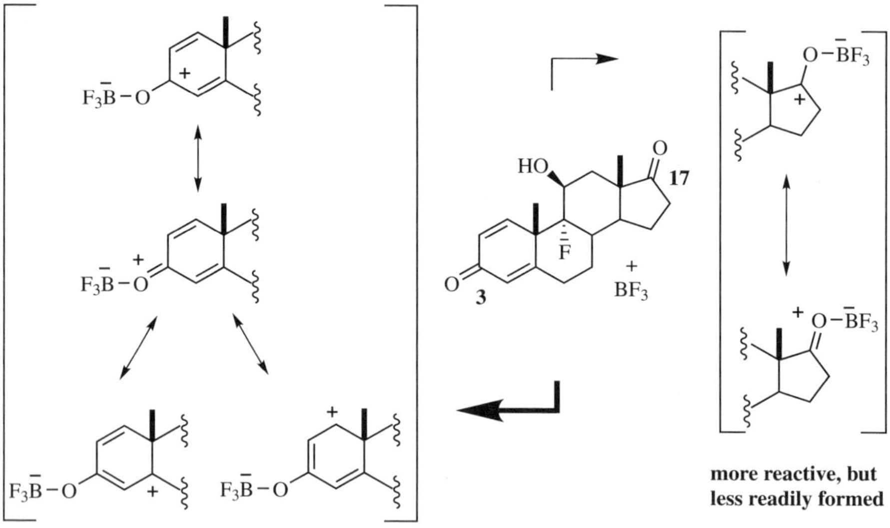

Figure 9.6. Comparison of 3- and 17-carbonyl activation.

adjacent double bonds, and two of the resulting resonance forms produce carbocations that are sterically encumbered by an adjacent quaternary center. Thus, addition of sufficient catalyst to ensure complexation of both centers leverages the greater reactivity of the activated 17-ketone, conferring a more rapid rate toward the desired product. The addition of excess thiol helped the dissolution of starting material; but counterintuitively, it also accentuated the kinetic advantage in conjunction with rapid product crystallization from the reaction mixture. Ultimately, a highly efficient, well-understood process was designed and scaled around these findings, the flow diagram of which is shown in Figure 9.7.

Although the personal side of the scientist would like to downplay the significance of the charging error and believe that knowledge and intuition played the dominant role in the successful design of the above process step, it is also important to recognize that serendipity frequently plays an important role in science. Ultimately, both elements are important. While chance favors the prepared mind, chance only favors those who court her.[2] Just as important, however, the potential for regioselectivity might also have been rationalized as a testable hypotheses *a priori*, thus allowing a manageable number of screening experiments to be targeted. It has been the authors' experience that taking such a "step back" is not necessarily limited by the interests and aptitudes of process scientists as by the extensive time required to carry out such experiments in the face of deadlines synchronous with the business of pharmaceutical development. Thus, tools that allow scientists to generate such information *more rapidly* should pay dividends in the form of better process technology and fewer development cycles.

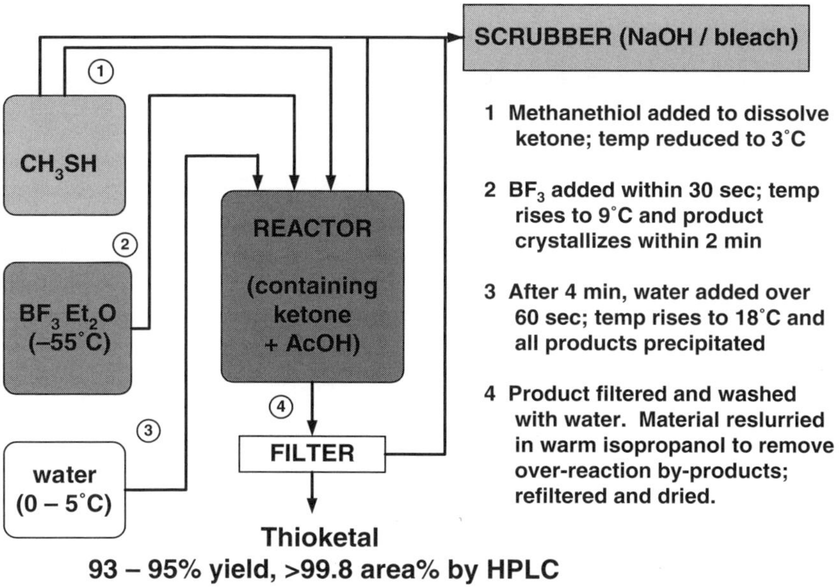

Figure 9.7. Final defined process.

If one could imagine starting development of the same process today without prior knowledge, what tools could enhance the rate at which key observations are made and built upon to reach the final process described in Figure 9.6 (or a better one)? The critical control parameters for the final process are shown in Figure 9.8, and it is worth emphasizing that the factors involved are all variables that could be identified through a systematic, structured experimental approach. Thus, if an experimentalist started with an appropriate analytical method to accurately measure responses for each experiment (assay for desired product, starting material, known impurities and overall mass balance), it is conceivable that those critical parameters could

- Use of high excesses of BF_3 (5 equivalents) and methanethiol (6 equivalents)
- Solvent / reagent / catalyst combination - provide solubility for starting material, insoluble conditions for product
- Controlled reaction temperature in the range of 0 to 10°C, absorption of heat generated during exothermic reaction
- Limited reaction time, and use of rapid quench

Figure 9.8. Critical process parameters identified.

have been reached from a structured approach in which all possible variables are tested blindly across wide ranges. Moreover, it would be reasonable to expect that the same critical parameters should be reached with still greater efficiency by limiting the experiments to short loops that test rational hypotheses (such as factors that could be expected to impact regioselectivity), to those that measure relative rates (i.e., directionality-indicating), or to those that measure solubilities so as to predict potential for a selective (and protective) crystallization of the desired product.

TOOLS AND APPROACHES

The term *automation* conjures up the image of a robot that operates unattended while working 24/7 on tedious work that would otherwise be carried out by human hand. Application of this concept is most applicable to highly repetitive operations that can be cleanly programmed; unfortunately, most process research and development activities do not fit this mold. The key concept is to build mechanisms that support scientists' capacity to carry out *parallel experimentation*. Since experimentation can vary widely, any mechanism chosen must necessarily also be sufficiently complex to allow *flexibility* in its application (analogous to the way that Microsoft Excel can be applied flexibly to a wide range of calculation problems). The following discussion focuses on a range of tools that have been developed within Bristol–Myers Squibb to foster the concepts of parallel experimentation:

Hardware

The simplest and most easily adopted innovation supporting parallel experimentation is the concept of a reactor block. Several designs have been created over the past 10 years, and a prototype developed within Bristol–Myers Squibb is shown in Figure 9.9. Built for both screening and controlled experiments, this tool allows for up to 24 independent experiments to be conducted in parallel using disposable

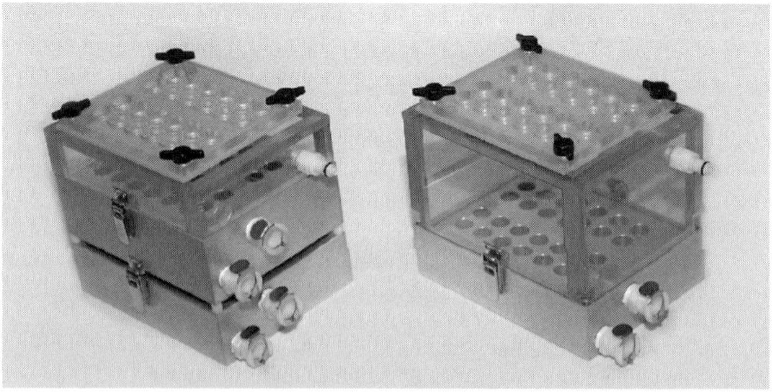

Figure 9.9. Early reactor block design.

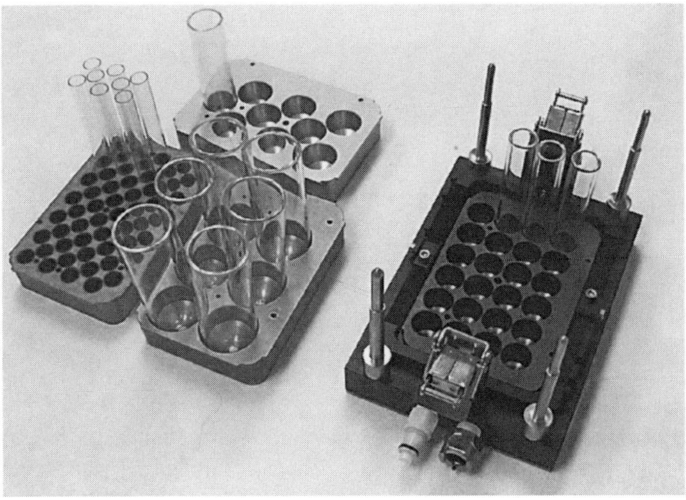

Figure 9.10. MiniBlock design: accommodates different size and number of reactors.

borosilicate test tubes. Use of aluminum as the block core permits ready circulation of a coolant or heating fluid around all tubes, while allowing for simultaneous stir bar agitation of all cells when the unit is appropriately placed on a standard stir plate. An optional upper block enables the circulation of a coolant near the upper areas of the tubes, thus permitting experiments to be performed under reflux conditions when the lower block is heated and the upper block is cooled. Finally, a plexiglass hood maintains all tubes under an inert atmosphere of nitrogen or argon, while Teflon-lined rubber septa fitted with the hood allows for the introduction of materials via syringe. When properly set up, the system has been demonstrated to retain >95% of an initial charge of methylene chloride held at reflux for 20 hours.

As the use of reactor blocks took hold among BMS process scientists a second generation system was initiated to allow greater range of flexibility based on the MiniBlock design developed within Bristol–Myers Squibb Drug Discovery. Starting with a frame that could heat or cool a common template of aluminum cores, a 48-well core was established to support screening work, while additional cores were created that could accommodate fewer, more carefully controlled experiments requiring larger volumes (Figure 9.10). Corresponding upper blocks enabled reflux conditions (Figure 9.11) and a more compact design to allow reactions to be run under inert atmosphere was established (Figures 9.12 and 9.13). The overall design was ultimately adopted by Mettler-Toledo Autochem, which now sells the system as the MiniBlock XT®.[3]

It would seem a major compromise of the above design that the entire block must be controlled at the same temperature. However, the gain of simplicity and reduced cost provides a positive trade-off against the desirability of temperature control for individual cells. As described below, experiments can usually be designed to test several specific temperatures in combination with other test

Figure 9.11. MiniBlock design: with and without condenser block.

variables, and the employment of three blocks to vary temperature in a given experimental design meets most experimental needs.

In many cases, the use of a reactor block with manual material charging is sufficient to support a specific set of parallel experiments. However, for more extensive designs or the screening of an array of conditions, an automated mechanism can frequently be advantageous. Ideally, the tool would perform a variety of functions under a set of instructions from the experimentalist, including the measurement and dispensing of starting materials and reagents, sampling of reaction mixtures, and timed addition of quench solutions. For this purpose, we frequently employ the Gilson 215 liquid handler, the sole function of which is to measure and move liquids according to the instructions of the experimentalist. Liquid handlers have good

Figure 9.12. MiniBlock design: teflon seal for reactors (upper right) allow delivery, sampling, and inerting.

Figure 9.13. MiniBlock design: reactors covered, sealed, and latched.

dynamic range (microliter to milliliter), show good reproducibility and accuracy, are cost-efficient, and, aside from some safety limitations (e.g., movement of pyrophoric liquids) are extremely flexible and applicable to a wide array of experiments. In fact, the most significant limitation of a liquid handler is the fact that it can only handle liquids, meaning that any solid starting materials or reagents to be transferred must be fully dissolved in solvents. Fortunately, process chemists prefer, whenever possible, to conduct chemistry in solution in order to achieve reproducible reaction times and reaction profiles, and process engineers prefer to handle solutions over solids from a material handling standpoint. Therefore, the limitations imposed are not highly restrictive. The deck of a Gilson 215 liquid handler (configured with reactor block, reagent and starting material solutions, and vials for HPLC sampling) is shown in Figure 9.14.

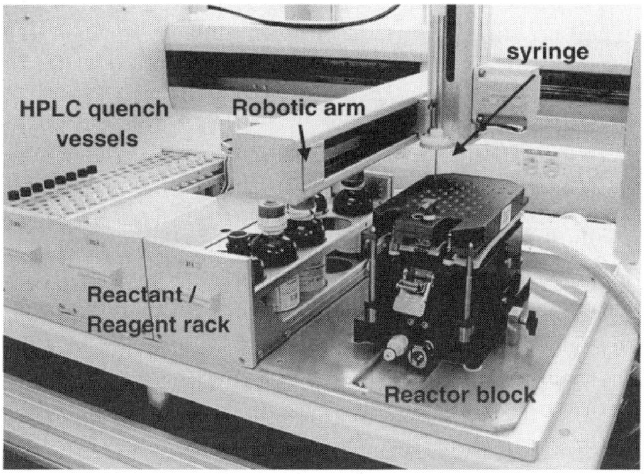

Figure 9.14. Gilson 215 liquid handler, set for experimentation.

The transfer of liquids to meet experimental design requirements has significant advantages. In addition to providing a means by which varying charges can be distributed among a variety of experiments, it allows all experiments to start from one stock solution, which in turn requires the experimentalist to make only one weighing of each starting material or reagent. In cases where one of the reagents or starting materials cannot be dissolved initially, the use of a "smart balance" can speed the weighing process. This technology (also developed within Bristol–Myers Squibb) includes software that allows the operator to set a target weighing range. As material is dispensed into the pre-tared vessel on the balance pan, the operator is given a visual cue on the monitor when the desired mass range is reached. With the click of the computer's ENTER key, the precise weight is stored in computer memory for future use. With practice, several weight measurements can be accomplished per minute. While exact amounts are not weighed, accurate amounts within a given range are provided. The accurate weights recorded can subsequently be used as the basis for calculating the solution charges of other starting materials and reagents by the liquid handler to reach desired molar equivalent ratios and reactant concentrations.

Analytics

The heart of any parallel experimentation system is the means by which responses of interest are measured, and new, unexpected information is captured. HPLC is the workhorse method used daily by process chemists throughout the pharmaceutical industry because of its proven reliability and its use to support regulatory registrations. However, the *development* of HPLC assay technology to support parallel experimentation is often time-consuming and can create a bottleneck. To address this problem, a system was developed within our Process R&D department for screening candidate HPLC columns and mobile phases [Multi-Dimensional Screening and Analysis (MDSA)] as an aid to method development.[4] Six diverse, but complementary, reverse-phase columns were chosen based on the past experiences, and these were matched up with four gradient mobile phase pairings to create a separation "gauntlet" (Figure 9.15). Use of a software-controlled column changer and a custom-designed program allows users to conduct automated "MDSA runs" unattended. Sequential injections of an analyte onto six columns in matrix with four gradient mixtures produce 24 chromatograms with a few mouse clicks.

Based on UV detection at two wavelengths, the concept can be applied not only toward HPLC method development, but also toward challenging purity assumptions of reaction mixtures or isolated products that would otherwise be based on a single assay method (it is not uncommon for significant synthetic byproduct impurities to co-elute with the desired product, or with other byproducts in a given HPLC system). Since process development in the pharmaceutical industry often involves reducing impurities below ICH guidelines of 0.15%,[5] such information is very important. A demonstration of the ability of the system to resolve one impurity from another for identification purposes is shown in Figure 9.16. Although measurements can be based on a variety of detection methods, the presence of

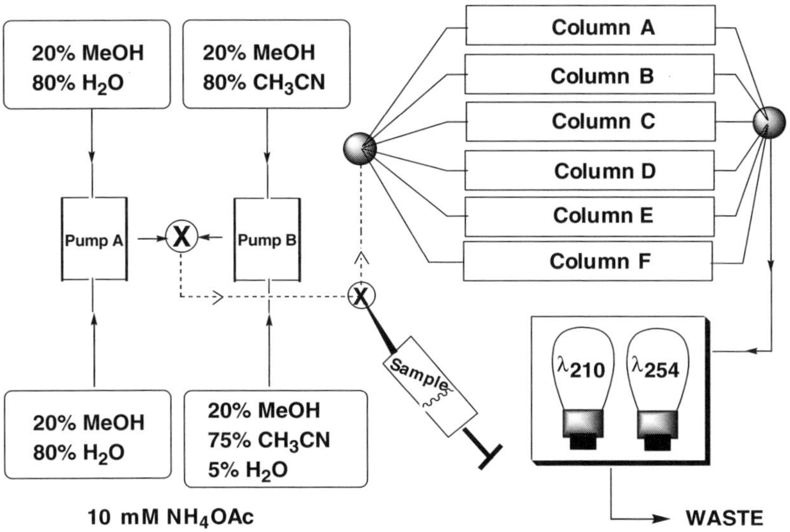

Figure 9.15. MDSA chromatography system.

UV-absorbing chromophores within the vast majority of drug molecules and their corresponding synthetic byproducts makes UV detection the most frequently used measurement option.

Just as important to identifying the presence of byproduct impurities is the elucidation of their structures. When the structure of an impurity is identified, it frequently provides immediate clues to the research scientist regarding why the impurity formed and how the impurity might be controlled or eliminated. Therefore, the rate at which impurity structures are identified is also of great importance. For this reason, the MDSA system employs only mobile phases containing volatile buffers that are compatible with existing LC/MS technology. LC/MS detectors are paired with several of our MDSA systems in a "walk-up" mode, allowing researchers to rapidly gain mass spectral information (frequently molecular ion data) on peaks of interest from a specific MDSA run. Similarly, with the emergence of methodology to collect isolates from analytical columns via solid phase extraction (SPE) and then backflush them with deuterated solvents into the cryo-flow probe of a 600-MHz NMR, impurity identification is becoming much more efficient.

Assuming appropriate stability of the impurity of interest, a chosen chromatographic condition from MDSA can also be transferred directly to automated semipreparative chromatography (systems that were developed within BMS drug discovery to purify libraries of compounds prepared via automated parallel synthesis). The availability of purified isolates is important for quantitative purposes because it allows relative response to be calculated against the desired product. Impurity isolates are also commonly put to use studying the reactivity of the impurity independent of the other components of the reaction of interest (e.g., studying rates of reversibility or further degradation).

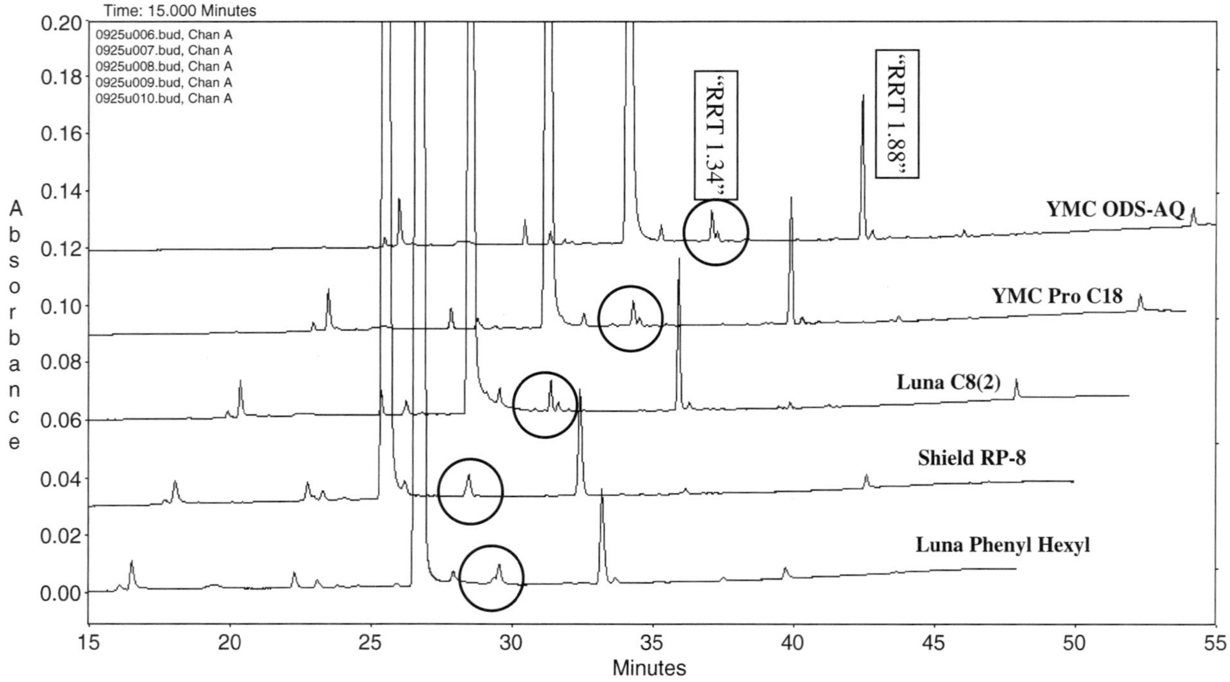

TFA Gradient: 254 nm

Figure 9.16. Separation of impurities using MDSA.

Experimental Design

The amount of knowledge gained from of each set of experiments ultimately depends on the creativity and resourcefulness of the experimentalist, who sets the flow of experiments based on theory or knowledge gained through past experiments. Quite often, however, optimization problems faced by researchers are sufficiently complex to require a "trial and error" approach as a starting point. Statistical design of experiments (DOE) is a structured approach that is particularly useful in those situations.

The basic methods for DOE were first developed by Sir Ronald Fisher in 1921 for studying the effect of various treatments (such as soil conditions, fertilizers, and watering rates) on crop yields. Further methods were subsequently refined by George Box, Genichi Taguchi, and others, and the techniques have been applied successfully in many industrial fields (including bulk chemicals) over the past 80 years. Surprisingly, DOE has seen limited use in pharmaceutical process R&D until recently. With the availability of tools such as those described above, however, many converts are being made among Pharma-based process scientists and the use of DOE is rapidly increasing.[6] A concise description of DOE and how automated methods can be used to harness its power is provided in the Chemistry/Process Examples section below.

Informatics and Systems Integration

The preceding discussion of tools and approaches illustrates that productivity gains made possible by parallel experimentation do not involve one specific tool, but a variety of tools working together. When vendors promote automation solutions to Pharma process R&D scientists, they typically focus on hardware enablement of experimental execution. The reality, however, is that there are a number of additional barriers to routine and systematic use of parallel experimentation concepts.

Cycles of parallel experimentation can be represented as shown in Figure 9.17, and while hardware components to support parallel experimentation are important,

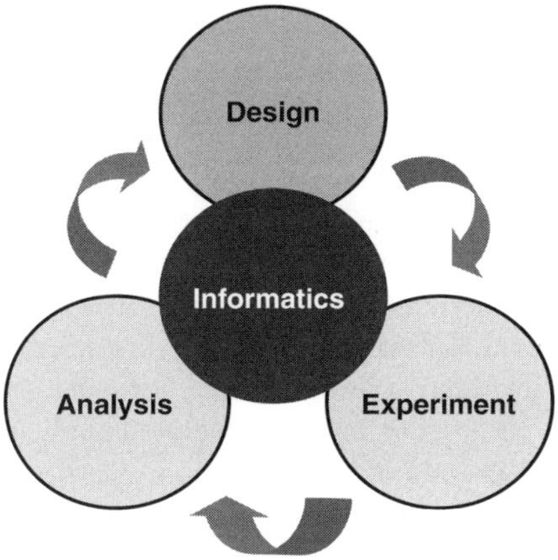

Figure 9.17. The cycle of automation-assisted parallel experimentation.

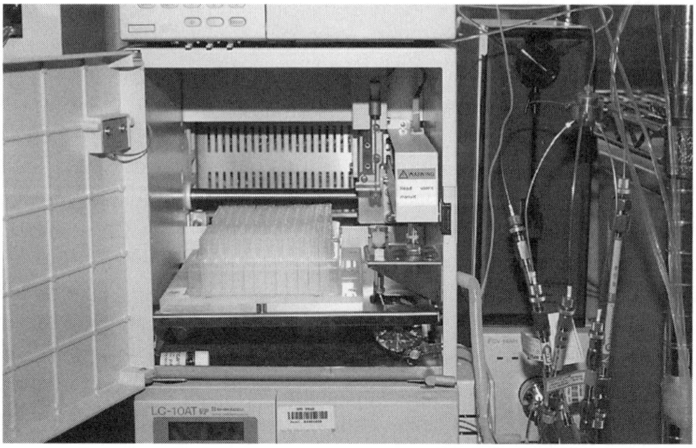

Figure 9.18. Direct transfer of 96-well plate from liquid handler to HPLC autosampler.

each component of the cycle needs to be developed and optimized to serve parallel experimentation. To the extent possible, systems should be integrated to allow transition of materials or information from one system to another. For instance, samples may need to be systematically withdrawn by a liquid handler and transferred to vials containing a quench/buffer system in preparation for HPLC or GC analysis. Given that off-line analysis is currently the norm,[7] the tray containing samples for analysis should be common to both the liquid handler and the analytical mechanism for auto-sampling. A standard 96-well plate, for example, can be readily used on the liquid handler, and direct transfer of the entire plate to the autosampling area of an HPLC or GC eliminates the need to laboriously hand transfer up to 96 samples into a second format (Figure 9.18).

The component linking design, experimentation, and analysis together is informatics, the importance of which has been largely unrecognized to date. As parallel experimentation systems gain popularity and complexity, the need for informatics will increase proportionately. Two important issues exist: (1) How can large numbers and types of experiments and their corresponding data sets be efficiently tracked, analyzed, and stored so that these activities don't become limiting, and (2) how can programs be designed to have intuitive, user-friendly interfaces to lower the energy-barriers for learning and use? [8] The former is particularly important in cases where a specific tool may not be needed frequently, and to the extent possible, software should be simplified and have the same "feel" across applications.

Within our department, lowering the barrier for scientists to learn and use a Gilson liquid handler involved writing custom software that translated codes for commands provided by the vendor into an integrated application that could be readily understood by nonexperts. Functions commonly carried out by the liquid handler were ultimately adapted to Microsoft Excel using Visual Basic for Applications (VBA), such that the scientist need only provide (1) a work list, which tells the instrument what to do and in what timing, and (2) a tray file that tells the instrument

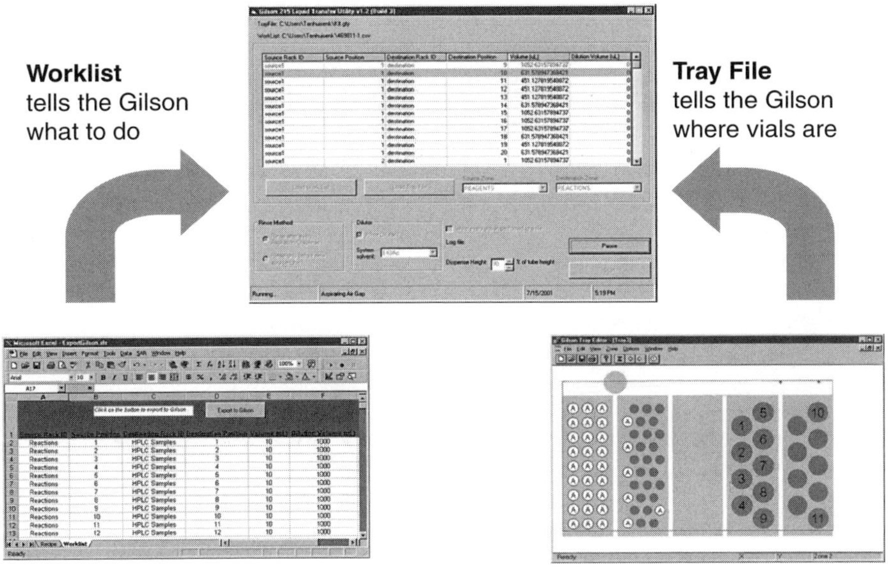

Figure 9.19. Simplified Gilson software interface.

where reagent containers, reactors, and sample vials are located on the instrument deck. A variety of tray templates are preprogrammed and can be flexibly chosen by the user as the experimental needs dictate (Figure 9.19). As relative equivalents and concentrations of reagents and reactants are defined in Excel by the user for each individual experiment, a volume of material is ultimately calculated for each. The work list used by the instrument is simply those volumes (in microliters) to be moved by the syringe-bearing robotic arm according to the Excel worksheet (Figure 9.20).

Lowering the barrier for learning and using HPLC tools (as described above) involved (a) collaborating with an HPLC vendor to define software tailored for

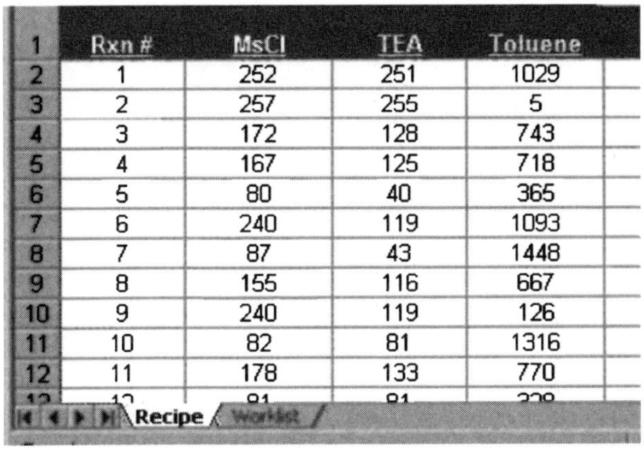

Rxn #	MsCl	TEA	Toluene
1	252	251	1029
2	257	255	5
3	172	128	743
4	167	125	718
5	80	40	365
6	240	119	1093
7	87	43	1448
8	155	116	667
9	240	119	126
10	82	81	1316
11	178	133	770

Figure 9.20. Example input file for Gilson (in microliters).

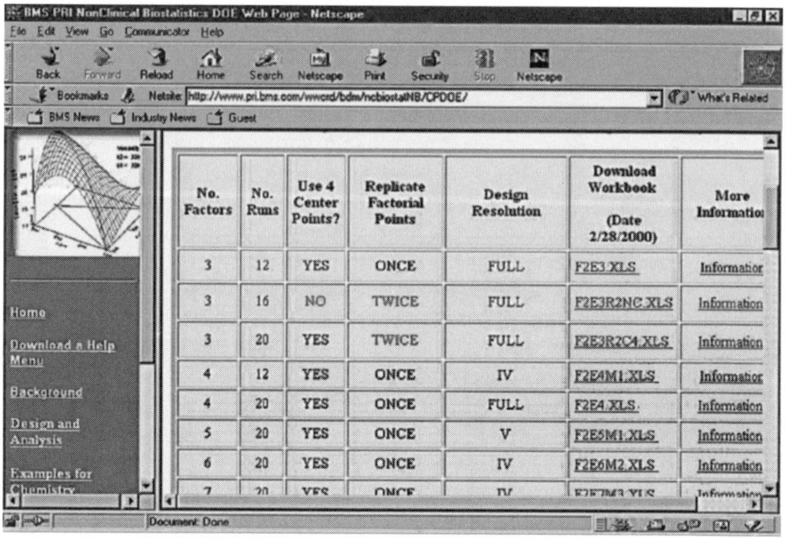

Figure 9.21. Web-download of spreadsheets to support DOE (structured experimentation).

the needs of process chemists (Process VP) and (b) defining user-friendly software to control the MDSA screening system. For alignment purposes, application of a standard set of volatile HPLC buffers in eluant gradients allows direct translation of HPLC methods to LC/MS systems (i.e., use of a specific array of volatile buffers). Similarly, direct translation of HPLC methods to semipreparative columns assists in the isolation of impurities for more thorough characterization.

While experimental design is historically a function of the creativity of the scientists, tools that enable the application of DOE concepts extend scientists' capabilities toward systematic screening and experimentation. Relatively few process scientists in the pharmaceutical industry have a strong grounding in the field of statistics, and this can prove daunting to those lack appropriate background knowledge. Simple, yet powerful, DOE applications can be programmed, however, to define specific designs for the nonexpert and that automatically calculate results based on responses measured and entered into the application. Lowering the barrier for learning and using DOE in our case involved (1) providing process scientists with training on the core fundamentals of statistics and (2) making available custom-made Excel worksheets that define an appropriate experimental design and calculate results based on response data entry. Such spreadsheets are made available to the scientists through an internal Web site (Figure 9.21).

CHEMISTRY/PROCESS EXAMPLES

Statistical design of experiments is a powerful technique, but tends to be tedious and requires rigor in both execution and tracking. As such, the tools described above that support parallel experimentation (i.e. reactor blocks, liquid handler, informatics supporting design, tracking and analysis) are very well adapted to support and expand its use.

Simply put, as applied to Process R&D, statistical DOE is a systematic approach to discovering trends and improving processing performance through the analysis of predefined *response variables* (e.g., yield, impurity level, %ee), by carrying out studies involving *class variables* (e.g., choice of metal, ligand, solvent, reagent) and a well-chosen set of *continuous variables* (e.g., reagent equivalents, time, temperature, concentration). Each continuous variable is assigned a "high" and a "low" value for use in the experimental protocol. The difference between high and low values should be sufficiently large that it will significantly exceed measurement error, but not so great that other factors might compromise the effect being studied. As a means of validating the relative error in conducting the experimentation, several experiments are added that employ all variable values held constant (and usually midway between the high and low values as "center-point runs"). If a protocol is carried out to study x variables and y center-points, in order to cover all permutations the number of experiments required would be $2^x + y$.

An example is provided in Figure 9.22. If only temperature is to be studied, only two experiments would be needed along with some center-point runs. As concentration, solvent, and catalyst are sequentially added, however, the number of experiments increases to 4, 8, and 16 experiments, respectively. This is known as a *full factorial design*. While such designs are easily carried out to study a few variables, the number of experiments expands exponentially as more variables are added. To gain almost equivalent information, a reduced number of experiments (i.e., a *fractional factorial design*) may be carried out by designing the study in such a way as to allow a statistical analysis of the responses. Based on statistics, DOE provides useful information about the variables that provide the most impact on the response (main effects). Importantly, it also provides information about *interactions* between

#	temperature	concentration	solvent	catalyst
1	high	high	MeOH	AcOH
2	low	low	MeOH	AcOH
3	high	low	MeOH	AcOH
4	low	high	MeOH	AcOH
5	high	high	DMF	AcOH
6	low	low	DMF	AcOH
7	high	low	DMF	AcOH
8	low	high	DMF	AcOH
9	high	high	MeOH	TFA
10	low	low	MeOH	TFA
11	high	low	MeOH	TFA
12	low	high	MeOH	TFA
13	high	high	DMF	TFA
14	low	low	DMF	TFA
15	high	low	DMF	TFA
16	low	high	DMF	TFA

Figure 9.22. Two-level factorial designs.

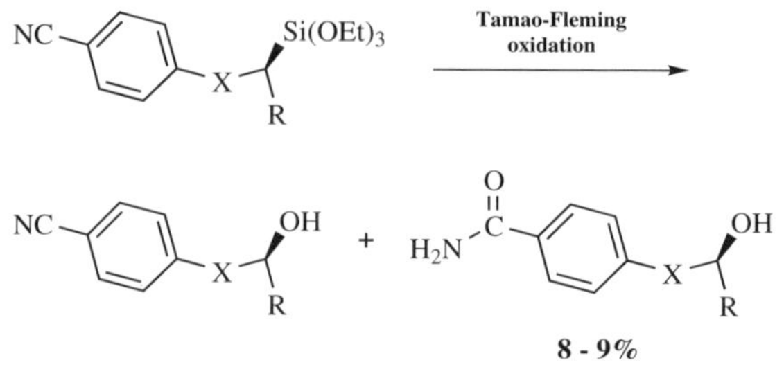

Factor	equiv. $KHCO_3$	equiv. KF	equiv. Urea H_2O_2	Volume (mL)	Temperature °C	Solvent Ratio: THF/MeOH
High	6.5	2	6.5	80	35	0.68
Centre	5	1.5	5	60	25	0.58
Low	3.5	1	3.5	40	15	0.48

Response Factor: Impurity area%

Figure 9.23. Use of DOE to determine factors that improve reaction selectivity.

variables (i.e., an indication that a variable is important, but that its importance is dependent on whether a second variable is in a "high" or a "low" state). Properly powered fractional factorial designs provide information about both main effects and two-way interactions. The trade-off is the resolution of higher-order (three-way or four-way) interactions that rarely are observed.

Example 1: Improvement in Reaction Selectivity

A description of a fractional factorial design from research carried out at Bristol–Myers Squibb is shown in Figure 9.23. In this case, during a Tamao–Fleming oxidation, the hydrolysis of a remote aromatic nitrile resulted in an undesired primary amide impurity to the extent of 8–9%. An HPLC system was employed to measure the relative formation of the primary amide as the response with the aim to minimize the impurity. Six variables were chosen for study, as shown. Rather than carrying out the study with 2^6 plus 4 center-point runs (68 experiments), a fractional factorial study was employed requiring only 20 runs (16 designed experiments plus 4 center-point validation runs).

To enhance understanding of how statistical methods can be used to gain information on the relative importance of variables, Figure 9.24 shows the responses (amide impurity levels) obtained in the above experiment, sorted according to temperature employed (15°C or 35°C) and subsorted according to the level of amide formed. Even a casual comparison of the values obtained for experiments carried out at 15°C (distributed among all other variable conditions) against corresponding values for experiments carried out at 35°C reveals no significant statistical difference among the two data sets. The range and distribution of results for each is similar, and if analysis of variance (ANOVA) were to be carried out with the two data sets, there would be an absence of statistical evidence indicating that they are from

RUN#	Equiv. KHCO 3	Equiv. KF	Equiv. urea-H2O2	Volume	Solvent Ratio: THF/MeOH	Temp °C	Impurity amount
2	5.0	1.5	5.0	60	0.58	25	8.65
8	5.0	1.5	5.0	60	0.58	25	8.57
11	5.0	1.5	5.0	60	0.58	25	8.24
18	5.0	1.5	5.0	60	0.58	25	8.17
5	6.5	1.0	3.5	80	0.68	15	2.86
16	3.5	2.0	6.5	80	0.68	15	3.29
13	6.5	1.0	6.5	40	0.68	15	3.93
7	3.5	2.0	3.5	40	0.68	15	4.07
15	3.5	1.0	6.5	80	0.48	15	5.72
17	3.5	1.0	3.5	40	0.48	15	7.35
6	6.5	2.0	6.5	40	0.48	15	8.35
9	6.5	2.0	3.5	80	0.48	15	9.87
3	3.5	1.0	3.5	80	0.68	35	2.94
14	6.5	2.0	6.5	80	0.68	35	3.18
1	6.5	2.0	3.5	40	0.68	35	3.52
10	6.5	1.0	6.5	80	0.48	35	4.75
4	3.5	1.0	6.5	40	0.68	35	5.22
20	3.5	2.0	3.5	80	0.48	35	6.17
19	6.5	1.0	3.5	40	0.48	35	6.54
12	3.5	2.0	6.5	40	0.48	35	12.33

Figure 9.24. Example of a variable showing little effect.

distinct pools. Thus temperature is not an important variable impacting the formation of this impurity. Note that the center-point runs reproduced four times provide consistent responses relative to the spread of responses in the experiments, thus providing validation for the interpretation of results.

Figure 9.25 shows a resorting of the same data, this time according to whether the ratio of solvents employed (THF to methanol) is 0.48 or 0.68. Now a scan of the values for each set uncovers a very different situation. Virtually all of the high data points arise from experiments employing the 0.48 ratio, while all but one of the lowest responses are present in the data set in which the 0.68 ratio was used. Statistical analysis would indicate that solvent ratio is a very important variable with regard to controlling the impurity (a main effect).

In Figure 9.26, the same data are sorted once more, this time according to high and low reaction volume, and subranked according to whether a 0.48 or a 0.68 solvent ratio is employed. Note that a comparison of all data points for the high dilution condition (80 ml/g) does not reveal a significant difference when compared against all data points for the low dilution condition (40 ml/g). However, subsorting according to whether a high or low solvent ratio is also employed uncovers with statistical significance that the lowest data points of the entire experiment set occur

RUN#	Equiv. KHCO3	Equiv. KF	Equiv. urea-H2O2	Volume	Solvent Ratio: THF/MeOH	Temp °C	Impurity amount
2	5.0	1.5	5.0	60	0.58	25	8.65
8	5.0	1.5	5.0	60	0.58	25	8.57
11	5.0	1.5	5.0	60	0.58	25	8.24
18	5.0	1.5	5.0	60	0.58	25	8.17
10	6.5	1.0	6.5	80	0.48	35	4.75
15	3.5	1.0	6.5	80	0.48	15	5.72
20	3.5	2.0	3.5	80	0.48	35	6.17
19	6.5	1.0	3.5	40	0.48	35	6.54
17	3.5	1.0	3.5	40	0.48	15	7.35
6	6.5	2.0	6.5	40	0.48	15	8.35
9	6.5	2.0	3.5	80	0.48	15	9.87
12	3.5	2.0	6.5	40	0.48	35	12.33
5	6.5	1.0	3.5	80	0.68	15	2.86
3	3.5	1.0	3.5	80	0.68	35	2.94
14	6.5	2.0	6.5	80	0.68	35	3.18
16	3.5	2.0	6.5	80	0.68	15	3.29
1	6.5	2.0	3.5	40	0.68	35	3.52
13	6.5	1.0	6.5	40	0.68	15	3.93
7	3.5	2.0	3.5	40	0.68	15	4.07
4	3.5	1.0	6.5	40	0.68	35	5.22

Figure 9.25. Example of a variable showing main effect.

when both the high volume and the high solvent ratio are present (compare boxed areas). This is an example of a two-way interaction. Solvent volume is important, but its effect depends on whether or not the solvent ratio variable is also high.

Example 2: Reaction Development

Some chemical transformations occur in multiple steps that are not easily dissected, and in these cases DOE can be a good starting point for development. Figure 9.27 shows such a multistep transformation recently employed within Bristol–Myers Squibb in which DOE was employed to gain knowledge that advanced development. At its initial stage of development, the desired product was only being isolated in 30–35% yield with incomplete reaction and low mass balance. The major byproduct present was the decarboxylated impurity shown (up to 25%).

It is important to note that more than one response can be gathered and analyzed for a given set of experiments. In this case, analytical techniques were developed to measure responses for overall yield of desired product (quantitation against a known standard), level of decarboxylated product, and level of starting material remaining after a set reaction period. High and low values studied are shown in

RUN#	Equiv. KHCO3	Equiv. KF	Equiv. urea-H2O2	Volume	Solvent Ratio: THF/MeOH	Temp °C	Impurity amount
2	5.0	1.5	5.0	60	0.58	25	8.65
8	5.0	1.5	5.0	60	0.58	25	8.57
11	5.0	1.5	5.0	60	0.58	25	8.24
18	5.0	1.5	5.0	60	0.58	25	8.17
1	6.5	2.0	3.5	40	0.68	35	3.52
13	6.5	1.0	6.5	40	0.68	15	3.93
7	3.5	2.0	3.5	40	0.68	15	4.07
4	3.5	1.0	6.5	40	0.68	35	5.22
19	6.5	1.0	3.5	40	0.48	35	6.54
17	3.5	1.0	3.5	40	0.48	15	7.35
6	6.5	2.0	6.5	40	0.48	15	8.35
12	3.5	2.0	6.5	40	0.48	35	12.33
5	6.5	1.0	3.5	80	0.68	15	2.86
3	3.5	1.0	3.5	80	0.68	35	2.94
14	6.5	2.0	6.5	80	0.68	35	3.18
16	3.5	2.0	6.5	80	0.68	15	3.29
10	6.5	1.0	6.5	80	0.48	35	4.75
15	3.5	1.0	6.5	80	0.48	15	5.72
20	3.5	2.0	3.5	80	0.48	35	6.17
9	6.5	2.0	3.5	80	0.48	15	9.87

Figure 9.26. Example of a variable showing an interaction.

the first and third rows of the table in Figure 9.27, while the validation conditions are listed in the center row. Results from the study are shown in Figure 9.28. Lightly shaded experiments represent the validation runs, which show good consistency across all three responses. While it is tempting to focus on the two experiments

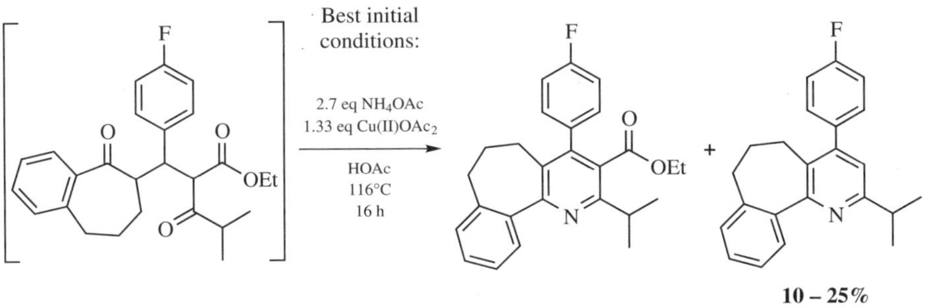

HOAc mL/g	Eq NH_4OAc	Eq Cu	Residual Water	Residual THF
4	1.35	1.1	0	0
5	2.7	1.33	0	0
7.5	4	2	5	10

Figure 9.27. DOE applied toward reaction development.

RUN#	mL HOAc / g S.M.	Eq NH_4OAc	Eq. Cu	Residual Water	Residual THF	Rel% completion	Rel% Decarb.	Yield (area%)
1	7.5	1.35	2.00	5	0	59.6	15.52	40.24
2	4.0	4.00	1.00	5	10	89.3	6.14	71.82
3	4.0	1.35	1.00	5	0	77.5	13.79	55.86
4	4.0	1.35	1.00	0	10	83.2	11.53	62.49
5	5.0	2.70	1.33	0	0	91.0	10.78	69.81
6	7.5	4.00	2.00	0	0	93.6	9.64	73.14
7	7.5	1.35	1.00	0	0	93.9	8.65	75.11
8	7.5	4.00	1.00	5	0	53.5	7.11	40.14
9	5.0	2.70	1.33	0	0	90.4	11.10	69.63
10	7.5	4.00	2.00	5	10	91.3	9.16	70.93
11	7.5	1.35	1.00	5	10	87.1	13.26	64.47
12	7.5	4.00	1.00	0	10	86.8	8.73	68.15
13	7.5	1.35	2.00	0	10	71.4	11.70	50.35
14	5.0	2.70	1.33	0	0	91.6	12.11	69.20
15	4.0	1.35	2.00	5	10	81.0	14.14	56.82
16	4.0	4.00	1.00	0	0	99.7	5.99	82.80
17	5.0	2.70	1.33	0	0	91.6	10.94	69.84
18	4.0	4.00	2.00	0	10	99.9	5.36	83.33
19	4.0	4.00	2.00	5	0	98.8	9.43	78.43
20	4.0	1.35	2.00	0	0	84.2	13.95	60.54

Figure 9.28. Screening DOE.

showing the best responses in all three categories (darker shaded rows), it is important to examine all statistical results since doing so may point the direction for subsequent experimentation. As provided by prearranged calculations in our custom-designed DOE spreadsheet and shown in Figure 9.29, the most important factors leading to higher conversions (represented in capital letters) involved low acetic acid volume (high concentration) and low moisture content, along with the higher number of equivalents of ammonium acetate. Main effects that were less significant

Factor	Low	High
HOAc mL/g S.M.	C, Q	
Eq NH4OAc		C, Q, D
Eq. Cu	d, q	c
Residual Water	C, Q, d	
Residual THF		c, q

Main effects favor
> Ammonium acetate,
< Volume,
< Residual water,
> THF.

C: Higher conversion to product
Q: Improved yield (area%)
D: Less decarboxylation

Effects of copper noted in two way interactions.

Figure 9.29. Statistical analysis of variable impact and two-way interactions.

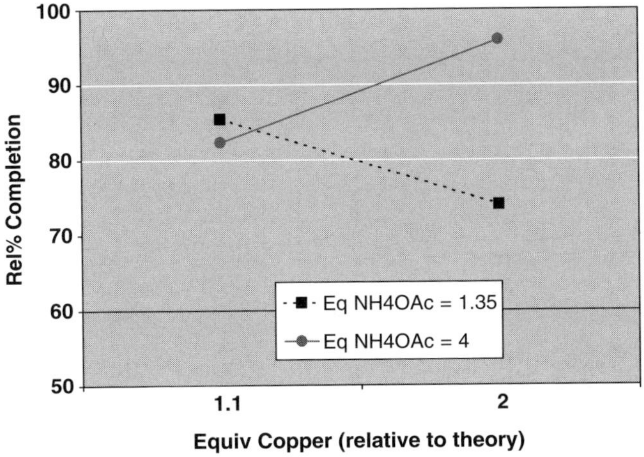

Figure 9.30. Interaction between ammonium acetate and copper with respect to percent reaction completion.

are shown in lowercase. For the response indicating consumption of starting material, an interaction was observed indicating that higher ammonium acetate levels are important, but only when a higher level of copper is added. When such an interaction is present, there is a reversal of the slope for the two variables when plotted as shown in Figure 9.30, and commonly the slopes cross.

Example 3: Reaction Condition Screening

Process R&D scientists commonly look for opportunities to eliminate chemical steps where possible, and it is worthwhile to screen for conditions where this can be accomplished. Such a case is shown in Figure 9.31, in which a controlled reduction of an ester to an aldehyde was sought to streamline a procedure that otherwise required overreduction to an alcohol followed by reoxidation to an aldehyde in separate operations. The ester was highly hindered in this case, and since the aldehyde can be expected to reduce more quickly than the ester, success in meeting this goal seemed unlikely. Experimental trials with the reducing agent RedAl-P (Figure 9.32) seemed worth exploring nonetheless, since the system was reported to involve one or more tetrahedral intermediates that are sufficiently long-lived to prevent substantial secondary reduction.[9] An experimental design was performed as provided in Figure 9.33, this time demonstrating that solvent choice (a class variable) provides a great deal of impact toward providing the desired selectivity. A key point to be made is that "hits" such this can often be overlooked unless a broad enough screen of variables is performed.

Example 4: Reaction Kinetics

In principle, process methodology could be developed purely on the basis of DOE experimentation. In practice, however, it is generally more efficient to use DOE to

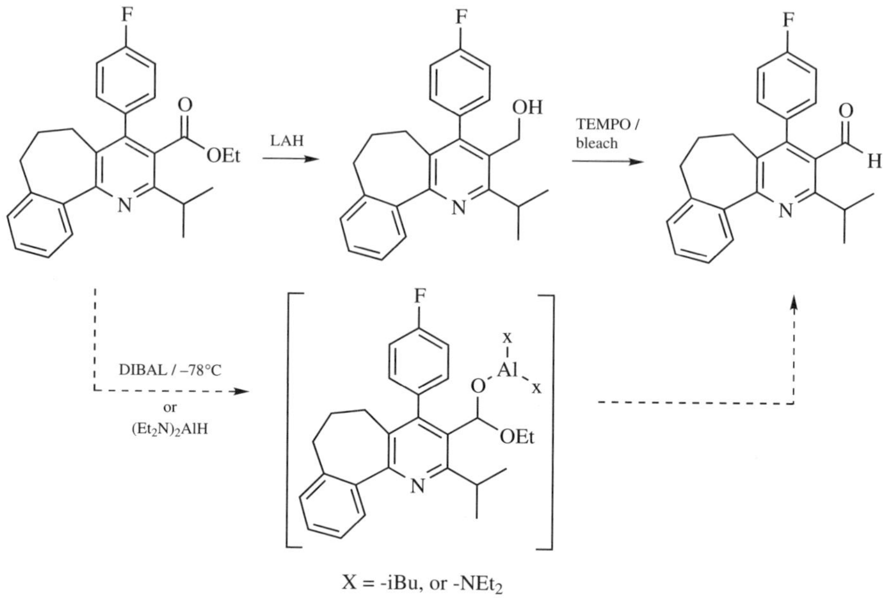

Figure 9.31. Example of reaction condition screening.

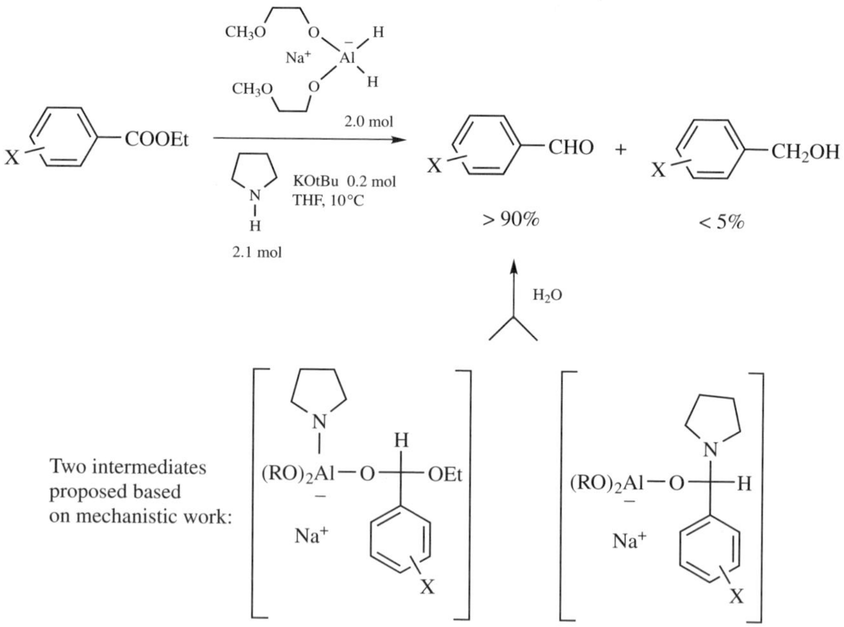

Abe, T., *et al.* Tetrahedron *57*, 2001, 2701-2710

Figure 9.32. Literature precedent for controlled reduction.

RUN#	Solvent	Temp	Eqs Pyrr	Eqs RedAl	Ratio aldehyde/alcohol
1	MTBE	20	2	4	4.69
2	THF	10	3	3	3.11
3	THF	0	2	2	1.57
4	Toluene	0	2	4	10.62
5	THF	20	2	4	1.10
6	MTBE	0	2	2	0.00
7	THF	10	3	3	2.95
8	Toluene	0	4	2	10.55
9	THF	10	3	3	3.41
10	MTBE	20	4	2	1.09
11	THF	0	4	4	4.47
12	Toluene	20	2	2	8.00
13	THF	10	3	3	3.37
14	MTBE	0	4	4	2.34
15	Toluene	20	4	4	10.88
16	THF	20	4	2	2.73

Figure 9.33. Results from DOE screen.

gain clues on the variables that are important, as well as their *directionality*. In order to achieve complete control of a synthetic process being developed, the process scientist should have a complete understanding of the dynamics of the chemical systems being used, and the clues provided by DOE should lead to experimentation that creates well-founded process knowledge, similar to what was described for the steroid intermediate in Figures 9.7 and 9.8.

The final example in this chapter will be used to illustrate how a flexible system designed to support parallel experimentation can be reconfigured to provide more detailed mechanistic data to create detailed process knowledge. The chemistry in question involved an amide coupling, as shown in Figure 9.34, which provided both the desired product and a small amount of the undesired epimer. By prior experimentation, the acid chloride starting material had been specifically shown to be subject to thermally induced racemization. Based on empirical observations, the suppression of this impurity was suggested to be effected by carrying out coupling at a low temperature (−30°C).

In order to establish the potential for an operating window at higher temperatures, a kinetics study was performed using the Gilson liquid handler, reconfigured as shown in Figure 9.35. In this case, samples of freshly prepared acid chloride were held at different temperatures in four consecutive experiments. Rates of racemization were measured at each temperature by having the liquid handler sample the acid chloride solution at a high rate and then quench each sample into a large excess of the amine substrate contained in a series of vials on the opposite side of the deck. The entire series of quenched samples were then transferred to an HPLC autosampler for analysis, where the relative amount of epimer was analyzed against the amount of amide formed. Data for the entire set were subsequently moved into

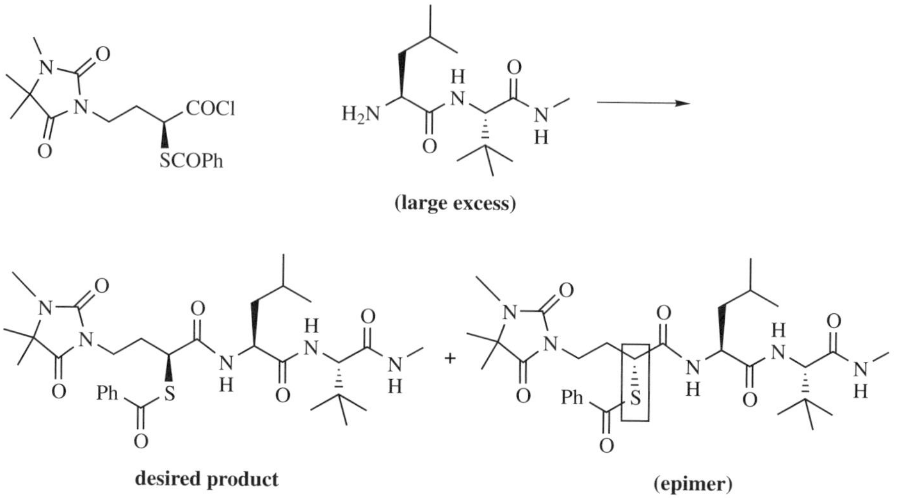

Figure 9.34. Use of kinetic modeling to design control of epimer formation during amide coupling.

Microsoft Excel, which provided reduction of data to the graph of reaction profiles shown in Figure 9.36.

While some scatter is evident among the data points, the consistency was still sufficient to provide curves that fit to first-order relationships with very high correlations (Figure 9.37). From the slopes of these curves, first-order rate constants

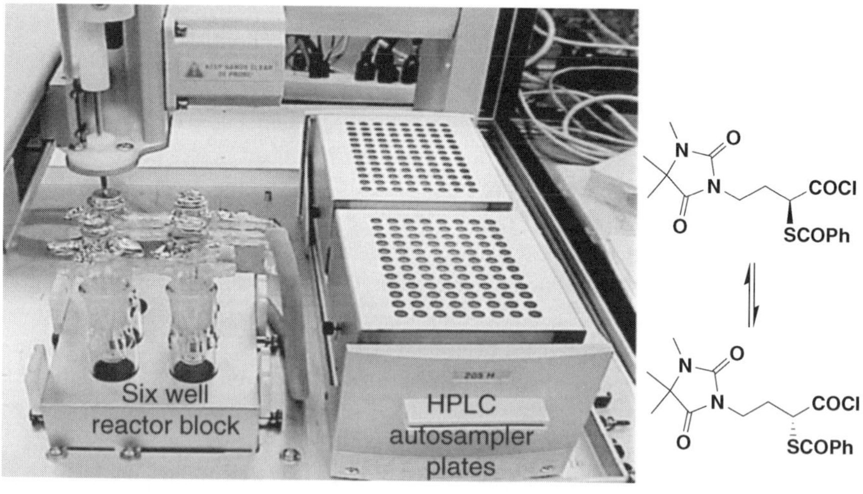

Figure 9.35. Gilson reconfigured to measure rates of acid chloride substrate epimerization as a function of temperature.

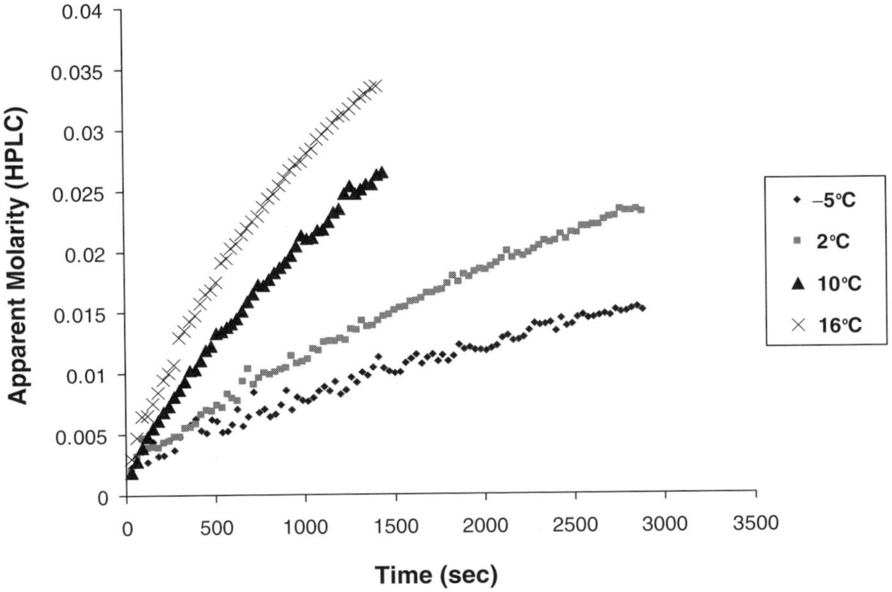

Figure 9.36. Results and data reduction of kinetic study.

were calculated and their natural logarithms were plotted against the inverse of temperatures at which they were measured (Arrhenius plot), thus providing a measured activation energy of 15.2 kcal/mole for the thermally induced racemization of the acid chloride (Figure 9.38). With such data, for any given set of conditions (time,

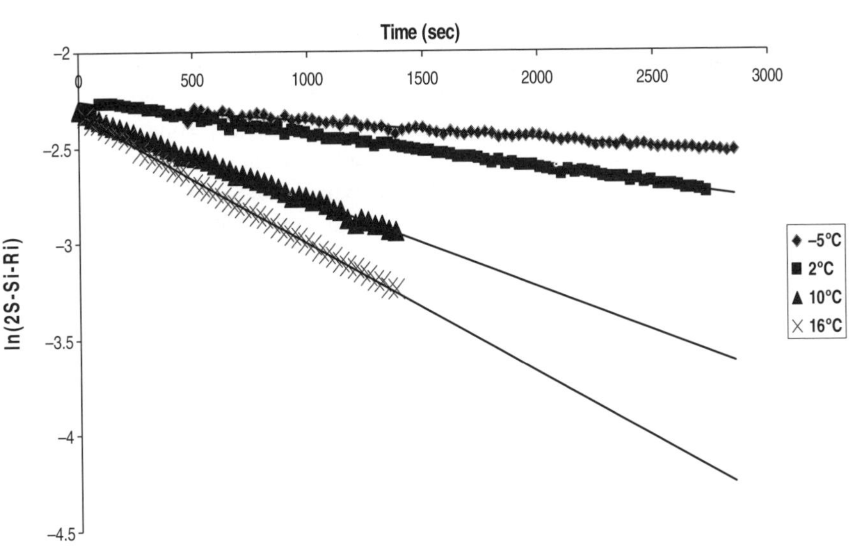

Figure 9.37. First-order plots.

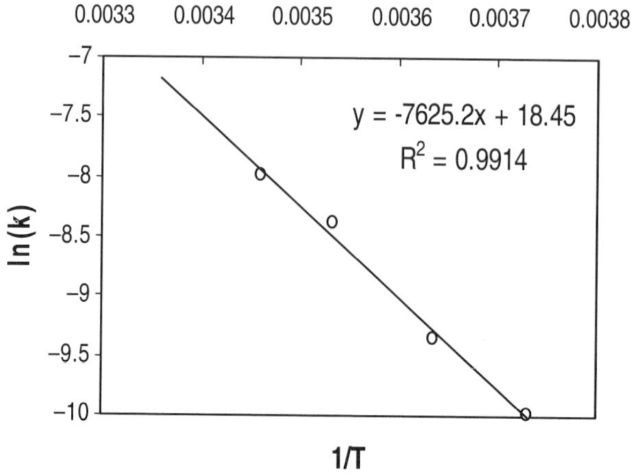

$$E_a = 7625.2 \times 8.314 \text{ J/mol K} = 63.4 \text{ kJ/mol} = \mathbf{15.2\ kcal/mol}$$

Figure 9.38. Arrhenius plot.

temperature, concentration) the extent of racemization can be predicted, allowing engineers to make appropriate trade-offs in designing the process equipment train (e.g., higher temperature can be used, as long as faster transfers can be effected; or, alternatively, the mixture can be formed and used at higher temperature under greater dilution).

SUMMARY AND CONCLUSION

The principles of automation, as applied to process research and development within the pharmaceutical industry, involve much more than setting up robotic systems to run parallel reactions or measurements. The key concept is the use of various automation and informatics technologies (either emerging or existing) that can support a wide array of activities involving *structured parallel experimentation*. The examples cited above bear witness to the fact that technologies currently available can enable experimental paradigms that could not be effectively supported in the past. With appropriate tools, statistically designed experiments that might have previously taken weeks to perform tediously can now be carried out efficiently in a fraction of the time. Kinetics experiments such as the one described above would not have been contemplated, because the time to carry them out would be considered prohibitive relative to the typical development timeframe. Yet, the experiments described above were completed easily within two days, with much of the experimentation carried out unattended.

Several examples of structured parallel experimentation have been provided in this chapter, and many others that support the generation of process knowledge are summarized in Figure 9.39. Determinations of stability and solubility, for example,

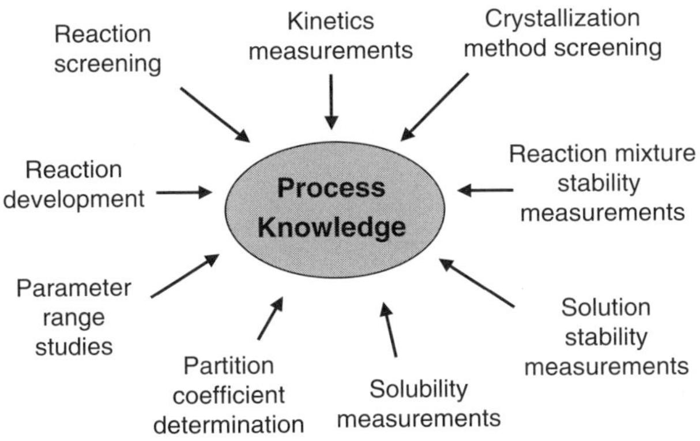

Figure 9.39. Applications of automation-assisted parallel experimentation.

represent areas where structured parallel experimentation can have profound impact on the efficient design of process workup and isolation conditions. Enabling hardware tools should be sufficiently flexible to allow a wide array of experimental work to be performed on one core system, while providing avenues for integration with others systems involved (e.g., with analysis and weighing). Similarly, to maximize uptake of such tools, enabling software applications will be needed to integrate and streamline all aspects of the experimental cycles—design, experimentation, and analysis—while being consistent in their design, intuitiveness, and feel.

While the examples provided in this chapter illustrate how tools and technologies can be used to improve the manner and efficiency with which process R&D can be carried out, the development of tools and technologies is far from rate-limiting. Much more important is the interest of process R&D scientists to learn and rely

	Past	→ *Future*
...basis of their field and work	intuition inspiration art	rational methodic analytical
...approach to route selection	linear approaches, of limited scope based on existing knowledge	parallel approaches, based on knowledge but poised for discovery
...approach to overall process design	handcrafted solutions, phenomenological basis	systematically-derived, integrated solutions, knowledge-based
...approach to scaleup	a test of the design	validation of what is known

Figure 9.40. Values/beliefs of process scientists.

upon the tools. At present, these technologies are still in an early state of development, which necessitates reliance upon a relatively few number of experts and early adopters. As the field matures, however, it will be best leveraged by accessibility and use by all process scientists.

Some resistance can be expected, because the concepts conflict with values and beliefs held by many. These values have been gradually shifting, as indicated in Figure 9.40, and will continue to shift as the tools evolve and improve. In the end, the tools described herein will not replace the scientific knowledge and creativity of individual process scientists. However, they should be expected to positively influence the rate at which scientists can bring their scientific knowledge and creativity to bear on the generation of process knowledge and its subsequent translation into highly reliable process technology.

ACKNOWLEDGMENTS

The authors thank James Pazden and James Bergum for their guidance in the application of DOE and for writing code that enables process scientists to carry out a powerful, yet simple, form of DOE unassisted.

REFERENCES

1. Pisano, G. *The Development Factory*, Harvard Business School Press, Boston, MA, 1997, pp. 36–48.
2. "Chance favors the prepared mind" is actually an abbreviation of the Louis Pasteur quotation "in matters of observation, chance only favors the prepared mind"; the quotation "chance only favors those who court her" is attributed to French physician and bacteriologist Charles Nicolle (1866–1936).
3. Weller, H. N., Rubin, A. E., Moshiri, B., Ruediger, W., Li, W.-J., Allen, J., Nolfo, J., Bertok, A., and Rosso, V. W. *J. Assoc. Lab. Automation*, in press.
4. Karcher, B. D., Davies, M. L., Venit, J. J., and Delaney, E. J. *Am. Pharm. Rev.* **2004**, *7(6)*, 62–66.
5. International Committee for Harmonzation (ICH) Guideline Q3A(R) can be found online at http//www.ich.org
6. An excellent series of articles covering automation-assisted applications of DOE toward pharmaceutical R&D was documented in issue 5, volume 7 (2001) of *Organic Process Research and Development*, including papers by Martin R. Owen, Chris Luscombe, Lai-Wah Lai, Sonya Godbert, Derek L. Crookes, and David Emiabata-Smith, pp. 308–323; Dennis Lendrem, Martin Owen, and Sonya Godbert, pp. 324–327; Victor W. Rosso, James L. Pazdan, and John J. Venit, pp. 294–298; and Paul Higginson, and Neal Sach, pp. 331–334.
7. Although liquid handlers can be configured to carry out at-line analysis, a single syringe arm servicing a sampling schedule can usually be expected to have timing conflicts if the arm is also being used to the deliver materials in setting up reactions. Work-arounds have

been devised, however (see Michael Harre, Neh, H., Schulz, C., Tilstam, U., Wessa, T., and Weinmann, H. *Org. Process Res. Dev.* **2001**, *5*, 335–339).

8. Higginson, P. D., and Sach, N. W. *Org. Process Res. Dev.* **2004**, *8*, 1009–1014 serves as a illustration of the importance of informatics to the success of automation-assisted parallel experimentation.
9. Abe. T. *Tetrahedron*, **2001**, *57*, 2701–2710.

10

LARGE-SCALE SYNTHESIS: AN ENGINEERING PERSPECTIVE

JOSEPH H. CHILDERS, JR.

Johnson & Johnson Pharmaceutical Research and Development, L.L.C. Chemical Development, Spring House, PA 19477

INTRODUCTION

The mass production of chemicals and pharmaceuticals is carried out in plants or production facilities capable of safely handling large quantities of chemicals, solvents, and reagents. The transition of chemical reactions from the small scale carried out in a laboratory to produce gram quantities of substances into large-scale chemical reactions capable of producing tens or hundreds of kilograms requires numerous technical modifications dealing with such operations as weighing, addition, stirring, heating, cooling, filtration, containment, and so on. It is the function of a chemical engineer to implement such modifications to chemical reactions and reproduce the results of the small-scale reaction on a large scale.

While the variety of processes and industries that benefit from the services of chemical engineers are numerous, the field is still not easy to define. The term *chemical engineer* describes what makes the field unique and different from the other branches of engineering. All types of engineers use their training and innovation to safely and economically overcome technical problems; only the chemical engineer utilizes the science of chemistry to resolve these problems. Chemical engineers deal with solving technical problems in achieving safe production of basic chemicals, complex molecules, polymers, explosives, radioactive isotopes, toxic substances, active pharmaceutical ingredients (APIs), and many more. The strong ties that

Fundamentals of Early Clinical Drug Development: From Synthesis Design to Formulation, edited by Ahmed F. Abdel-Magid and Stéphane Caron

bind chemistry and chemical engineering are unique in the fields of science and technology. This close relationship has been beneficial to both fields. The collaborative efforts of chemists and chemical engineers in transitioning small-scale reactions into large-scale processes is usually referred to as chemical process research.

Many age-old human activities such as brewing and soap-making employed some early versions of chemical engineering. Modern chemical engineering came about at the same time as the Industrial Revolution, which saw the rise of modern chemistry and industrial chemical processes after the seventeenth and eighteenth centuries.[1] Chemical engineering was first conceptualized in England over a century ago; however, its primary evolution, both educationally and industrially, has occurred in the United States. The outbreak of World War I in 1914 provided a primary impetus for the initiation of this evolution. At the time, Germany was the world leader in organic chemistry and chemical technology. The war forced the United States to develop the training and innovation necessary to improve chemical processes for handling many new challenges. The most immediate challenge was to produce the materials necessary to wage war. With industry cooperation, the United States built ammonia plants that produced the explosives (and fertilizers) that helped win the war efforts.

World conflicts continued to motivate chemical engineering innovation. At the beginning of World War II, the United States was again dependent on other countries for essential raw materials and medicines. After Japan captured many of the rubber-producing lands that supplied 90% of America's natural rubber, innovation was required to produce synthetic rubber. The collaboration between the United States government and the chemical industry resulted in the increase of synthetic rubber production by over a hundredfold. A similar cooperation between pharmaceutical manufacturers and the U.S. government helped produce new medicines. For example, the pharmaceutical manufacturers worked together with the government to boost the production of penicillin from 425 million units in all of 1943 to 646 billion units in June 1945 alone.[2] It was the unique collaboration between scientists and chemical engineers that made these advances possible. All of these advances started with inventions in the laboratory and required an engineering ingenuity to develop the technical means for the scale-up to the final production process. Many engineering inventions were developed during this period, such as the production of high-octane gasoline, plastics, the automobile, synthetic rubber, nuclear fission, television, insulin, antibiotics, anti-toxins, and other pharmaceuticals.

MODERN CHEMICAL ENGINEERING

The advances in chemical engineering during World War II helped spur the postwar expansion in the pharmaceutical industry. The companies funded the research efforts that led to the discovery, development, and launch of many novel drugs. The period from 1950 to the early 1990s witnessed an unprecedented growth of the pharmaceutical industry and consequently the advancement of chemical engineering to handle the development and manufacturing of the new drugs. During this period, chemical process scale-up from laboratory to pilot plant was a very deliberate process. After extensive process research and repetitive reactions, the chemical process was optimized in the laboratory. The process was then gradually scaled-up to pilot plant equipment. Scale-up factors were rarely greater than 10. The challenge for the engineers was to scale-up the process while maintaining the properties and purity of the product as developed at the laboratory scale. This was no simple task because on scale-up a range of factors from reaction time, concentration effects, and mixing of raw materials could affect the outcome of a chemical reaction or produce safety hazards that were not observed on the small scale. In addition to making initial test material, the pilot plant was used for process development, where most challenges were observed before the process moved into manufacturing. Valuable hands on experience was acquired with this approach and frequently resulted in refined chemical procedures ready for manufacturing-scale production. However, after all this time and effort, too frequently the project did not advance to market.

In the 1990s the pharmaceutical R&D strategy shifted to higher throughput. Large numbers of compounds are screened for activity. Promising compounds are quickly synthesized and tested. The principal role of the pilot plant is the supply of the initial batch of active pharmaceutical ingredient (API); it is no longer a full-scale process development. The goal is to quickly eliminate compounds that are not going to market and advance those with market potential. Most compounds do not advance past the initial scale-up operation, so why waste time refining the chemical process?

This strategy has affected the approach to scale-up of chemical procedures. Process development chemists are looking to *quickly* turn over a safe process for scale-up. Scale-up factors have increased to 20, 50, and even 100+ for many chemical steps. This scale increase from bench to pilot plant is now typically the largest the process will experience. These changes, coupled with the increasingly complex multistep chemistry, make today's chemical scale-up processes even more challenging.

The first scale-up batch of material is often the most difficult to prepare because the process is not optimized. The pilot batch preparation will establish scale-up feasibility. For success the engineer must focus on the batch and unit operations required to scale-up these synthetic procedures into typical chemical reactor systems ranging in size from 20 liters to 1000 liters. The ultimate goal is to produce the first 5 to 15 Kg batch of GMP API. The primary use for this material is toxicity testing, stability studies, and first-in-human (FIH) or initial clinical trials. In addition, a significant amount of this batch is used for early dosage form development.

The chemical engineer must focus on the chemical scale-up processes that turn-key raw materials or process intermediates into valuable APIs. Because of the great number, variety, and complexity of these modern processes, it is more practical to break down the overall process into a series of steps, called unit operations.[3] Unit operations occur in process after process and employ common techniques based on similar scientific principles. Unit operations are used to carry out the predominantly physical steps of preparing the reactants, separating and purifying the products, and controlling the energy transfer into or out of the chemical reactor. For example, in most processes, solids and fluids must be moved, heat or other forms of energy must be transferred from one substance to another, and tasks like distillation or evaporation, extraction, filtration, and drying must be performed. This chapter will attempt to contrast how these operations are handled on the pilot scale versus the laboratory scale.

Since this material will be used in human testing it, must be prepared using Good Manufacturing Practices (GMP). The establishment of the Food, Drug, and Cosmetic Act in 1938 provided guidelines and procedures for GMP production that added to the challenges of chemical engineers to comply with these procedures and guidelines.[4]

SCALE-UP PROCEDURES AND PROCESSES

The starting point for a successful scale-up is a detailed laboratory procedure or Laboratory Report. The Laboratory Report includes the following important background information:

- The reaction scheme with starting materials, solvents, reaction conditions, known byproducts, and the laboratory apparatus.
- Detailed procedure including any observations or comments by the process research chemist.
- A reaction safety evaluation including Material Safety Data Sheets (MSDS), reaction calorimetry evaluation, and differential scanning calorimetry (DSC).[5]
- Analytical methods.

A project team (includes process development chemists, process safety personnel, and the chemical engineer responsible for the scale-up) is assembled to review the laboratory procedure and determine the scalability of each synthetic step. Once the team agrees that the process can be safely scaled, the assigned chemical engineer takes the lead. Working together with process chemists, a scale-up strategy is developed. Based upon reaction volume requirements and other process considerations, a work center is assigned. With this information the chemical engineer prepares a batch record detailing step-by-step the intended process operations and operating parameters for the synthetic process. Preparation of this document requires careful evaluation of each process or unit operation. After the intended process is detailed in a batch record format, a reaction safety review (RSR) of the process is performed.

SAFETY REVIEW

Whole books have been written about the reaction safety review or the Hazards and Operability Study (Haz-Op).[6,7] For the initial scale-up, a full Haz-Op is not normally conducted. This could take many days or weeks to complete and is normally performed in a manufacturing setting where the larger scale presents a much greater potential hazard. During manufacturing, the vessel size will range from hundreds to thousands of gallons in size. At this earlier stage of the project, volumes and risks are not of the scale seen in manufacturing. However, as stated previously, this initial scale-up is normally the greatest jump in scale and requires careful safety assessment prior to proceeding.

The RSR discussion focuses on the chemical process equipment, vessel charging and general material handling, reaction calorimety results and process operating limits, materials compatibility, hazardous byproducts, reaction sampling, in-process controls, possible stopping and holding points, and personnel protective equipment. Throughout the discussion the participants should consider what could possibly go wrong and consider the ramifications of this upset to the process. This is considered the "what if" scenario. Each participant should ask the question "What could go wrong?" rather than later asking the question "What went wrong?" The key objective is to understand and manage the scale-up risk. Once the reaction safety review is complete the process is ready for scale-up.

PROJECT WORK CENTER

For most steps a work center or batch reactor equipment train includes two jacketed reactor systems. The reactor systems are equipped with overhead condensers, feed vessels for liquid feeds via metering pump, receivers for liquid collection, and a slurry filter for collecting product. Other support equipment is incorporated into the system as required. Normally, this type equipment is portable and includes items such as metering pumps, receivers, liquid-phase separators, and various type filters.

MATERIAL HANDLING

Raw materials weighing and charging is the first processing step. Solvents are weighed and charged directly into reactor systems using diaphragm pumps or vacuum transfer. Potent or toxic reagents are weighed in fume hoods, diluted with solvent, and then transferred to the reactor system by vacuum. Using the correct handling techniques, air flow control and personnel protection these materials can be handled safely. The most difficult materials to handle and charge into reactors during scale-up operations are solids or powders. These are often unique chemical intermediates with limited or no toxicity data. Several kilograms are normally required, especially in the early stages of a synthesis campaign. The old process of donning a respirator, opening the reactor, and scooping the powder from a bulk container directly into the reactor is no longer acceptable—not only from an industrial hygiene perspective, but also to minimize cross-contamination. During such an uncontrolled operation the concentration of airborne particles can exceed 10,000 micrograms per cubic meter ($\mu g/m^3$).

Pharmaceutical industrial hygiene personnel have established guidelines for assessing the risk of handling chemicals, intermediates, and APIs. The operator exposure levels (OELs) are categorized by their potential hazards in airborne concentrations based on physical properties such as toxicity and therapeutic category. Typical OEL categories[8] include:

- Category I: $>100\ \mu g/m^3$
- Category II: $20 < 100\ \mu g/m^3$
- Category III: $1 < 20\ \mu g/m^3$
- Category IV: $<1\ \mu g/m^3$

To reduce operator exposure to these airborne particles, the typical engineering solution is airflow control. Conventional uncontrolled powder scooping and weighing operations using a local exhaust air pickup would maintain operator exposure levels (OELs) between 1000 to 10,000 $\mu g/m^3$. This is acceptable for handling small quantities of nonhazardous (Category 1) solids. A common industry trend is the incorporation of laminar-directional airflow (LDA) systems for handling powders. The LDA system is a dedicated containment booth or room (Figure 10.1).

Laminar down-flow booths provide vertical airflow from ceiling panels over the operator's head and upper body, alleviating the rise of dust into the breathing zone. Airflow inside the booth achieves up to 850 air changes per hour of HEPA-filtered air (High-efficiency particulate air). Using proper work procedures, chemical

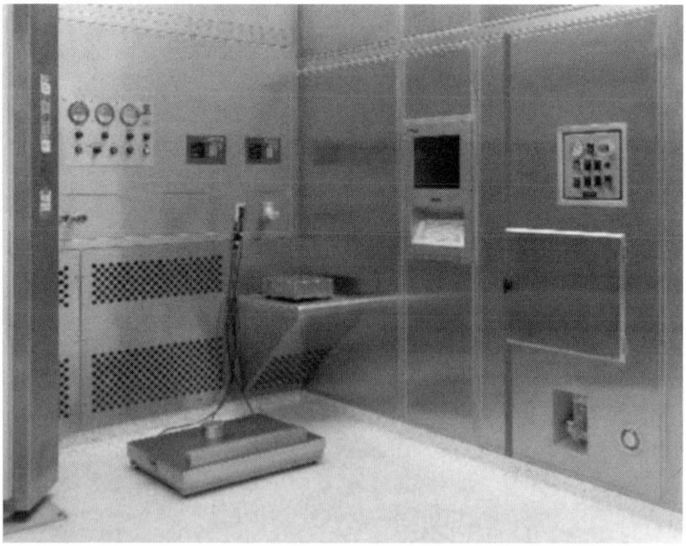

Figure 10.1. Extract laminar down-flow booth. Courtesy of CPS PHARMA.

Figure 10.2. High containment barrier isolator.

powders, ethical pharmaceutical intermediates, and API powders can be handled inside the laminar down-flow booth and maintain OELs of below 20 $\mu g/m^3$. This is acceptable for handling larger quantities of Category 1, all of Category 2, and relatively low-toxicity Category 3 chemicals.

For weighing or handling higher-toxicity Category 3 materials, the typical engineering solution is a high containment barrier isolator. This is a completely enclosed work area with built-in gloves that completely isolates the operator from the hazardous material (Figure 10.2). Handling Category 4 materials may require robotic systems.

The transfer of solids/powders into the reactor is another issue. One technique is to dissolve or slurry the powder in an appropriate solvent and then charge this mixture to the reactor. This is effective for materials that form a uniform mixture for easy transfer or exhibit good solubility in the reaction solvent. Another technique involves the use of a split butterfly valve system[9] (Figure 10.3). The powder is weighed inside a laminar down-flow booth directly into charge bottles or containers that are equipped with the passive portion of the split butterfly valve. The active portion of the valve is mounted on the reactor charge point. The passive and active valve halves are mated, and then the valve is opened as a single unit and the powder flows from the charge bottle into the reactor (Figure 10.4). Once the charge is complete, the valve halves are separated without exposing the process area to significant contamination. Another approach employs a high containment barrier isolator chamber that is docked directly to the reactor opening during powder charging operations.

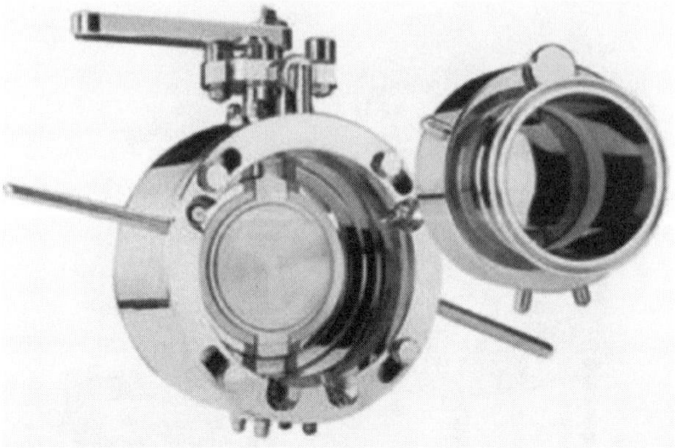

Figure 10.3. Charge point transfer coupling (CTC) valve. Courtesy of Powder Systems Limited (PSL).

Figure 10.4. ChargePoint connection for powder/solids addition. Courtesy of Powder Systems Limited (PSL).

REACTOR SYSTEMS

The scale-up of high-value pharmaceutical API's requires the use of multiuse process equipment. The agitator-equipped vessels with jackets for heating or cooling (chemical reactors) are the core system and workhorse for chemical scale-up. It represents the greatest flexibility among reactor types. The straight-sided cylindrical batch stirred-tank reactor with a centrally positioned vertical mixing shaft on a dished head is the most common configuration. However, conical bottom configurations, useful for extraction cuts and low-volume processing, are often used. Typically, resistance temperature detector (RTD) sensors are located in the mixing baffle and the bottom reactor valve. RTD sensors measure temperature based on the principle that the electrical resistance of a metal increases with temperature. The temperature can be determined by passing a small current through a thin metal wire and measuring the resistance. The most common type currently used in pilot plant facilities is the 100-ohm platinum RTD.

An important fact of scale-up is that heat transfer gets worse with larger vessels because the effective heat transfer area is reduced. A 1000-gallon reactor has only 10% of the heat removal area of a 1-gallon reactor. Height-to-diameter ratios between 1 and 1.5 are normal. Making the vessel taller can increase the heat transfer area, but bulk mixing may be compromised. The rate of heat transfer to or from an agitated liquid mass in a vessel is a function of the physical properties of that liquid and of the heating or cooling medium, the vessel geometry, and the degree of agitation. Once installed, the vessel geometry is fixed. The rate of agitation and jacket temperature can be varied.

For reactors in sizes ranging from 20 liters to 1000 liters, the most common material of construction is glass-lined steel because of its excellent chemical resistance and ease of cleaning. The inside surfaces are coated with multiple layers of high-temperature-fired glass enamel to produce a uniform smooth surface finish with a thickness of 1–2.25 mm. Internal operating pressures typically range from full vacuum to 100 or 150 psig. The overall heat transfer coefficient for a glass-lined carbon steel reactor is about 30 BTU/hr-°F-sq. ft.

There are some cautionary notes regarding the use of glass-lined reactors. They should not be used with hot concentrated alkali. They are not recommended for very high or very low temperatures, and thermal shock from large temperature differentials (ΔT) may cause the glass to crack. The maximum recommended ΔT between the reactor jacket and the process is around 50°C. Metal reactors do not have these temperature limitations. The overall heat transfer coefficient for a standard metal stirred reactor is about 40 BTU/hr-°F-sq. ft.

The higher thermal conductivity of metal reactors results in superior heat transfer. Therefore for extremely high and low temperatures, metal reactors are preferred. Although expensive, the highly chemically resistant Hastelloy C metal reactor has become the choice over stainless steel reactors in the pharmaceutical industry. Linked with the proper heat transfer system using a silicone polymer heat transfer fluid, Hastelloy reactor systems can operate at temperatures from −100°C to +260°C.[10]

For some reactions, cold/cryogenic conditions result in greater selectivity, yield, and purity. Commonly employed on the laboratory scale using low-temperature dry-ice baths to attain very cold reaction conditions, only more recently have these conditions become more common on scale-up. The additional time and resources required to alter the synthesis route has made the low-temperature process more competitive for the preparation of the initial scale-up batch of API. Some low-temperature organometallic reactions require reactor heat transfer systems that cool the process as low as −80°C. When using a high-performance silicone polymer heat transfer fluid like Syltherm XLT™ designed for use at lower temperatures, a low-temperature heat exchanger circuit can easily be incorporated into a heat transfer system (HTS). To obtain these low temperatures, liquid nitrogen is used as an indirect coolant for the heat transfer fluid because injecting liquid nitrogen directly into the reaction mixture poses difficult control problems. The HTS system is cooled down in the first stage to about −20°C using a cold glycol loop, and then low-pressure (1.5 bar) liquid nitrogen is injected into a second-stage (low-temperature) spiral heat exchanger to obtain reactor jacket temperatures down to −90°C.

Figure 10.5. Modular reactor system. Courtesy of Sentenel Process Systems.

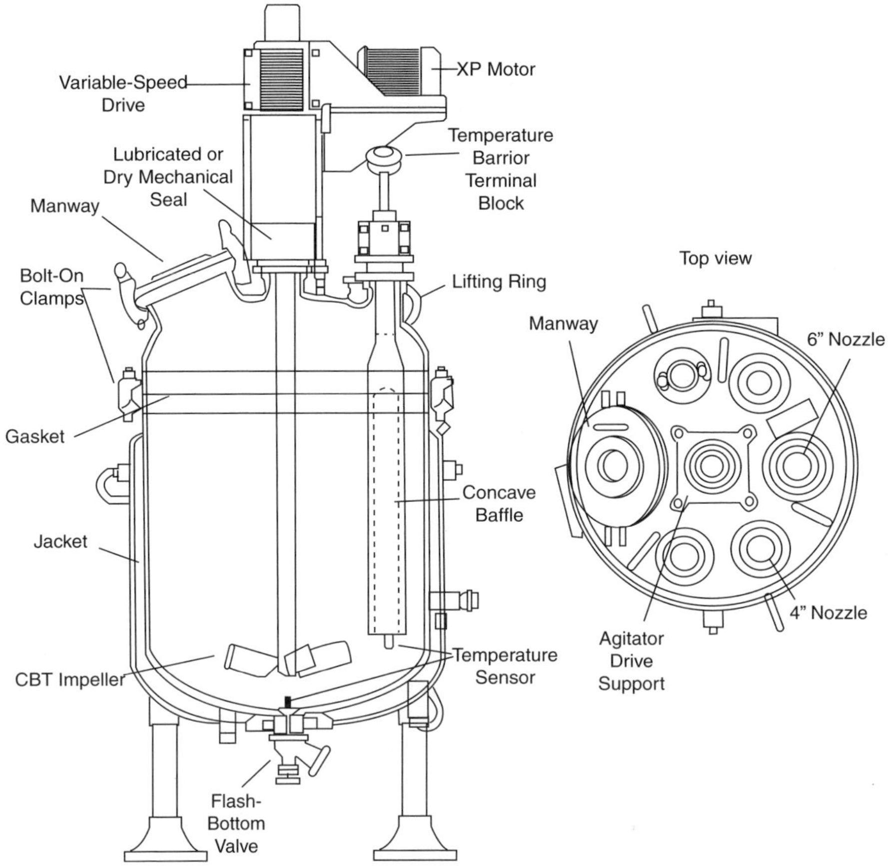

Figure 10.6. Reactor P&ID. Courtesy of Pfaudler Reactor Systems.

INSTRUMENTATION AND PROCESS CONTROL

The development of automatic control systems in the past 75 years is considered by some to be a second industrial revolution. Rapid innovation in this field of engineering continues to this day. The simple manual recording and control of processes has largely been abandoned. Today, sophisticated instrumentation and control systems are common in most pilot plant facilities. The primary application of this equipment is focused on the actual reaction step of the process. The reaction step of the synthetic process is the most sensitive to temperature (and sometimes pressure) control. Even though this is probably the most optimized step in the laboratory, the scale-up operation requires careful attention and precise control systems to monitor and maintain the desired reaction conditions.

In the laboratory heating is often controlled by adjusting a rheostat connected to a heating mantel. Immersing the reaction vessel in an ice bath may be used to control cooling. On scale-up more precise temperature control is required. A simple

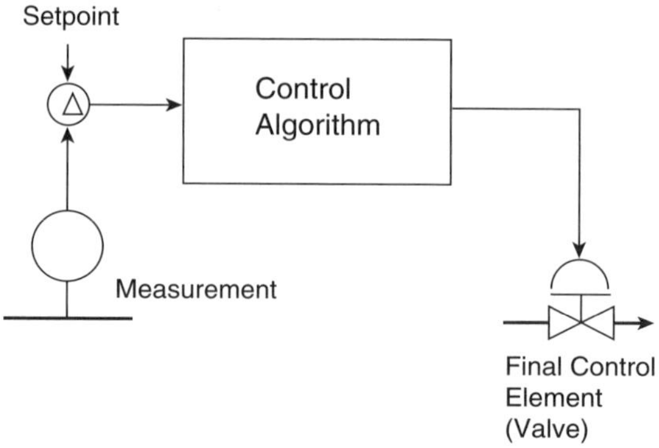

Figure 10.7. Simple control loop.

temperature control loop (Figure 10.7) for a chemical reactor includes an RTD with temperature transmitter, a PID-controller, and a control valve.[11] The three devices form a closed-loop or feedback control loop.[12] Closed-loop control is a method in which a real-time measurement of the process being controlled is constantly being fed back to the controlling device to ensure that the value, which is desired, is being achieved. The RTD constantly transmits a temperature to the controller and then the controller sends a signal to the control valve to adjust its position to increase or decrease flow to the reactor jacket. The mission of the controlling device is to make the measured value equal to the desired value. The best way of accomplishing this task is with the use of a control algorithm know as PID. In its basic form, PID involves three mathematical control functions working together: proportional, integral, and derivative.

Controllers are designed to eliminate the need for continuous operator attention. Cruise control in a car and a house thermostat are common examples of how controllers are used to automatically adjust some variable to hold the measurement (or process variable) at the set-point. The set-point is where you would like the measurement to be. Controller error is the difference between the set-point and the actual measurement. The output of the PID controller will change in response to a change in the measurement or set-point. To optimize control the PID variables are adjusted to fit the particular process, a technique called control loop tuning. Control loop tuning is the art of selecting values for the tuning parameters so that the controller will be able to eliminate an error. How quickly the error is eliminated is not only a function of the tuned control loop but also a function of the system under control. It may be possible to heat one gallon of water to boiling in several minutes, but the same is not possible for 10,000 gallons of water. Again consider the cruise control in a car. It can accelerate the car to a desired cruising speed, but not instantaneously. The car's inertia causes a delay between the time that the controller engages the accelerator and the time that the car's speed reaches the set-point. This same situation applies to heating a reactor: Heat transfer causes a delay between the time that the controller signals the heating control valve to open and the time that the reactor temperature reaches set-point. How well a PID controller performs depends in large part on such lags.

- The proportional (P) parameter determines the magnitude of the difference between the set-point and the process variable (known as error), and then it applies or directs appropriate proportional changes to the control variable to eliminate the error (proportional control).
- The integral (I) parameter looks at the offset of set-point and the process variable over time and when required corrects it (integral control).
- The derivative (D) parameter monitors the rate of change of the process variable and accordingly makes changes to the output variable to adjust for unusual changes (derivative control).

The process of determining the values of these parameters is known as PID tuning. Fundamental to developing a well-designed control loop is a thorough understanding of the process and what affect adjusting the tuning parameters will have on its control. Often, dramatic improvements in control performance can be achieved by breaking the overall process into its pieces and building appropriate control circuits around the pieces. The technique of cascade control is one way of doing this (Figure 10.8). Most batch reactor systems employ cascade temperature control. The cascade control technique breaks the overall process into two parts connected in series: a primary process and a secondary process. The secondary process (reactor jacket temperature) feeds the primary process (reactor internal temperature). With this technique of control, disturbances in the secondary process (reactor jacket) are corrected by the secondary loop before they can significantly influence the primary process (reactor temperature). The result is more precise temperature control of the process.

With modern distributed PLC (programmable logic controller) networks and distributed control systems (DCS), data logging, trending, and operation have never

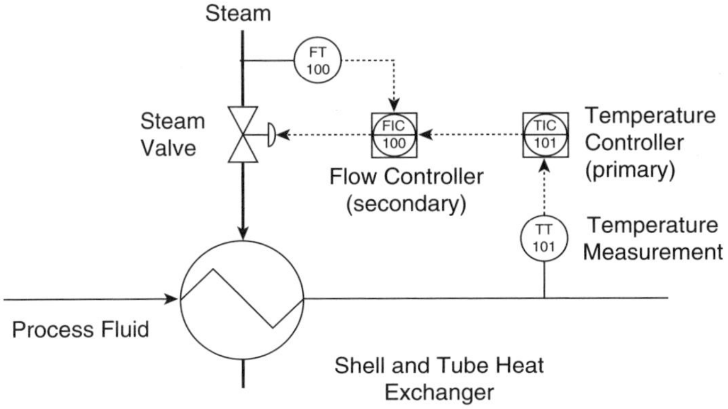

Figure 10.8. Cascade control loop.

been easier. These type advances make the periodic investment in instrumentation upgrades an important aspect of maintaining a modern scale-up facility.

REACTOR SAMPLING

In order to determine completeness of a reaction, an in-process control or assay sample of the reaction mixture is required. The challenge on scale-up is the collection of a representative process sample from the reactor vessel. The need to know the condition of a process in a reactor has always presented problems since it is not always possible to stop the reaction process and/or open the reactor to take a sample. Some processes may allow the use of local exhaust and manual sampling through the reactor opening. Frequently, this is not acceptable because of operator

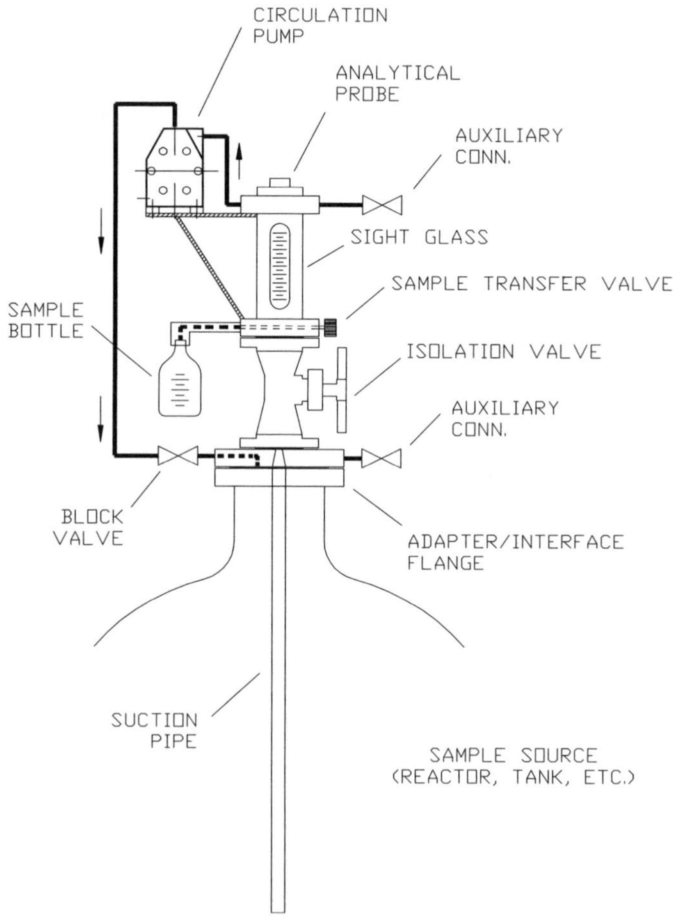

Figure 10.9. Sampler schematic.

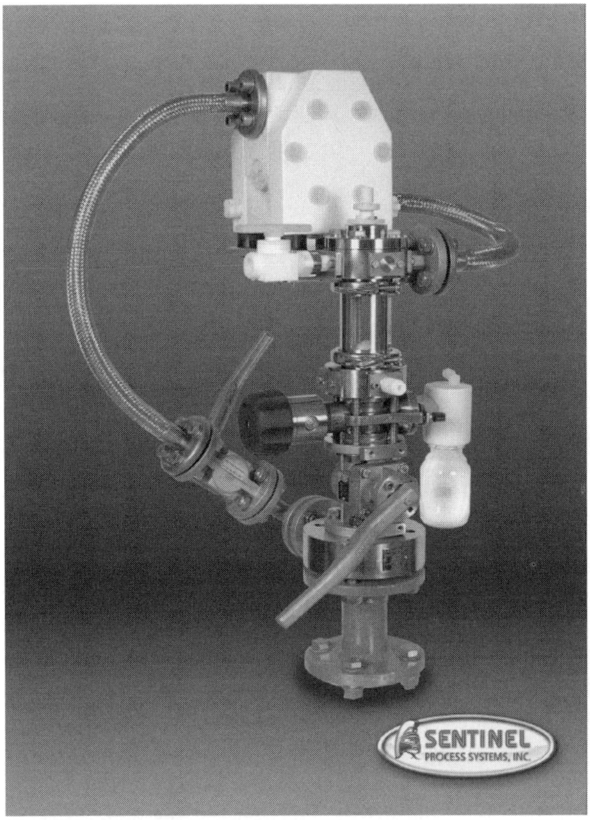

Figure 10.10. Reactor sampler.

safety issues and environmental concerns. It is becoming more common for reactor systems to be equipped with a sampling apparatus that provides safety for the operator and the environment. The process vessel sampling is done with closed-loop systems that make it possible to lift a sample (liquid) from the reactor under process conditions without stopping the process or opening the reactor. (Figures 10.9 and 10.10).

REACTION WORKUP

Typically after the reaction is determined to be complete, the reaction mixture is subjected to further processing referred to as the reaction workup. The workup may be as simple as collecting a solid product directly by filtration. It may include an aqueous quench to dilute or neutralize the reaction mixture followed by one or several extractions to remove components such as byproducts, impurities, or unreacted starting material. It may also include distillation of solvents or volatile components to facilitate isolation of products. These routine laboratory procedures are usually simple and fast on the lab scale but often pose major challenges during scale-up. During scale-up, aqueous quench solutions must be added slowly with good agitation and cooling. The quench addition may result in evolution of a gas byproduct, which can cause serious foaming problems and pressure buildup or it may result in an exothermic reaction that requires efficient cooling and pose a potential safety hazard.

Liquid–liquid extraction involves the transfer of mass from one liquid phase into a second immiscible liquid phase. The process can be carried out in many different ways, but the most common scale-up batch operation is the addition of a second immiscible phase into a reactor vessel holding the reaction mixture. The selection of the second immiscible phase is based on its ability to pick up (extract) the desired component in the batch. It could be water to extract some inorganic material or a particular solvent to extract some impurity in the batch. After a period of agitation, the layers are allowed to separate and the two phases are split between two vessels. An inline sight-glass is normally installed between the two vessels to observe the interphase between layers. This process may be repeated several times. Pilot scale extraction processes are often time-consuming and labor-intensive and require additional vessels or equipment. Continuous liquid–liquid extractors are available as packaged units; however, for early scale-up the batch process is favored because it requires less time to optimize in the laboratory, although an aqueous batch extraction procedure that worked well in the laboratory may turn into an emulsion during scale-up. In the pilot plant, good mixing is required to generate efficient two-phase contact, but too much turbulence may cause an emulsion. To minimize the risk of an emulsion on scale-up, it is normal to agitate at a low speed for longer time. Numerous low-volume batch extractions yield the most efficient mass transfer, but the time and effort required during scale-up is much greater. A compromise between efficiency and labor (time) is required. Rarely observed in the laboratory is a dark interfacial layer that typically results from insoluble matter contained in raw materials. This layer is usually discarded.

Another key advantage of the batch extraction process is that it is generally very effective at ambient temperature and pressure. If greater efficiency is required, temperature is an easy parameter to alter during scale-up. Also, changing temperature can sometimes help break an emulsion. After the extraction process is complete, a very common laboratory technique for the dehydration of the organic phase is the addition of solid drying agents such as magnesium sulfate, sodium sulfate, or molecular sieves. Occasionally, color is removed from the solution by the addition of powdered activated charcoal. These procedures may seem routine in the laboratory but on scale-up can pose major problems. Once charged to the reactor, the insoluble additive must be removed by filtration. This additional processing step may result in product losses and vessel cleaning problems. Removal of residual water by azeotropic distillation drying is often a desired alternative to the addition of solid drying agents. Another alternative is the use of filter cartridges with the drying agent or decolorizing agent impregnated within the cartridge material. With this option the batch can be passed or circulated through the cartridge without contaminating the reaction vessel with an additive.

Batch distillation operations for volume reduction or solvent exchange in the laboratory are normally done at reduced pressure using a rotary evaporator because it is fast and simple to operate. During scale-up, this operation is commonly performed in reactor systems. To remove higher boiling solvents, prevent the degradation of a heat-sensitive compound or the formation of impurities, vacuum distillation is a frequently used technique. In the laboratory, these operations may take minutes to hours to complete. In pilot plant equipment, the same operation will require hours to days to complete. To improve throughput or to handle heat-sensitive compounds that cannot tolerate the extended residence time of a large-scale batch distillation, one possible solution involves using a thin-film or wiped-film evaporator. Film-type evaporation or distillation involves mechanical generation of a thin product (using wiper blades or rollers) over the heat transfer surface to promote minimal resistance to heat transfer and thus dramatically reduce product residence times. Thin/wiped film evaporator systems (Figure 10.11), while not essential

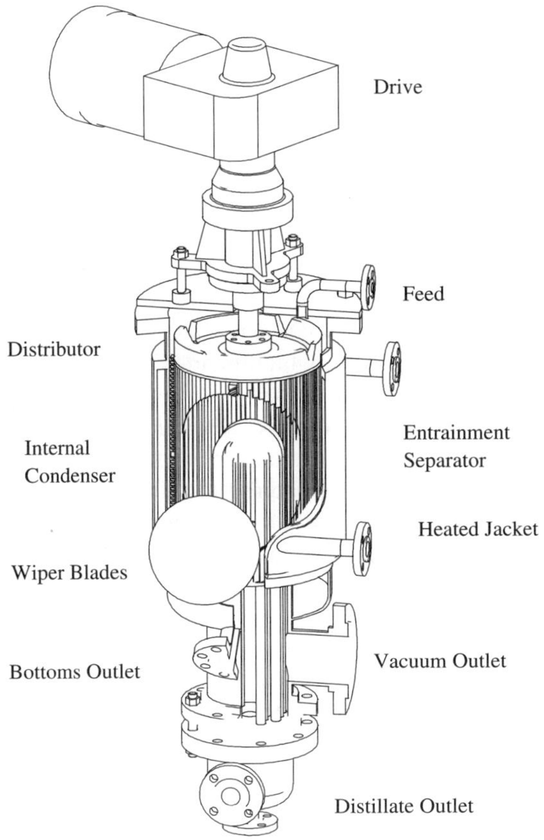

Figure 10.11. Wiped-film evaporator (WFE). Courtesy of Pfaudler Engineered Systems.

scale-up equipment, are becoming more common additions to pilot plant facilities, especially for rapid solvent stripping/concentration operations. A thin/wiped-film evaporator consists of a jacketed cylindrical heating section and a top vapor section, a top cover with mechanical seal housing and roller bearings, and a bottom conical section housing the lower bearing. An internal revolving rotor equipped with either wipers or blades provides internal distribution and rapid transport of the film over the heated area or surface. Evaporation takes place under vacuum, and the vapors are condensed externally using existing facilities process condensers and vessels.

Frequently, the rationale for distillation/stripping operations is solvent exchange. It is common for one of the final batch processing steps to involve the switch from the reaction solvent to another solvent more appropriate for the final product crystallization. Once the solvent exchange is complete and prior to crystallization, a clarification or polish filtration of the batch is required. The normal method is to pump the solution through a fine-pored (5 μm or less) cartridge filter into a separate precleaned vessel or crystallizer. This will remove any undissolved particulates that may have been introduced in the raw materials or from other sources during normal processing. A polish filtration step is an absolute requirement prior to crystallization of the final API.

CRYSTALLIZATION

The formation of crystals from a homogeneous solution is essentially another solid–liquid separation technique. It is a very important operation and probably the most difficult to scale-up. Batch crystallization has been and continues to be the most widely practiced form of crystallization in the laboratory and on scale-up. The equipment used is the same stirred vessels used for performing the reaction steps (round-bottom flasks or reactor vessels). A batch crystallization process can be left unattended (overnight) if necessary, which is an advantage for smaller operations using a single-shift workforce.

In order for crystallization to take place, a solution must be supersaturated. Primary nucleation is the first step in crystallization. It is the growth of first crystals. On scale-up a large driving force is required to initiate primary nucleation. Rapid cooling can provide this driving force and produce the instantaneous formation of many nuclei. However, in this case, crystallization may occur too fast and the nuclei formation is often accompanied by "crashing out" of the solid from solution. This process is difficult to scale-up and often results in the occlusion of solvent or impurities in the crystal lattice. It also may cause problems in stirring and filtration. In many cases, gradual cooling and stirring is enough to start nucleation and crystallization of the solid at a reasonable rate to avoid these problems. The next or chief mechanism in crystallization is secondary nucleation. Secondary nucleation requires "seeds" or existing crystals (Figure 10.12) to initiate crystal growth at lower supersaturation where crystal growth is optimal.

Control of the batch crystallization process is difficult and in the past was rarely practiced in the laboratory or during initial scale-up. The process was considered as

Figure 10.12. SeedPoint valve. Courtesy of Powder Systems Limited (PSL).

much of an art form as a science. Today even the very first batch of API undergoes rigorous quality testing to ensure that it exhibits consistent polymorph and crystal shape characteristics. To properly control consistently the crystallization process, a minimum requirement is knowledge of the solubility of the product versus temperature (solubility curve) and when the initial solution is supersaturated. Seed crystals are introduced at the proper concentration and temperature, and then a slow rate of cooling is applied to the batch. While cooling is the chief method for controlling crystallization, other factors including mixing, evaporation, or co-solvents may be employed. After crystallization, the product must be isolated. Therefore, in addition to meeting final product specifications, ideally we want to produce a free-flowing slurry (good slurry density) or a slurry that is easily pumped from the crystallization vessel to a filtration unit.

Regarding pumps, they are so prevalent in almost every aspect of scale-up that the subject cannot be totally overlooked. "Only the sail can contend with the pump for the title of the earliest invention for the conversion of natural energy into useful work, and it is doubtful that the sail takes precedence. Since the sail cannot, in any event, be classified as a machine, the pump stands essentially unchallenged as the earliest form of machine, which substituted natural energy for human muscular effort. Perhaps most interesting, however, is the fact that with all the technological development that has occurred since ancient times, including the transformation from water power through other forms of energy all the way to nuclear fission, the pump remains probably the second most common machine in use, exceeded in numbers only by the electric motor."[13] Since pumps have existed so long and are in such widespread use, it is hardly surprising that they are produced in a seemingly endless variety of sizes and types and are applied to an apparently equally endless variety of services. In the laboratory, solvents are poured, reagents are added by addition funnel, and slurries are transferred to filter by pouring. On scale-up, pumps (positive displacement diaphragm or piston pumps) are used for each of these applications. In addition, pumps are used to supply virtually every utility service in the pilot plant from vacuum to purified water to conditioned air.

FILTRATION

The removal of crystalline product from a solid–liquid slurry in the laboratory is normally accomplished in something simple like an open Buchner funnel. During scale-up the pilot plant equipment becomes a bit more sophisticated. While open Nutsches are sometimes used, the trend is toward more containment for minimization or elimination of cross-contamination. Present-day chemical processing requires a variety of solid–liquid separation techniques. A success story of the last decade is the introduction of agitated pressure Nutsche filter dryers into pilot plant scale operations (Figure 10.13). These units have evolved from simple pressure/vacuum filter units into sophisticated pieces of equipment capable of performing many process operations including filtration, displacement washing, reslurry, smoothing and compression, drying, and product discharge and collection. All these operations are done under containment. Combining these operations into one unit minimizes exposure of potent products to operators and the environment. The installation of a split butterfly valve (Figure 10.14) or high containment barrier on the discharge virtually eliminates exposure. Product centrifuges offer another scale-up option for solid–liquid separation. They can generate centrifugal forces of around 1000 times gravity. This force can push more solvent out of the cake and reduce drying time.

DRYING

For the final product drying operation, vacuum tray dryers are still used for small batches. For larger batches, tray drying is usually a slow process, yields clumpy product, results is operator exposure during tray handling operations, and is very labor-intensive. Still most facilities maintain tray drying as an option for some materials. The tray dryer unit can be installed inside a laminar down-flow booth or workstation to control operator exposure during tray handling operations. For drying large pilot scale batches, potent APIs, or hazardous materials, the pressure filter dryer is gradually replacing the tray dryer as the industry trends toward the

Figure 10.13. FiltroDry, an enclosed agitated filter dryer. Courtesy of 3V Cogeim.

Figure 10.14. Contained dryer off-loading system. Courtesy of Powder Systems Limited (PSL).

consolidation of several process operations into one piece of equipment and the elimination of operator exposure and cross contamination.

ENVIRONMENTAL

Tremendous scientific, political, and social changes have impacted air quality and environmental regulations. This promotes new initiatives for controlling volatile organic compound (VOC) and hazardous air pollutant (HAP) emissions. The main focus of these initiatives was on larger manufacturing facilities, but pilot plant scale operations are now coming under ever-increasing scrutiny. A variety of technologies are used to reduce emissions from chemical facilities.[14]

Absorption or scrubbing (using a packed tower or wet scrubber) is most commonly used for controlling emissions of acidic or inorganic gases, such as HCl. Scrubbers are especially useful when the pollutant is soluble in water. When sized properly, efficiency removals to 99.9+% are easily obtained, making scrubbers the primary selection for trapping of toxic chemicals that can be neutralized. A packed tower scrubber is a column filled with a depth of packing material, a sump at the bottom, a spray header at the top, and a pump for recirculating the scrubber liquid. In some applications an exhaust fan is required to provide the pressure to force the gas through the column and out the stack. A venturi scrubber is similar to a packed tower, except that the pumped liquid provides the motive force for the gas (similar to a water aspirator used in the laboratory). Additionally, part of the neutralizing reaction takes place in the venturi nozzle, reducing the packing depth. Frequently, a venturi is coupled with a packed tower for increased efficiency.

If neutralizing doesn't work, the chemical is most likely a volatile organic compound (VOC). Separation via condensation using refrigerated condensers at atmospheric pressure is commonly used to remove nonhalogenated and halogenated VOCs/HAPs. If condensation does not control VOCs within permissible limits, other technologies are employed to control emissions. For large facilities, a thermal oxidizer (TO) may be installed in the process exhaust stream containing the VOC laden gases from the process. The gases pass through the TO and are then oxidized. Thermal oxidation, or thermal incineration, is the process of oxidizing combustible materials by raising their temperature above the autoignition point in the presence of oxygen and maintaining it at a high temperature for sufficient time to complete combustion to carbon dioxide and water. A TO system requires a burner and operates around 1500°F. TO-equipped pilot plants are unusual.

Another practical and cost-efficient technology for containment in a pilot plant that runs intermittently with relatively low levels of emission is carbon absorption. The process exhaust stream is passed through a series of carbon canisters or a bed of carbon to absorb the VOCs. As the removal efficiency of the carbon diminishes, the canisters or bed is replaced and regenerated for later use.

During scale-up, aqueous waste is pH adjusted and held in double-contained storage tanks for later disposal. Organic solvent waste is packaged in drums, shipped off-site, and often burned for energy recovery.

EQUIPMENT CLEANING

In the laboratory, dirty glass equipment is rinsed with suitable solvents and sent to the glass wash area for final cleaning. On scale-up, the batch process operations/equipment used to prepare a single API can leave residual product that could contaminate subsequent batches. This is not acceptable for human pharmaceutical products. Therefore the equipment cleaning procedure is a key-processing step that gets incorporated directly into the process batch record. Cleaning procedures vary based on the process, but it typically starts with a pressurized water rinse of the system. A detergent and solvent wash follow this. As a rule, the final step is a solvent boil-up or reflux. Proper cleaning may require some equipment dismantling and reassembly after cleaning. Clean-In-Place (CIP) is a method designed to automatically clean process equipment without the time and labor requirement of manually dismantling equipment components. CIP is more a design method than a cleaning method. When properly designed, CIP is more consistent than conventional manual cleaning methods. The CIP design eliminates places where residue can accumulate and incorporates spray balls into the process equipment, used for cleaning the equipment internally. Today most new process equipment or reactor systems include CIP in the design. Existing equipment can be retrofitted with spray balls for cleaning (Figure 10.15).

Proper integration of CIP into the process requires a centralized CIP System Skid, which consists of cleaning and rinsing storage tanks, pumps, a detergent feeder which makes the cleaning solutions, a controller for variable CIP delivery, a heat exchanger for heating the wash solution, and integrated piping to supply

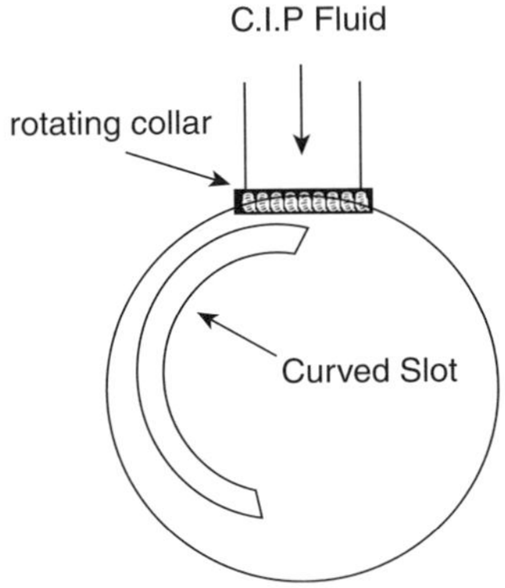

Figure 10.15. Slotted spray ball.

the wash solution to the satellite spray balls. The variables of temperature, velocity/pressure, detergent concentration, and exposure time are precisely controlled by the CIP System Skid.

For smaller parts or components an alternative to manual cleaning is an automated Cabinet Washer also known as a Clean-Out-of-Place (COP) system. A COP System Skid consists of similar components as a CIP Skid but it also includes a cabinet or enclosure with an integrated rack for stacking dirty equipment. Small equipment or parts are placed on the cabinet rack for automated cleaning similar to a home dishwasher.

CONCLUSION

This chapter discussed the basic and key processing operations used during large-scale synthesis from an engineering perspective, starting with the initial evaluation of the laboratory procedure and concluding with process equipment cleaning. All of the operations discussed could be expanded and still tell only part of the story. Maintaining successful scale-up of modern chemical processes requires the continual modernization of the scale-up facility. In order to remain competitive, sustain a safe operation, maintain ever-increasing quality requirements, and meet more aggressive timelines, modern systems and procedures must be routinely introduced. The pilot plant must continually evolve with time to meet these new challenges. Many pilot scale facilities are in a perpetual state of upgrade.

Today's sophisticated and integrated instrumentation systems provide process monitoring and data collection capabilities that further expand the scale-up potential of modern pilot scale facilities. Current heat transfer systems offer a wide range of temperature capabilities in a single package that expand processing capabilities. Process Analytical Technology (PAT) techniques provide noninvasive methods for monitoring process operations and enhancing the scalability of the process. PAT methods like Near-Infrared Spectroscopy (NIR) are being used to detect the endpoint of a reaction. This is becoming a useful tool on scale-up because often by knowing when the reaction is complete not only maximizes the yield but also protects the product from being overheated, which may lead to decomposition and forming impurities. Modern PAT techniques for early scale-up permit the accurate monitoring of the crystallization process. The most common tool for monitoring crystal or particle growth is focused beam reflectance measurement (FBRM). In FBRM a probe is inserted into the reactor and a spinning light beam (laser) bounces off the particles suspended in the slurry solvent. This provides a measurement of the crystal cord length and overall particle size distribution of the slurry. The particle size distribution within the crystallization can be adjusted by altering heating or cooling rates to fit the conditions optimized in the laboratory. PAT techniques can speed the transition from the laboratory to scale-up and provide a more reliable scale-up process.

These techniques represent only a fraction of the innovations taking place within the industry today. The successful implementation of these new techniques requires the close collaboration between process development chemists and chemical engineers. This collaboration is by far the most important factor to successful large-scale synthesis.

REFERENCES AND NOTES

The discussion presented in this chapter is focused on basic information around chemical engineering and fundamentals of scale up operations. For general references, the author suggests the following sources:

Perry, R., and Green, D. *Perry's Chemical Engineers' Handbook*, 7th edition, McGraw-Hill, New York, 1997.

McConville, F. X., *The Pilot Plant Real Book: A Unique Handbook for the Chemical Process Industry*, FXM Engineering and Design, Worchester, MA, 2002.

1. www.pafko.com/history.edu, History of Chemical Engineering.
2. Helfand, W. H., Woodruff, H. B., Coleman, K. M. H., Cowen, D. L. Wartime Industrial Penicillin in the United States, in *The History of Antibiotics: A Symposium*, Parascandola, J., Ed., American Institute for the History of Pharmacy, Madison, WI, 1980, pp. 31–56.
3. McCabe, W. L., and Smith, J. C. *Unit Operations of Chemical Engineering*, 3rd edition, McGraw-Hill, New York, 1976.
4. The FDA began as the Division of Chemistry and then (after July 1901) the Bureau of Chemistry, and the modern era of the FDA dates to 1906 with the passage of the Federal Food and Drugs Act; this added regulatory functions to the agency's scientific mission. The Bureau of Chemistry's name changed to the Food, Drug, and Insecticide Administration in July 1927, when the nonregulatory research functions of the bureau were transferred elsewhere in the department. In July 1930 the name was shortened to the present version. FDA remained under the Department of Agriculture until June 1940, when the agency was moved to the new Federal Security Agency. In April 1953 the agency again was transferred, to the Department of Health, Education, and Welfare (HEW). Fifteen years later, FDA became part of the Public Health Service within HEW, and in May 1980 the education function was removed from HEW to create the Department of Health and Human Services, FDA's current home. To understand the development of this agency is to understand the laws it regulates, how the FDA has administered these laws, how the courts have interpreted the legislation, and how major events have driven all three. For more information on the history of the FDA see www.fda.gov/oc/history, FDA History.
5. Wedlich, R. C. Tame Process Thermal Hazards, *Chem. Eng. Prog.* **1999**, September, 71–77.
6. Lees, F. P. *Loss Prevention in the Process Industries*, Vols. I and II, Butterworth-Heinemann Ltd., Linacre House, Jordan Hill, Oxford, 1992.
7. Kletz, T. A. *What Went Wrong? Case Histories of Process Plant Disasters*, 2nd edition, Gulf Publishing Company, Houston, TX, 1988.
8. Fay, T., and Nicholas, P. Safe Handling of Potent Compounds, *Chem. Eng.* **2002**, *April*, 62–68; **2002**, *May*, 68–72.
9. DiMarco, C., Johnson, B., Izzo, B., and Dienemann, E. Implementation and characterization of a solids charging System in a Pharmaceutical Pilot Plant, AIChE Annual Meeting, November 2003.

10. Syltherm XLT, Product Technical Data, The Dow Chemical Company.
11. Liptak, B. G., and Venczel, K. *Process Control—Instrument Engineers Handbook*, revised edition, Chilton Book Company, Radnor, PA, 1985.
12. Liptak, B. G., and Venczel, K. *Process Measurement—Instrument Engineers Handbook*, revised edition, Chilton Book Company, Radnor, PA, 1982.
13. Karassik, I., Krutzsch, W., Fraser, W., and Messina, J. *Pump Handbook*, 2nd edition, McGraw-Hill, New York, 1986.
14. American Institute of Chemical Engineers, *Practical Solutions for Reducing and Controlling Volatile Organic Compounds and Hazardous Air Pollutants*, AIChE Center for Waste Reduction Technologies, New York, 2001.

11

SYNTHESIS AND APPLICATION OF RADIOISOTOPES IN PHARMACEUTICAL RESEARCH AND DEVELOPMENT

LARRY E. WEANER AND DAVID C. HOERR
Johnson & Johnson Pharmaceutical Research and Development, L.L.C., Spring House, PA 19477

INTRODUCTION

The importance and impact of radioisotopes on the advancement of the life sciences is well known and documented in the literature.[1–3] When incorporated into a molecular structure for use as a radiotracer, they can provide rapid *in situ* detection and quantification in animals and complex biological matrices with high sensitivity.[4,5] Pharmaceutical and biomedical organizations routinely use radiolabeled compounds to study the metabolism and pharmacokinetics of drug candidates[6] and to develop the supporting data required to obtain regulatory approval to market a new drug.[7]

Numerous radionuclides are available for incorporation as radiotracers. The most commonly used include hydrogen-3 (tritium), carbon-11, carbon-14, sulfur-35, phosphorus-32, phosphorus-33, iodine-125, iodine-131, and fluorine-18.

Of these, ^{11}C, ^{14}C, and ^{3}H are the radionuclides of choice for pharmaceutical and biomedical research studies because they are isotopes of the common elements found in the molecular structures to be studied. They are used to prepare radiotracers with physical and chemical properties identical to those of the unlabeled compounds, and their incorporation rarely affects biological activity or results in changes in metabolic pathways.

Fundamentals of Early Clinical Drug Development: From Synthesis Design to Formulation, edited by Ahmed F. Abdel-Magid and Stéphane Caron

Radiosyntheses are usually designed to incorporate the label(s) in the least number of synthetic steps and as late in the synthetic scheme as possible in order to minimize handling and costs. In many cases, the radiolabeled compound can be prepared using the same chemical sequence as that used for the preparation of the unlabeled analogue.

CHEMISTRY OF RADIONUCLIDES

Chemical reactions for the most part take place between the orbiting electrons of atoms while nuclear transformations occur primarily in the nuclei of the atom. Because of this, radionuclides, undergo the same chemical reactions as nonradioactive nuclides, and the radioactive properties of a radionuclide are almost completely unaffected by its chemical form. Therefore, a labeled compound can usually be prepared employing the same chemical reactions as those used for preparing nonradioactive substances. Additionally, the radioactive properties of the labeled atoms will be unaffected by chemical state except for those radionuclides that decay by processes involving the orbital electrons (e.g., electron capture). For those radionuclides, there is a very small change in their average lifetime, but these effects are not detectable when used for tracer applications.

There may, however, be measurable differences in the rates of chemical reactions between labeled and unlabeled substances because heavier atoms form stronger, more stable bonds.[8] This behavior is directly attributable to atomic mass differences and is known as an *isotope effect*. For many radionuclides, the isotope effect is small or not observed, particularly for larger radionuclides. However, when large mass differences exist between the two isotopes, chemical behavior is affected.[9] For example, the mass difference between the carbon isotopes, ^{12}C and ^{14}C, is small; consequently, only small isotope effects are usually observed. However, the mass differences between the hydrogen isotopes are significantly larger: deuterium (D) is twice as heavy as protium (H), and tritium (T) is three times heavier than H. When a bond to tritium (e.g., a C–T bond) is being broken or formed in a rate-limiting step, large isotope effects are observed (primary hydrogen isotope effect). Isotope effects are of course not observed in radioisotopically labeled molecules when the bond to a radionuclide is not involved and reaction occurs at any of the other unlabeled atoms in the molecule.

RADIONUCLIDE SELECTION AND LABELING POSITION IN RADIOTRACERS

For most radiotracers, selection of which position to place the label is determined by factors such as the type of experiment to be performed, the ease of synthesis, half-life of the radionuclide, amount of activity required, stability, and a number of other factors. Because only the radioactive atoms are traced, knowledge of the specific location of the radioactive label(s) within the molecular structure is generally required to ensure that accurate experimental results and valid conclusions are obtained. For this reason, most radiolabeled compounds used in experiments such as metabolism and reaction mechanism studies involve a rational synthesis in which the label is introduced into a precise location based on the synthetic route and chemistry involved. Since labeled compounds can undergo biotranformations that may result in cleavage and loss of the radioactive label when used in *in vivo* studies, careful consideration must be given to the position selected for incorporation.

Single or multiple positions may be labeled with one or more different radionuclides in a radiotracer. Selection of the radionuclide to be used is determined by the type of study being performed.[10] Most of the commonly used radionuclides undergo nuclear transformations that result in the release of radiation in the form of particles and/or X rays or gamma rays. The rate at which a radioactive substance decays is designated by its half-life, which is the average time required for one-half of the radioactivity present in a sample to decay away. The shorter the half-life, the more disintegrations per unit time. Additionally, each particle or photon that is released has a different amount of energy associated with it, and the ease of detecting and measuring that emission varies depending on the radionuclide (Table 11.1). Radionuclides with higher emission energies and high disintegration rates are easier to detect. The ideal radionuclide has a half-life that is long enough to allow for synthesis of the radiotracer and completion of planned experimental work. Both the decay rate and resulting radioactive emission must be sufficient to be easily measured in high sensitivity in the matrix being studied. Additionally, once incorporated, the physical properties and chemical behavior of the tracer molecule should be indistinguishable from the unlabeled material. However, in

TABLE 11.1. Properties of Commonly Used Radionuclides[11]

Radionuclide[a,b]	Mode of Decay[c]	Half-Life	Maximum Energy (keV)	Maximum Specific Activity (Ci/mmol)
^{3}H	β^-	12.32 years	18.6	29.0
^{11}C	β^+	20.39 min	1982	9.22×10^6
^{14}C	β^-	5700 years	156	0.0627
^{18}F	β^+	109.77 min	1655	1.71×10^6
^{32}P	β^-	14.26 days	1711	9150
^{33}P	β^-	25.34 days	249	5150
^{35}S	β^-	87.51 days	167	1491
^{125}I	EC, γ	59.4 days	186	2197
^{131}I	β^-	8.02 days	971	1.627×10^4

[a]The terms *radionuclide* and *radioisotope* can be used interchangeably, where *radionuclide* applies to atoms that undergo a radioactive transformation, while *radioisotope* refers to the radioactive species of an isotopic group. Therefore, for the element carbon, carbon-14 is a radioisotope for the isotopes of carbon-11, carbon-12, and carbon-13, as well as the other isotopes of carbon.

[b]Radionuclides are represented by a notation written in the form ^{A}X, where A is the mass number of the atom and X represents the chemical symbol. The notation for carbon-14 is therefore written as ^{14}C and read as "C fourteen."

[c]β^-, negative beta decay; β^+, positive beta decay; EC, electron capture; γ, gamma emission.

some cases it may be necessary to incorporate a radionuclide that provides an analogue or adduct of the molecule of interest.

Tracer compounds are generally not prepared at 100% isotopic abundance, and an atom denoted in a molecular structure as "labeled" with a particular radionuclide is in most cases a mixture of the radionuclide and its unlabeled isotope. Thus, a ^{14}C-labeled compound prepared with a specific activity lower than 62.7 mCi/mmol (2.32 GBq/mmol, Table 11.1) will contain some percentage of ^{12}C in the position denoted in a structural formula as being "^{14}C-labeled." Additionally, regardless of how little isotope is present, the compound will always be named with both the radionuclide and its location in the molecule indicated. The preferred way to name labeled compounds is by using the "square-brackets-preceding" system: for example, [^{14}C]carbon dioxide or [1,2-^{14}C]acetic acid. For quantitative studies, however, accurate knowledge of the amount of radioactivity incorporated in the labeled compound is required. For radiotracers, this is expressed as the "specific activity" of the compound, where specific activity is the amount of radioactivity per unit mass of a substance containing a radionuclide. The greater the amount of incorporation of a radionuclide into a compound, the higher the specific activity. Examples of units for specific activity are kBq/mmol or kBq/mg, where the International System unit for radioactivity, the becquerel (abbreviated Bq), is equal to 1 disintegration per second (dps). Specific activity is also frequently expressed using an older but still commonly used unit, the Curie (Ci), where 1 Ci is equal to 3.7×10^{10} Bq; examples with this unit are Ci/mg, mCi/mmol, and Ci/mmol. For example, 100% isotopic abundance of ^{14}C in a single position provides material having a specific activity of 2.32 GBq/mmol or 62.7 mCi/mmol.

SYNTHESIS OF RADIOLABELED COMPOUNDS

Tritium (^{3}H) and ^{14}C are the radioisotopes most widely used in pharmaceutical research. Both radionuclides emit low-energy beta particles (energetic electrons that are emitted from the nucleus) that are easily detected and pose no external radiation hazard to the workers who handle them. Each isotope can be readily incorporated using traditional synthetic chemistry without altering the molecular structure of the compound being labeled. Additionally, because of their relatively long half-lives, 12.32 years for ^{3}H and 5700 years for ^{14}C, correction for loss of activity due to natural radioactive decay is not required for most uses. Detection and quantitation of ^{3}H and ^{14}C radiotracers is generally accomplished using liquid scintillation counting, which accurately counts the number of ionizing particles emitted from a radioactive sample. Other techniques, such as accelerator mass spectroscopy, quantitative autoradiography, or quantitative imaging techniques are also utilized.

Synthesis with Carbon-14

Carbon-14-labeled compounds are primarily prepared using synthetic methods and to a lesser extent by biosynthetic methods. Incorporation of the ^{14}C isotope generally requires multiple synthetic steps starting from an organic intermediate labeled with ^{14}C. Classical organic synthesis methods and reactions are employed; however, the stoichiometry is adjusted to maximize the radiochemical yield of the labeled starting intermediate. For most applications, synthesis is completed so that the position of the radionuclide in the radiotracer is known based on the synthetic scheme employed. Labeling may require incorporation into single or multiple positions and with more than one radionuclide. However, dual labeling a compound with, for example, both ^{3}H and ^{14}C is rare. When required, dual-labeled compounds generally are produced by individually preparing the ^{14}C- and ^{3}H-labeled compounds and then mixing them.

Suitable radiolabeled intermediates can be purchased from any number of commercial radiolabeling companies or prepared on site from a basic inorganic compound such as $Ba^{14}CO_3$ or $K^{14}CN$. For a single position, the maximum incorporation of ^{14}C possible is 62.7 mCi/mmol (2.32 GBq/mmol). Most compounds, however, are prepared at specific activities in the range of 5–55 mCi/mmol. [^{14}C]Barium carbonate, one of the primary raw materials used for ^{14}C synthesis, is produced using a nuclear reactor and is routinely available with specific activities in the 50- to 58-mCi/mmol range.

Synthesis is generally completed on a 0.1-g to several-gram scale and can require millicurie to curie (megabecquerel to gigabecquerel) quantities of radioactivity. Because many of the starting intermediates are volatile, such as $^{14}CO_2$ and [1,2-^{14}C]acetylene, the initial steps of a radiosynthesis may involve containment by manipulating the materials in a closed vacuum manifold. Additionally, special glassware may be required for gases and difficult chemical transfers, while ensuring containment of the radioactive components. For example, during the synthesis of [1,2-^{14}C]chloral hydrate,[12] shown in Figure 11.1, the chlorination of [1,2-^{14}C]acetaldehyde required a special apparatus that allowed chlorine gas and radiolabeled acetaldehyde to be simultaneously added to the reaction at 95°C.

Carbon-14-labeled compounds and intermediates are also produced by biosynthetic methods that include the use of photosynthesis with [^{14}C]carbon dioxide and plants,[13] as well as by incubation/fermentation methods using small labeled molecules with enzymes, microorganisms, algae,[14] cells, and culture media. Chiral intermediates and compounds such as natural amino acids and carbohydrates are prepared and isolated with the carbon-14 label uniformly distributed throughout the molecular structure. For example, sugars can be prepared uniformly labeled (U) with ^{14}C in all

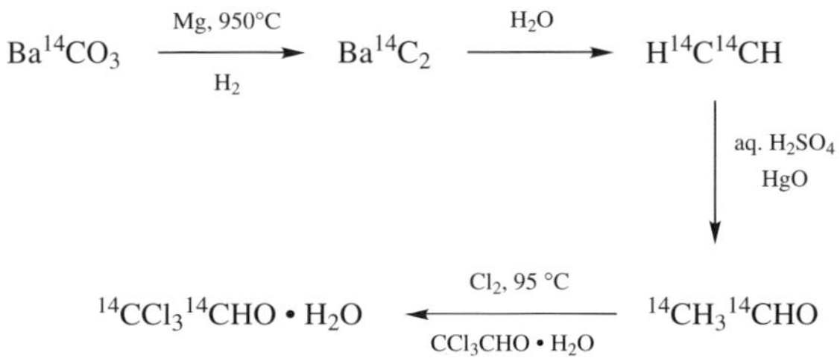

Figure 11.1. Preparation of [1,2-^{14}C]chloral hydrate.

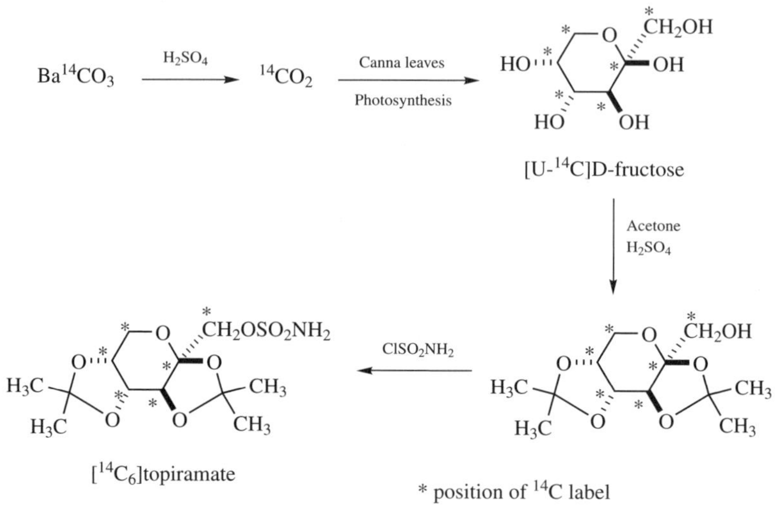

Figure 11.2. Synthesis of [^{14}C]Topiramate.

positions by photosynthetic $^{14}CO_2$ fixation into plant leaves. For the preparation of the antiepileptic drug, [^{14}C]topiramate, [U-^{14}C]D-fructose was photosynthetically prepared and extracted from the leaf of a Canna plant[15] (Figure 11.2).

In many cases it is necessary to label a compound in several different positions in order to understand its mechanism of action and metabolic fate. For the compound sodium palmoxirate, an orally active inhibitor of long-chain fatty acid oxidation, it was necessary to prepare samples with ^{14}C in several positions.[16] To determine the extent and rate of decarboxylation that occurred due to oxidative cleavage *in vivo*, material was synthesized with a ^{14}C label in the carboxyl group. The synthetic route used is shown in Figure 11.3.

A second sample was prepared with a more metabolically stable label for *in vivo* studies to follow the absorption, distribution, metabolism, and excretion of the

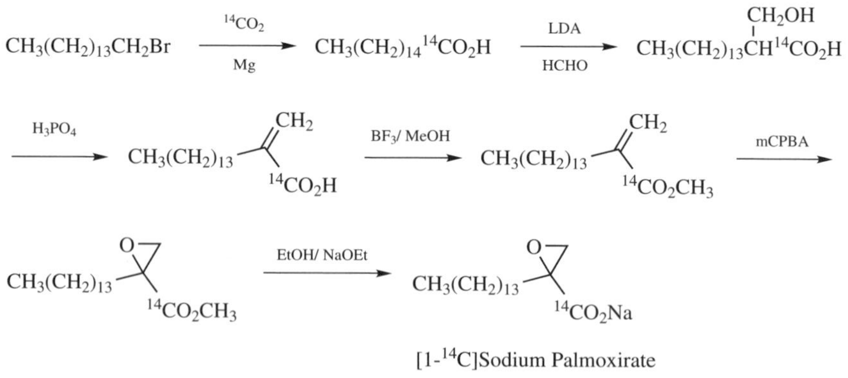

Figure 11.3. Synthesis of sodium [1-^{14}C]palmoxirate.

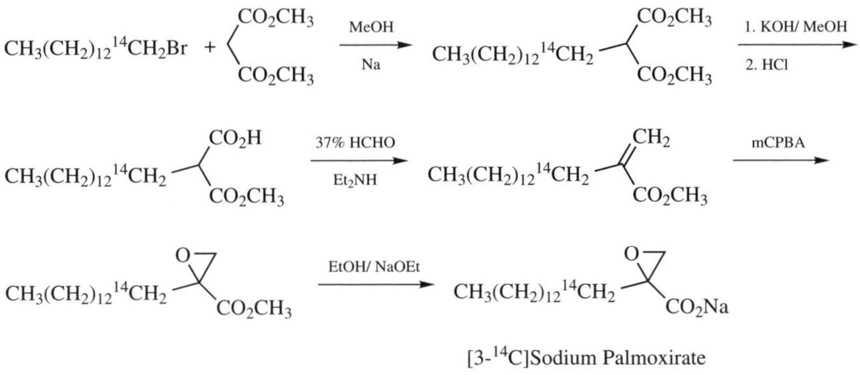

Figure 11.4. Synthesis of sodium [3-^{14}C]palmoxirate.

compound (Figure 11.4). Additionally, one of the biologically active metabolites, the coenzyme A ester of TDGA, was enzymatically synthesized for *in vitro* mechanism studies (Figure 11.5).

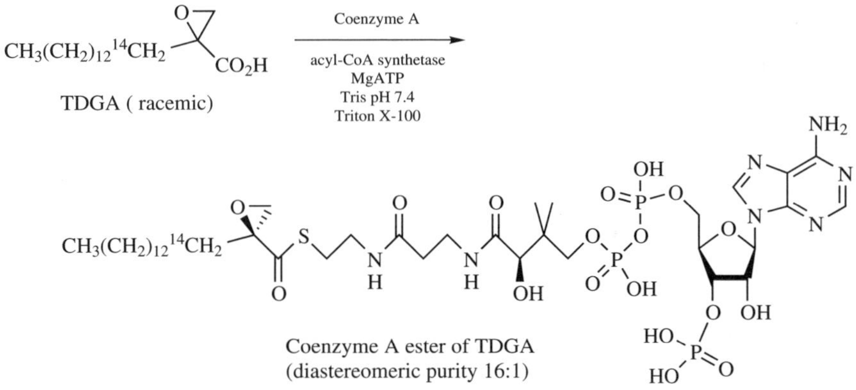

Figure 11.5. Enzymatic synthesis of the coenzyme ester of sodium [3-^{14}C]palmoxirate.

Synthesis with Tritium (^{3}H)

When very small amounts of a tracer compound need to be tracked and quantified, tritium is the isotope of choice because of its much higher specific activity. It can be introduced into a target molecule at specific activities that are hundreds of times those obtained with ^{14}C, which provides greatly enhanced detection and sensitivity. The general approaches for tritium labeling are by chemical synthesis and by hydrogen isotope exchange reactions in which selected hydrogen atoms in the compound to be labeled are directly exchanged with tritium atoms. Because of the ease of incorporation, isotope exchange is preferred for labeling more complex molecular structures. However, there are two main disadvantages associated with tritium exchange labeling: First, the position of the label may not be specifically known based on the chemistry used; second, there is a potential that the label will be lost when used in metabolic experiments, which causes release usually in the form of tritiated water.

How a tritium-labeled compound is to be used in a tracer study determines the amount of radioactivity that must be incorporated. For many applications, material having a very high specific activity is required, and it is not uncommon to require material with between two and four tritium atoms incorporated in specific positions in a single molecule. Since the theoretical amount of radioactivity that can be incorporated into one tritium atom is 29 Ci/mmol (1.07 TBq/mmol), radiotracers having specific activities of 50–110 Ci/mmol (1.85–4.07 TBq/mmol) are regularly prepared by incorporation of 2–4 tritium atoms. Both synthetic and isotope exchange reactions with tritium use higher amounts of radioactivity than is used in ^{14}C syntheses. Because of the small scale and high amounts of radioactivity involved, tritium is normally introduced in one or two synthetic steps from the final product.

Typically, tritium gas or a tritiated solvent such as [^{3}H]water is employed for introduction of the label(s). The most common tritium-labeling reactions involve the metal-catalyzed reactions of dehalogenation, hydrogenation of double and triple bonds, and isotopic exchange with tritium gas.[17] Reactions are completed in closed manifold systems generally using up to ten of curies (37–370+ GBq) of activity. The catalysts used most often are platinum and palladium metals supported on carbon. To obtain maximum incorporation, reactions are completed in solvents with non-exchangeable protons such as anhydrous 1,4-dioxane, tetrahydrofuran, and *N,N*-dimethylformamide. Trace quantities of water or functional groups that contain hydrogen atoms that are readily exchangeable (called "labile" protons), such as in —OH, —SH, —NH, and —COOH, will exchange directly with the tritium gas and give products having lower specific activities. Any tritium that exchanges into labile positions is readily exchanged and removed during HPLC purification of the product.

Tritium labeling in a specific position of a complex molecule requires preparation of 5–100 mg of a suitable unlabeled precursor. Direct iodination or bromination of the target compound is preferred when possible, in order to avoid a lengthy synthesis. In the case of Atosiban, an *O*-ethyl-tyrosyl analogue of oxytocin, a specific tritium label was introduced in two steps by direct iodination and dehalogenation with tritium gas and a catalyst[18] (Figure 11.6). The specific activity of the final compound was 16 Ci/mmol (592 GBq/mmol), and specificity of the label position was confirmed by tritium-NMR.

Alternatively, tritiated water can be used and in the case of the alpha-2-adrenergic high-affinity ligand **1**,[19] tritiation was accomplished by iodination, metalation with *tert*-butyllithium, and, finally, quenching with tritiated water to give material having a specific activity of 50 Ci/mmol (1.85 TBq/mmol, Figure 11.7). It should be noted that attempts to introduce the label(s) by direct tritiodehalogenation using a palladium catalyst and tritium gas gave material with specific activities that ranged from 4 to 10 Ci/mmol (148 to 370 GBq).

A variety of intermediates such as tritiated iodomethane or metal tritides (e.g., $NaBT_4$) are commercially available or can be prepared on-site. These allow incorporation, for example, by reduction of a starting intermediate in the reaction sequence or by

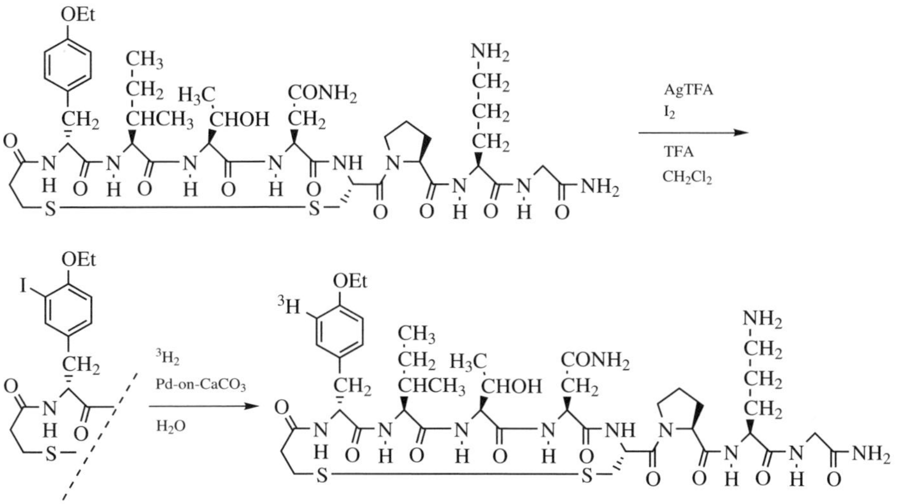

[^{3}H]Atosiban

Figure 11.6. Tritium labeling of *N*-protected amino acids and peptides containing a tyrosyl residue.

reaction with a tritiated starting material. Tritiated iodomethane is frequently used when radioligands labeled at high specific activity are required. Reaction is completed in a single step and provides material with specific activities in the range of 80–85 Ci/mmol (2.96–3.15 TBq/mmol). For preparation of high-specific-activity tritium-labeled aroyl aminoacyl pyrrole **2**, required for receptor site binding studies, CT_3I was used to rapidly introduce the label via the des-methyl precursor[20] (Figure 11.8).

Tritium is also frequently introduced by direct isotopic exchange and may incorporate into one or more sites within a molecule with unequal isotopic abundance at the different sites. It can be exchanged with hydrogen in a molecule in both the solid state at high temperature and in solution with a catalyst using tritium gas, ^3H_2O, or by transfer from a tritiated reagent such as [^{3}H]formic acid. Catalysts used are generally metallic, but acids and bases are also employed. Aqueous

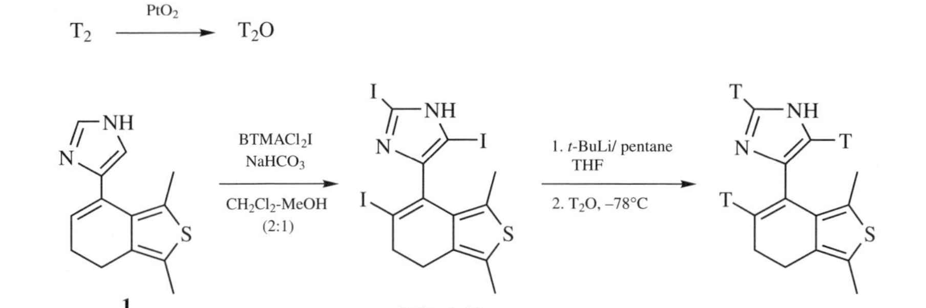

1

60% yield

Figure 11.7. Introduction of tritium by hydrolysis of an organolithium intermediate with T_2O.

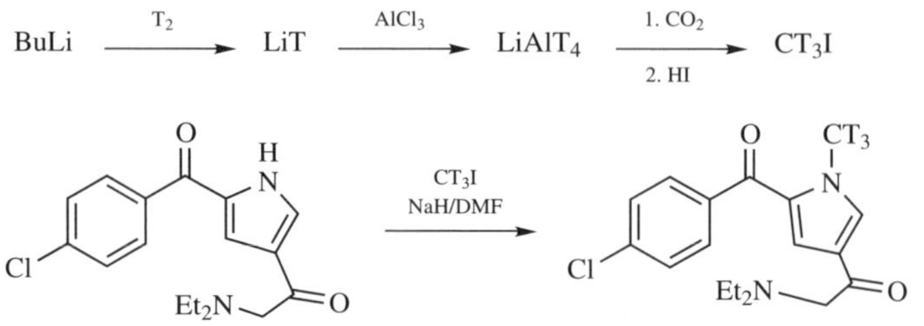

2

80 Ci/mmol (2.96 TBq/mmol)

Figure 11.8. Preparation of high-specific-activity tritium-labeled aroyl aminoacyl pyrrole **2**, using tritiated methyl iodide.

exchange reactions usually require large amounts of radioactivity (5–20 Ci), and the resulting specific activities obtained are generally lower than those obtained following a synthetic approach. Side reactions can occur, as well as radiolytic decomposition at high temperatures and in the presence of acids, bases, and high amounts of radioactivity. Because tritium can be exchanged into different positions under the reaction or exchange conditions used, tritium NMR is used to determine the distribution and location of tritium atoms in the final labeled product.

For pharmaceutical compounds, the exchange method of choice is the use of homogeneous iridium complexes with T_2 gas based on the work of Crabtree,[21] Heys,[22,23] Hesk,[24] and others. The method allows exchange, in most cases, into the hydrogens in the ortho-ring positions of acetanilide, aryl and heterocyclic amides, esters, and ketones. Substitution is generally regioselective, and tritium is incorporated at high specific activity. It can be used for labeling a target molecule without the need for synthesis or preparation of a suitable precursor. One disadvantage of the exchange conditions, however, is that the reaction works well in only a limited number of solvents with methylene chloride being the solvent of choice.

Synthesis with Sulfur-35, ^{32}P, and ^{33}P

Sulfur-35, ^{32}P, and ^{33}P are radionuclides that are used less frequently in pharmaceutical drug development. The β^- particles emitted by ^{35}S and ^{33}P are only slightly more energetic than those emitted by ^{14}C and offer the same ease of detection without the need for shielding from external radiation. The β^- emission of ^{32}P has a significantly higher energy and requires the use of shielding with low-density material (e.g., plastic) during handling. The half-lives of these isotopes are short enough that correction for loss of activity due to natural radioactive decay must be made during use. However, they offer the advantage of having significantly higher maximum specific activities, which provide highly increased detection sensitivity compared to that obtainable with ^{3}H and ^{14}C.

Sulfur-labeled compounds are used at lower specific activity for metabolism studies to study the fate of the sulfur moiety while radiotracers having high specific

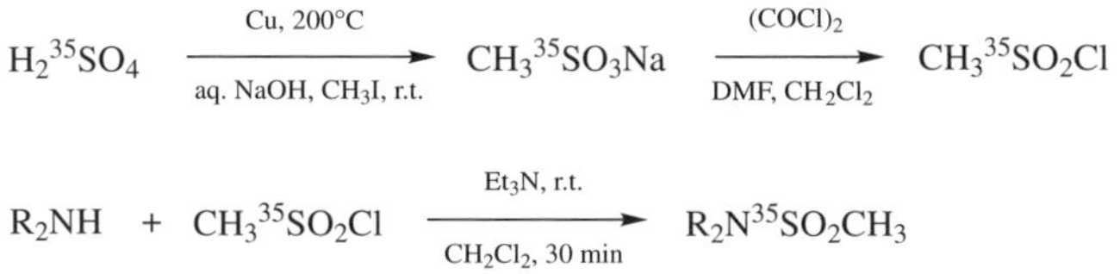

Figure 11.9. Synthesis of [^{35}S]sulfonamide radioligands at high specific activity.

activity are employed for applications such as receptor site binding studies. For a single position, the maximum incorporation of ^{35}S possible is 1491 Ci/mmol (55.17 TBq/mmol); however, most compounds are prepared at specific activities that range from several mCi/mmol to >1200 Ci/mmol.

Both ^{35}S starting materials and methods for incorporation are limited. For pharmaceutical applications, ^{35}S-labeled compounds are primarily prepared using synthetic and isotopic exchange reactions. Incorporation of the ^{35}S label generally starts from organic and inorganic intermediates such as elemental [^{35}S]sulfur, $^{35}SO_2$, $Na_2{}^{35}S_2$, $H_2{}^{35}S$, $H_2{}^{35}SO_4$, and [^{35}S]thiourea. For example, for the preparation of a high-specific-activity ^{35}S-labeled sulfonamide radioligand,[25] methane [^{35}S]sulfonyl chloride was used (Figure 11.9). The method is useful for preparing ^{35}S derivatives of ligands containing primary and secondary amines. The ^{35}S-radioligand was prepared having specific activities that ranged from 700 to >1200 Ci/mmol (25.9 to >44.4 TBq/mmol) for various preparations.

Isotopic exchange reactions with a suitable ^{35}S starting intermediate are also routinely used. Figure 11.10 shows a typical example in which sulfur exchange in a thiocarbonyl group is completed with elemental [^{35}S]sulfur.[26]

Although readily exchanged, numerous examples of ^{35}S-labled compounds undergoing self-decomposition following preparation have been reported in the literature.[27] Especially unstable are proteins labeled with either [^{35}S]methionine or [^{35}S]cysteine which have been shown to degrade on storage and release volatile components identified as ^{35}S-labeled methanethiol, hydrogen sulfide, and sulfur dioxide.[28]

The ^{32}P and ^{33}P radionuclides are routinely employed in molecular biology studies for labeling proteins, nucleic acids, and macromolecules of interest, and to a lesser extent for pharmaceutical compounds. When very high specific activity is required, ^{32}P is used. Incorporation into macromolecules is normally completed using enzymes and a ^{32}P/^{33}P-labeled nucleotide. Typical specific activities in the range of 800–6000 Ci/mmol (29.6–222 TBq/mmol) are obtained for ^{32}P-labeled compounds. For drug candidates, syntheses have been reported starting from labeled intermediates such as $^{32}POCl_3$, [^{32}P]orthophosphoric acid, and $^{32}PCl_3$.[29] Unlike ^{35}S, radiotracers labeled with ^{32}P do not give off volatile radioactive decomposition products.

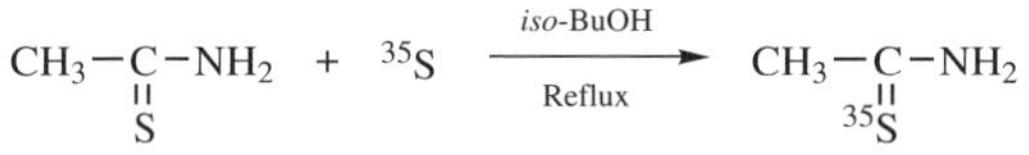

Figure 11.10. Synthesis of [^{35}S]Sulfonamide by sulfur exchange.

Synthesis with Iodine

Compounds labeled with radioiodine are used in pharmaceutical research, and the radionuclides used most frequently are ^{125}I, ^{131}I, and ^{123}I. Because of its longer half-life (59.4 days), ^{125}I is the isotope of choice for most *in vitro* studies. Unlike ^{14}C and ^{3}H, which emit β particles, the isotopes of iodine can be traced by their characteristic gamma emissions (γ). Gamma emissions offer significant detection and measurement advantages over β emitters because they can easily be counted directly without the need for special sample preparation. When external imaging is required for *in vivo* studies, ^{123}I with a half-life of 13.27 hours is employed because of its readily detected γ emission (159 keV). The high specific activity attainable by introduction of radioiodine provides increased sensitivity over tritium and to a lesser extent, ^{35}S. It is particularly useful for applications where picomole (10^{-12} mole) and smaller quantities need to be quantitated in biological matrixes or for use in receptor-site binding studies. For ^{125}I-labeled compounds, specific activities in the range of 2000 Ci/mmol (74 TBq/mmol) are obtained.

Since iodine is not an atom usually found in pharmaceutical entities, its introduction provides a mimetic of the parent compound. Radioiodine is usually substituted for another halogen atom in the molecule without alteration of the molecular structure or introduced into a position that does not affect biological activity or the physical properties of the compound. Radioiodination is the most efficient method for labeling large peptides and proteins due to the presence of tyrosyl and histidyl amino acid residues, which can readily incorporate one or two iodine atoms usually without alteration of biological activity.

Radioiodine is most often introduced directly by electrophilic substitution with $^{125}I^{+}$, which is commercially available in reagent form (e.g., as [^{125}I]iodine or [^{125}I]iodine monochloride) or generated *in situ* by oxidation of $Na^{125}I$ with an oxidizing agent such as hydrogen peroxide, peracetic acid, chloramine-T,[30] or iodogen. Other methods of radioiodination include direct halogen exchange[31,32] or incorporation by replacement of a leaving group such as a suitable organo-stannyl,[33] -lithium,[34] and -boron,[35] or -silicon[36] derivative by radioiodine. Figure 11.11 shows the procedure used by Kampfer and co-workers for the preparation of no-carrier-added, ^{123}I-labeled, (*R*,*R*)[^{123}I](iodo-3-quinuclidinylbenzilate using a Cu-catalyzed halogen exchange reaction.[31]

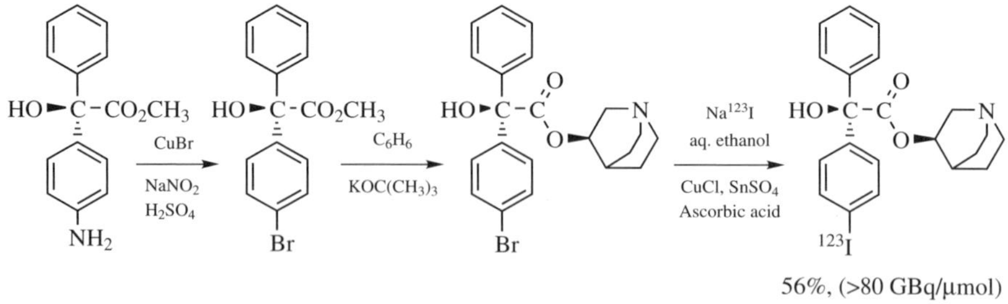

Figure 11.11. Radioiodination by halogen exchange.

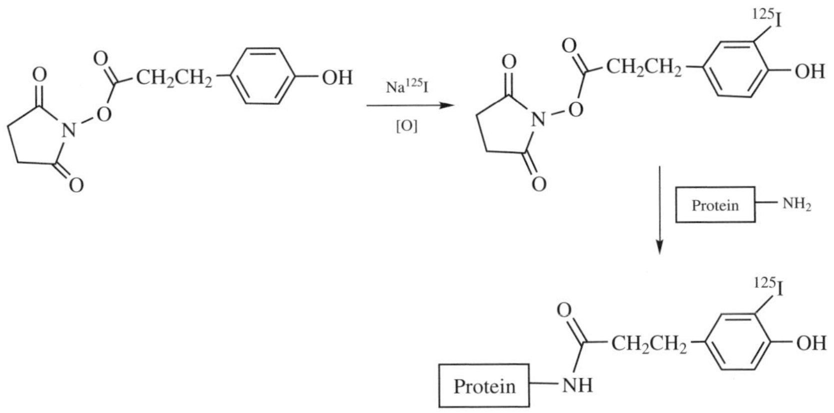

Figure 11.12. Radioiodination using the Bolton–Hunter reagent.

When iodine cannot be introduced directly, a procedure that produces a radioiodinated conjugate with amino groups in the compound of interest may be employed.[37] This involves radioiodination of a substrate that contains an activated aromatic group, such as a phenolic moiety, followed by coupling to an amine group on the molecule of interest to form a covalent bond. The most commonly used reagent for this purpose is the Bolton–Hunter reagent,[38] *N*-succinimidyl 3-(4-hydroxyphenyl)propanoate. The reagent can be radioiodinated either before or after formation of the conjugate. As depicted in Figure 11.12, an amino group on the protein to be labeled is coupled with the iodinated reagent to give the Bolton–Hunter adduct. Depending on the conditions used, either mono- or diiodinated product may be prepared.

Iodinations with ^{125}I are conducted on very small scale, generally with 1–50 mCi (37 MBq to 1.85 GBq) of radioiodine in small volumes of solvent. Reactions are rapid and generally completed in reaction times of 1–10 minutes. Unreacted radioiodine must be removed at the end of the reaction, and this is accomplished by the addition of a reducing agent that converts $^{125}I^{+}$ to $^{125}I^{-}$, which is then removed by chromatography. Iodinated products are stored refrigerated in solution to improve radiochemical stability. Loss of the radioiodine label over time, either as radioiodide or radioiodine, has been widely reported and needs to be taken into account when using any of the iodine radionuclides.[39,40] It occurs during *in vitro* and *in vivo* experimental use, as well as on storage.[41,42] Samples that lose the radioiodine label can be purified to remove any impurities formed; however, the specific activity of the recovered sample will be lower.

PURIFICATION, ANALYSIS, AND RADIOLYTIC SELF-DECOMPOSITION

Radiotracers are purified using traditional purification methods; these include HPLC, extraction, and recrystallization procedures. Because the mass prepared is usually less than a few milligrams for tritiated compounds and less for radionuclides such as ^{35}S and radioiodine, they are normally purified by HPLC and are handled and stored in solution. An HPLC chromatographic isotope effect is frequently observed for compounds containing three or more positions labeled with tritium—for example, fully tritiated *N*-methyl groups.[43] When observed, the multi-labeled tritium compound is more polar and elutes before an unlabeled standard on a reversed-phase HPLC column.[44] Separation is generally sufficient to allow chromatographic fractionation of the labeled molecules providing isotopic enrichment with higher molar specific activity.[45]

The purity required for a radiotracer is dependent on its intended use. Chemical and radiochemical purities in the 97–99+ % range are usually required for pharmaceutical applications. Purified samples are typically analyzed for identity, chemical purity, radiochemical purity, strength, and specific activity, as well as for other requirements or properties such as the presence of labile tritium, chiral purity, sterility, or having a particular polymorphic form. For radiotracers, radiochemical purity is defined as the fraction of radioactivity in the desired radiolabeled compound relative to the total amount of radioactivity in the sample. It may be determined by HPLC, GC, or TLC with a radioactive detector. For ^{3}H and ^{14}C, analysis is most commonly performed by passing the effluent from a high-performance liquid chromatograph through a radioactive flow detector. Because the energy of the β-particles emitted is so low, the effluent needs to be exposed to a scintillator for detection. This involves mixing the HPLC effluent stream in a detector flow cell with a solution that contains an organic molecule that fluoresces when exposed to β-particles. Flow cells containing a solid scintillator may also be used. In either case, detection of the light pulses produced using photomultiplier tubes allows measurement of the radioactivity present. Peak areas observed are directly proportional to the amount of radioactivity associated with the product and any radiochemical impurities present in the sample.

Pharmaceutical compounds frequently need to be prepared at specific activities that match the active drug dose with a desired quantity of radioactivity. To accomplish this, unlabeled drug substance called "carrier" is added to the radiolabeled compound until the desired drug dose and radioactive dose are obtained.

Once prepared, radiotracers need to be properly stored and periodically reanalyzed to ensure that they have not degraded. Degradation occurs as radiation released by radionuclides interact with nearby molecules, associated water, and any solvent if present.[28,46] Radiation energy may cause the formation of ions, free radicals, and other reactive species that may react and cause the radiolabeled compound to undergo degradation. This effect is known as *radiolytic self-decomposition* and rapidly accelerates with increasing amounts of radioactivity, radioactive particle energy, and time. Compounds having high specific activities are particularly susceptible to accelerated rates of decomposition. The degradation occurs both during synthesis and with the final compound during storage. Following synthesis, crude reaction mixtures need to be purified as rapidly as possible to avoid accelerated decomposition in the presence of excess radioactivity used in the radiolabeling procedure. To reduce the rate of decomposition, purified radiolabeled compounds are

$$^{14}C \longrightarrow {}^{14}N + \beta^- + \text{antineutrino}$$

Figure 11.13. Radioactive decay of ^{14}C, producing a ^{14}N atom.

generally stored refrigerated at diluted radioactive concentrations in a solvent such as ethanol. Under these conditions, the solvent preferentially absorbs and interacts with the released radiation and extends the useful working time of the radiotracer before purification is required. Intermediates sensitive to polymerization, rearrangement, or decomposition must be prepared at lower specific activities to avoid rapid loss of chemical and radiochemical purities.[47]

In addition to radiolytic self-decomposition, the radionuclides in the tracers also undergo natural radioactive decay that results in the formation of a new atom. For example, tritium transmutes to a ^{3}He atom and, when present in a C–T bond, a $[C{-}^{3}He]^{+}$ species is formed, which immediately dissociates, leaving a carbocation that reacts with any available nucleophile to produce a new chemical entity. In the case of ^{14}C, radioactive decay produces a ^{14}N atom (Figure 11.13).

Natural radioactive decay alters the structure of the labeled compound and both radioactive and nonradioactive impurities can result. The rate of formation of new impurities due to natural decay is approximately 0.5% per month for compounds labeled with tritium in a single position. The rate is much faster for compounds with multiple labels where radioactive decay at any of the labeled positions usually results in formation of new compounds, which are both chemically different and radioactive. For ^{14}C the effect of natural radioactive decay on purity is small due to its long half-life. However, when one of the ^{14}C-labeled atoms in [1,2-^{14}C]ethane decays, [^{14}C]methylamine is detected (Figure 11.14).[48]

$$^{14}CH_3{}^{14}CH_3 \longrightarrow {}^{14}CH_3NH_2$$

Figure 11.14. Decay of one of the ^{14}C-labled atoms in [1,2-^{14}C]ethane, producing [^{14}C]methylamine.

POSITRON EMISSION TOMOGRAPHY (PET)

Positron emission tomography (PET) is an imaging technique that uses labeled compounds containing short-lived radionuclides that emit positrons as they undergo radioactive decay (positrons are particles that have the same mass as an electron but are positively charged, with the symbol β^+ or e^+). The most commonly used radionuclides are ^{11}C, ^{18}F, and ^{13}N because they have sufficiently long half-lives and can usually be incorporated without alteration of biological activity. Following radioactive decay, the positron emissions immediately react with surrounding electrons to produce gamma rays. These are easily detected using a PET imaging detector and can be used to monitor and map in live time the exact location (within 4–8 mm) of the labeled compounds as they are absorbed and interact with receptors and enzymes in the tissues and organs of living animals and humans. Furthermore, because the amount of radioactivity can be quantitatively measured, it allows for dynamic *in vivo* measurement of specifically bound radioligands in target tissues.[49] It has been proven to be an accurate, noninvasive method of detecting and evaluating cancers[50] as well as for use in cardiac and brain imaging.[51] Additionally, it can be used to study disease states in humans for which no animal models exist. The technique, therefore, provides a direct, noninvasive way to elucidate the disposition and local interactions of new drug candidates *in vivo*.

The radionuclides most used to prepare PET radiotracers are ^{11}C and ^{18}F, which undergo nuclear decay with half-lives of 20.39 and 109.77 min, respectively. As a consequence, synthesis, analysis, and purification procedures are extremely time-limited and must be completed in 3 to 4 half-lives from radionuclide production in a cyclotron. Reaction schemes need to be as short as possible, and radiochemical yields need to be fully optimized within the timeframe available. Supplies of the radionuclides must be produced either nearby or on-site in a cyclotron and rapidly transported to the synthesis facility. Syntheses are completed using remote-controlled reaction equipment, which is shielded to protect workers from radiation exposure.

For pharmaceutical studies, the ^{11}C radionuclide can frequently be incorporated in a few synthetic steps to provide an isotopically labeled form of the target compound without loss of biological activity. The radionuclide and positions available for labeling may, however, be limited since the labeling position(s) needs to be selected so that the label is not lost by metabolic biotranformations during the study. If the target compound cannot be readily synthesized using ^{11}C, then labeling using a suitable analogue may be considered. An acceptable alternative is ^{18}F, if fluorine is present in the molecular structure or if introduction provides a suitable structural analogue—for instance, by ^{18}F substitution for hydrogen or another functional group.[52] The longer half-life of ^{18}F makes it the isotope of choice when synthesis time cannot be appropriately shortened or when a longer *in vivo* study is required. Additionally, because of the higher specific activities obtainable with ^{18}F radiotracers, better detection and spatial resolution is obtained than those observed with ^{11}C.

Incorporation of the label into the target molecule is normally completed using molecular intermediates containing the PET radionuclide. The primary starting materials used for ^{11}C labeling are [^{11}C]carbon dioxide and, to a lesser extent, [^{11}C]methane. Syntheses are completed using traditional synthetic methods and reactions, such as carbonations and the use of organometallic derivatives, to give a variety of useful intermediates (^{11}CO, $^{11}CH_3I$, $^{11}CH_3Li$, $^{11}COCl_2$, $Pr^{11}CH_2I$, etc.; Figure 11.15).

For ^{18}F-labeled compounds, high-specific activity intermediates are prepared by electrophilic and nucleophilic radiofluorinations from either $^{18}F_2$ or $Na^{18}F$, which are facilely incorporated to give fluorinated alkyl and aryl derivatives. Depending

$$^{14}N(p,\alpha)^{11}C \xrightarrow[\text{Trace } O_2]{N_2} {}^{11}CO_2 \xrightarrow{\text{RMgX}} R^{11}CO_2H \xrightarrow[\text{2. HI}]{\text{1. LAH}} R^{11}CH_2I$$

Figure 11.15. Preparation of ^{11}C-labeled alkyliodides.

on the radionuclide used, radiotracers containing a single label are prepared on a 0.5- to 6-pmol scale with specific activities as high as possible.

When used to study receptor–radioligand binding, PET imaging works best with drug candidates that bind strongly. This provides high concentrations of localized radioligand that can be easily distinguished from circulating and nonspecifically bound radioactivity that may interfere with the sensitivity of the measurement by raising the background activity levels. Dilution with carrier is not appropriate with PET isotopes, because unlabeled ligand will compete at the site of interest and decrease detection sensitivity. To determine if a compound is suitable for a receptor–radioligand binding study in humans, a tritium-labeled analogue is prepared first to study the compound's distribution in animals, usually a rodent, as a model. This is required to determine binding efficiency in the tissue of interest, the overall metabolic distribution in tissue and target organs, and the excretion rate. Compounds found to be highly metabolized may not be suitable because PET cannot distinguish between radioactivity emitted by the parent radiotracer being studied and any labeled metabolites that may form. Additionally, selecting the appropriate compound to label can be difficult, and in many cases poor pharmacokinetic properties may prevent accurate *in vivo* measurements of the number of binding sites or enzymes present.[53]

RADIOACTIVE TRACERS IN DRUG DISCOVERY

Radiolabeled compounds are routinely used throughout the various stages of the drug discovery process. In the very early stages, they are employed to investigate enzyme and protein receptor–drug interactions and can be used for the measurement of the binding affinities of new chemical entities with target receptors. They can also be employed for developing an operating model for screening compound libraries for biological activity and lead compound identification. For identified lead compounds, radiotracers are employed for structure–activity optimization and for use in early metabolic studies. When labeled with PET radionuclides, they can be used to study disease models in animals to determine if a target compound is having an intended interventional effect.[49]

Enzyme and Protein Receptor–Drug Interactions

To be effective, a drug must selectively interact with a specific receptor on a target protein or enzyme. Identifying and characterizing the specific recognition site and the distinct types of molecular structures that interact at that site are critical for discovering new and novel medicinal compounds. Identification of drug–receptor pairs is frequently explored with radiolabeled compounds with very high specific activities, referred to as "radioligands." Because the drug–receptor interaction involves only a single drug molecule at each binding site and there are only a very low number of receptors available for binding, radioligands must be used to facilitate detection. The radionuclides normally used in radioligands are ^{3}H, ^{35}S, or ^{125}I. Tritium is generally the isotope of choice because it has a long half-life and can be incorporated into the molecular structure at high specific activity (30–110 Ci/mmol) without loss of biological activity. Sulfur-35 (1000–1200 Ci/mmol) and ^{125}I (2000–2200 Ci/mmol) can be incorporated at significantly higher specific activity and are used in cases where much higher sensitivity is required and incorporation does not affect biological activity. In practice, the radioligand must have high radiochemical purity or it might not be possible to distinguish if binding by a competing radiochemical impurity is occurring.

Binding to receptors is studied either (a) directly by labeling the compound of interest and measuring the amount of radioactivity that binds to a particular site or (b) indirectly by using a radioligand that is known to specifically bind to the target receptor site and measuring the amount of radioactivity that is displaced when an unlabeled probe compound is added. Both the number of receptor sites available and the binding affinity of a target compound or series of compounds can be measured. When receptor–drug interaction is studied directly, the compound of interest is radiolabeled and incubated in increasing concentrations with either (a) scintillant-coated beads to which receptors have been bound[54] (scintillation proximity assay, SPA) or (b) biological preparations containing the suspected receptors. The resulting radioligand–receptor complex is then measured directly in the case of SPA or through an isolation process in which any excess, unbound radioligand is first separated and the remaining amount of radioactivity bound in the complex measured. The process generally involves an equilibration step with excess unlabeled ligand to remove any radioligand that may have complexed with low-affinity, nonspecific sites in the biological mixture. Physiologically relevant receptor sites are identified as sites that bind specifically and show appropriate saturation kinetics upon the addition of increasing amounts of radioligand. Once a ligand is identified to preferentially bind at a particular receptor site, a screen can be developed to evaluate other structural analogues and their relative potency.

Another technique used for exploring drug–receptor site interactions and molecular recognition mechanisms with radiotracers is through the use of "photoaffinity labeling."[55–57] In this method, radiolabeling of the target receptor is accomplished by allowing a radioligand containing a photoreactive group to form a reversible complex with the receptor of interest in the dark. The complex is then irradiated with UV light to initiate a photochemical reaction to form a reactive intermediate that bonds covalently either on or nearby the receptor site (Figure 11.16). The

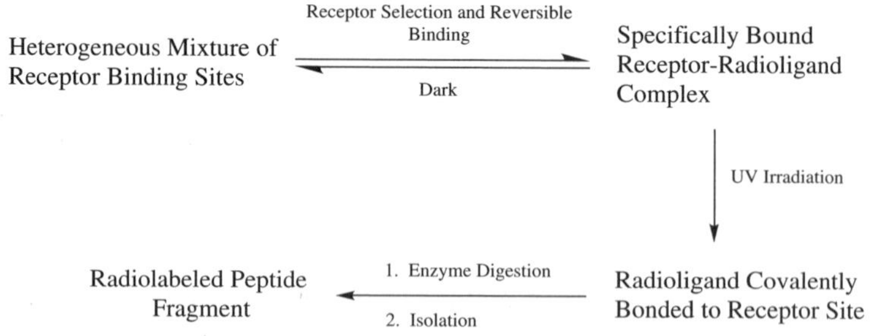

Figure 11.16. Illustration of photoaffinity labeling at target receptor sites.

method provides a covalently bonded radiolabeled derivative, which can be formed in a complex mixture of binding sites and used to isolate and study the structural domain of the receptor.

Radiolabeled photoaffinity ligands are prepared by incorporation of both a radionuclide, such as ^{3}H, ^{32}P, ^{35}S, or ^{125}I, and a photosensitive group that when irradiated produces a reactive intermediate that reacts with functional groups on the receptor to form the covalent bond. Photoreactive groups reported include azides, diazoketones, diazirines, benzophenones, and disulfides. Occasionally, parent drugs[58,59] that are capable of generating free radicals or other reactive intermediates and covalently bonding to the receptors are utilized. The most commonly employed group is the azido group which, when activated with UV light, produces a nitrene intermediate. Reaction of the nitrene by insertion into a C–H or other bond on the active receptor forms a covalent bond between the radioligand and receptor. An example was reported by Chudziak et al.[60] in which a radioiodinated azido-analogue of glibenclamide was prepared to study binding with a sulfonylurea receptor. Figure 11.17 shows the synthesis used and illustrates the photoaffinity-labeling concept with an azide group.

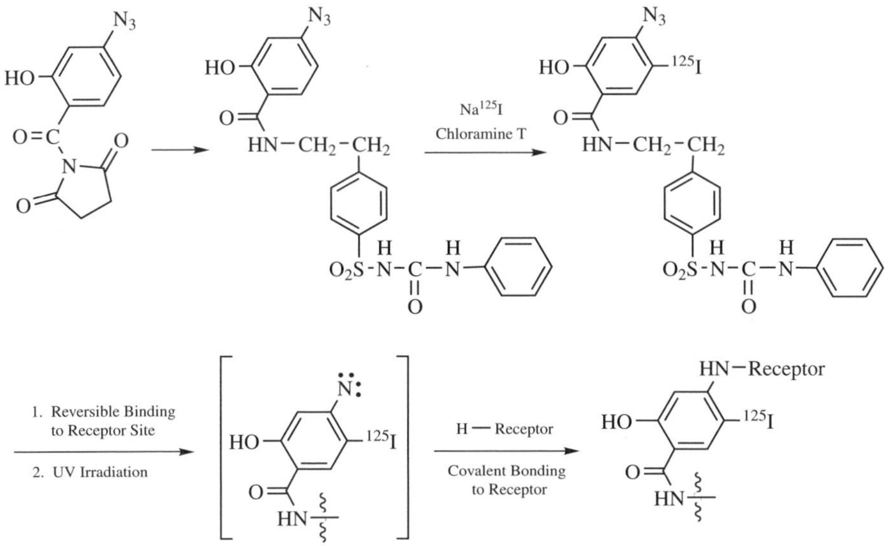

Figure 11.17. Covalent binding to a receptor using a radiolabeled photoaffinity ligand with an azido-photoreactive group by insertio into a C–H or other bond.

Drug Metabolism of New Drug Candidates

Radiotracers are increasingly being used in drug discovery research to rapidly assess the metabolism and pharmacokinetics of new chemical entities, as well as for understanding structure–pharmacologic action and structure–toxicity relationships for lead compound optimization. Early metabolic screening identifies and eliminates potential drug candidates that have undesirable metabolic, pharmacokinetic, or toxicity profiles. Such screening has increased significantly because of the large number of new drug entities being generated and their high rate of failure in the early stages of research.

Both ^{14}C- and ^{3}H-labeled compounds are routinely used to measure drug distribution and concentrations in various tissues and organs. The two most commonly used methods for measurement involve either whole-body autoradiography (WBA) or manual collection of the different tissues followed by radioactive measurement using liquid scintillation counting. Basic information on the distribution and penetration of the drug into specific tissues (e.g., brain tissue), as well as on the specific locations,[61] routes of elimination, and clearance rates, can be obtained.

For quantitative assessment of the radioactive distribution in tissues and organs, WBA is the preferred technique. It provides comprehensive visualization of the major organs and tissues and can detect radioactive uptake into areas that might not have been sampled by manual harvesting of tissues. In a WBA study, several animals (rodents) are dosed with the radiolabeled compound of interest, and the amount and location of radioactivity at various time points are quantitatively measured by scanning prepared cross sections of the sacrificed animals using an imaging scanner. Information about the time course and areas of penetration can be used to evaluate and compare a series of analogue compounds. Limitations of the technique are that pure radiolabeled sample (greater than 97% radiochemical purity) needs to be prepared, which can be time-consuming, and that it does not distinguish between radioactivity associated with the parent drug and that of any metabolites formed during the study.

Labeled compounds are also used to study toxicity issues when new drug candidates fail. Of particular interest are compounds that undergo metabolism to produce metabolites that further react with proteins and other macromolecules in the body and result in unwanted biochemical changes and toxicity. Radiotracers can be used to provide information about the types of proteins affected, adducts formed, concentration levels, and biological half-lives in high sensitivity. This information, coupled with structural identification of the metabolites, aids in the design of modified molecular entities that either minimize or eliminate the unwanted toxic effects.

Compounds labeled with ^{14}C and ^{3}H are the radionuclides most frequently employed to study covalent binding of the active metabolites produced. The radiolabeled samples must be prepared with the label incorporated into positions of the molecule that do not interfere with the metabolite formation process and are not lost during metabolic transformations. To assess whether metabolism and covalent binding of one or more of the metabolites is excessive and likely to result in toxicity, *in vitro* studies are conducted with the radiolabeled drug candidate in liver microsomes from humans and animals. The radiolabeled compound (20–50 μCi) is incubated in the microsomal mixture, and the amounts of radiolabeled metabolites that bind to the proteins in the mixture are measured. The measurement requires precipitation of the complexed proteins from solution, followed by dissolving the precipitate in a matrix suitable for radioactive counting using a liquid scintillation counter. The amount of radioactivity present provides a measure of the extent of total binding that occurred following administration of the drug. Depending on the extent of metabolism, a drug candidate may need to be labeled in more than one position to fully examine all of the molecular moieties that may form.

RADIOACTIVE TRACERS IN DRUG DEVELOPMENT

In the later stages of drug development a metabolic profile of the drug in humans is required and the use of radiolabeled compounds is essential for obtaining accurate and quantitative data. This includes data about the absorption, distribution, metabolism, and excretion (ADME) of the drug as well as about its plasma and tissue protein binding properties. Early information about the metabolism of the drug in humans is normally completed with nonradioactive material, and the rates of formation and major metabolites formed are identified by LC/MS/MS and proton NMR. Although modern analytical technologies can be used for many of these studies, radiolabeled compounds are preferred because they permit complete quantification in animals and humans in low quantities. Amounts of parent drug and metabolites formed following administration can be quantitated without the use of standards, using simple sample workup procedures and minimum development time. Carbon-14 is generally the isotope of choice for ADME studies because the ^{14}C label can be incorporated directly into the backbone of the molecule and its radioactive β^- emission is both easy to detect and measure. Tritium may also be used for metabolism experiments, and it is the isotope of choice for high-potency drugs with a very low pharmacologically active dose. It is also useful when high drug–protein binding is occurring and high sensitivity is needed to measure very small amounts of unbound drug or to follow distribution into blood components that can dramatically affect the pharmacokinetic properties of the drug.

To obtain accurate and quantitative metabolic data, a labeled molecule must have the same chemical and physical properties as the unlabeled compound. Additionally, incorporation of the isotopic atom(s) must be placed in a site that is not involved in metabolic bond breaking to avoid loss of label and/or isotopic effects. Numerous examples are reported in which metabolic pathways are "switched" from those normally observed with the unlabeled compound to alternative pathways when an isotopic atom is incorporated into a position that undergoes a metabolic transformation.[62,63] When this occurs, the metabolic breakdown of the drug is changed or inhibited and both the metabolic profile and corresponding metabolite levels are altered. In many cases the effect may not be detected and a distorted profile of the metabolism of the compound is obtained. The potential for an isotope effect and loss of label is generally much higher for tritium than for ^{14}C. The stability of the tritium label(s) to loss by exchange or degradation must be verified prior to use. This requires an excretion study be conducted, usually in a rodent, to confirm that loss of activity is not observed or, if it occurs, is within acceptable parameters of the intended study.

For ADME absorption studies, most radiolabeled drugs that involve small molecules are administered orally in water or in a pharmaceutical formulation. For less soluble compounds, administration may be made by capsule or as a suspension. However, the labeled sample needs to be prepared in the desired final form—for example, polymorphic form, salt form, and particle size. Intravenous injection in humans is not commonly used, but required for PET-labeled drugs and large biomolecules. These require filtration for sterilization and testing for sterility and pyrogenicity.

The radioactive dose that can be administered to a human subject is dependent on both (a) the length of time it takes the drug to be eliminated from the body and (b) whether the drug concentrates in a particular tissue or organ. To ensure that only a minimum of radioactive material is used, regulatory agencies require that a biodistribution study first be performed in animals to measure drug concentrations and the potential duration and exposure of major tissues and organs to the drug. These are normally performed in rats or mice; however, dogs and primates may also be used. The results are then used to calculate and ensure that the amount of radioactivity to be dosed in a clinical study is safe and in line with regulatory guidelines, which require that subjects receive the lowest possible exposure to radiation without hindering or compromising the quality of the data obtained from a study.

The amount of radioactivity administered varies widely depending on the analytical technology being used and the length of time required to complete the study. The most common method for measuring radioactivity is by liquid scintillation counting (LSC), which measures the number of atoms that undergo nuclear decay per minute. Atoms that are present but do not decay are not measured. More recently, accelerator mass spectrometry (AMS), a mass spectrometric method for quantifying the amount of radionuclide present in biological samples, has also started to be used. It is a more sensitive technique, since it directly measures all of the radionuclide present in the sample at once. Its detection limit is about 10^6 times lower than LSC.[64] When LSC is used, the radioactive dose for a ^{14}C-labeled drug given to humans is typically between 10 and 300 μCi per subject, with 100 μCi being the average for a single dose. Because the tritium β^- emission has lower energy and is more difficult to detect than that emitted by ^{14}C, ^{3}H-labeled drugs are dosed with higher amounts of radioactivity; this ranges from 50 to 1000 μCi, with the average dose being 250–300 μCi. For AMS the radioactive dose is in the range of 10–100 nCi per subject, which from a radiation exposure perspective is considered negligible.

The much higher sensitivity of AMS also offers the advantage of lowering the total drug dose used in early clinical studies to obtain information about the absorption and metabolism of investigational new drugs. The use of significantly reduced drug doses allows humans to safely participate in very early studies and reduces the number of animal studies needed, as well as eliminating the need to extrapolate from animal test data to humans.[65] Drug doses can typically be lowered down to the 1- to 100-μg dose range, about 1/1000 of the amount used in conventional studies, without loss of sensitivity. The technique requires the test compound to be radiolabeled, and most studies are completed using ^{14}C. The radiolabeled drug product needs to have high radiochemical purity for human use and samples prepared in the 1- to 10-μg range present analytical challenges for both final analyses and for establishing drug stability of the formulated product. While a number of studies have demonstrated the concept of using low subtherapeutic doses ("microdosing") to study drug disposition and pharmacokinetics with AMS, concern remains that microdosing may not be applicable to all drug classes and that misleading data may be obtained for candidate molecules that display nonlinear pharmacokinetics.[66,67]

For human mass balance studies, required to demonstrate that the entire dose administered is completely excreted, the use of radiolabeled drug substance is essential. Excretion studies require an accurate knowledge of the amount of radioactivity present in the dose. Collection of all of the excreta (urine, feces, etc.) and demonstration of a quantitative recovery of the radioactivity is necessary to meet regulatory requirements. Measurements of the collected samples are done by LSC of the urine samples and by combustion and LSC of the collected fecal samples. Additional studies, such as breath analysis for expired $^{14}CO_2$ or activity lost through sweat, may be required to attain mass balance. The labeled compound used must have a high radiochemical purity (>98+%) since labeled impurities can leave traces of radioactive residues in tissue. Additionally, if chemical decomposition or loss of the label occurs during the study, radioactive metabolic fragments may be incorporated into the tissues, resulting in invalid conclusions about tissue concentrations.[68]

SUMMARY

Radiotracers are widely employed in all areas of pharmaceutical research and development. The most widely used radionuclides are ^{14}C, ^{3}H, and ^{125}I for metabolism and receptor binding studies, while ^{11}C and ^{18}F are the most employed for noninvasive PET imaging studies. Numerous other radionuclides are available, and multiple strategies and techniques can be used to obtain specific and quantitative information that is not easily obtained using alternative analytical technologies. Technological advances in areas such as PET imaging, AMS, and imaging detectors have led to new applications, which require radiotracers to be prepared with high specificity and with high specific activity. To meet the challenge, new synthetic labeling methods, such as iridium-catalyzed tritium exchange reactions, have evolved that allow for faster and more efficient preparation of labeled compounds. Over the last several years, radiotracer use has dramatically increased in the early stages of drug discovery as researchers try to screen the large numbers of new drug candidates being generated and rapidly identify compounds with undesirable metabolic, pharmacokinetic, and toxicity profiles. The use of PET isotopes has also become invaluable in studying and understanding a variety of disease states in humans, including some for which no animal models exist. Finally, radiotracers continue to be essential for generating reliable ADME data to support regulatory filing requirements for bringing new drugs to the market.

REFERENCES

1. Stocklin, G., Qaim, S. M., and Roesch, F. *Radiochimica Acta* **1995**, *70/71*, 249–272.
2. Dalvie, D. *Curr. Pharm. Des.* **2000**, *6*, 1009–1028.
3. Tubiana, M. *J. Soc. Francaise Radiother. Oncol.* **1999**, *3*, 11–12.
4. Nexoff, D. *Handb. Radiopharm.* **2003**, 283–305.
5. Jonker, A., de Boer, P. A., van den Hoff, M. J., Lamers, W. H., and Moorman, A. F. *J. Histochem. Cytochem.* **1997**, *45*, 413–423.
6. Baillie, T. A., and Kassahun, K. *Adv. Exp. Med. Biol.* **2001**, *500*, 45–51.
7. Barthe, L., Woodley, J., Lavit, M., Przybylski, C., Philibert, C., and Houin, G. *Arzneimittel-Forschung* **2004**, *54*, 286–292.
8. Fang, Y.-R., Gao, Y., Ryberg, P., Eriksson, J., Kolodziejska-Huben, M., Dybala-Defratyka, A., Madhavan, S., Danielsson, R., Paneth, P., Matsson, O., and Westaway, K. C. *Chem. Eur. J.* **2003**, *9*, 2696–2709.
9. Lewis, B. E., and Schramm, V. L. *J. Am. Chem. Soc.* **2003**, *125*, 7872–7877.
10. Volkert, W. A., and Hoffman, T. J. *Chem. Rev.* **1999**, *99*, 2269–2292.
11. The National Nuclear Data Center, Brookhaven National Laboratory; www.nndc.bnl.gov, May 16, 2005.
12. Weaner, L. E., Burghard, G. L., Blackburn, D. W., and Mendelson, W. L. *J. Labelled Compd. Radiopharm.* **1977**, *13*, 141–146.
13. Yokoshima, T., and Hasegawa, M. *Radioisotopes (Tokyo)* **1961**, *10*, 435–442.
14. Barakat, M. F., Farag, A. N., Ragab, M. T., El-Fouly, M. M., and El-Baz, F. K. *Isotopenpraxis* **1990**, *26*, 311–317.
15. Laufer, L., Gutcho, S., Castro, T., and Grennen, R. *Biotechnol. Bioeng.* **1964**, *6*, 127–146.
16. Weaner, L. E., and Hoerr, D. C. *J. Labelled Compd. Radiopharm.* **1986**, *23*, 355–369.
17. Evans, E. A. *Tritium and Its Compounds*, Butterworths: London, 1974.
18. Weaner, L. E., Yim, N. C. F., and Hoerr, D. C. *Synthesis and Applications of Isotopically Labelled Compounds 1994, Proceedings of the International Symposium, 5th, Strasbourg, June 20–24, 1994*, pp. 137–140.
19. Ross, T. M., Jetter, M. C., McDonnell, M. E., Boyd, R. E., Connelly, C. D., Martinez, R. P., Lewis, M. A., Codd, E. E., Raffa, R. B., and Reitz, A. B. *J. Med. Chem.* **2000**, *43*, 1423–1426.
20. Carson, J. R., and Pitis, P. M. *PCT Int. Appl.*, **2002**.
21. Crabtree, R. H., Lavin, M., and Bonneviot, L. *J. Am. Chem. Soc.* **1986**, *108*, 4032–4037.
22. Shu, A. Y. L., Saunders, D., Levinson, S. H., Landvatter, S. W., Mahoney, A., Senderoff, S. G., Mack, J. F., and Heys, J. R. *J. Labelled Compd. Radiopharm.* **1999**, *42*, 797–807.
23. Chen, W., Garnes, K. T., Levinson, S. H., Saunders, D., Senderoff, S. G., Shu, A. Y. L., Villani, A. J., and Heys, J. R. *J. Labelled Compd. Radiopharm.* **1997**, *39*, 291–298.
24. Hesk, D., Das, P. R., and Evans, B. *J. Labelled Compd. Radiopharm.* **1995**, *36*, 497–502.
25. Dean, D. C., Nargund, R. P., Pong, S.-S., Chaung, L.-Y. P., Griffin, P., Melillo, D. G., Ellsworth, R. L., Van Der Ploeg, L. H. T., Patchett, A. A., and Smith, R. G. *J. Med. Chem.* **1996**, *39*, 1767–1770.
26. Chiotan, C., Zamfir, I., Szabo, M., and Turcanu, C. N. *Nov. Metody Poluch. Radioaktiv. Prep., Sb. Dokl. Simp.* **1970**, 386–401.
27. Meisenhelder, J., and Hunter, T. *Nature.* **1988**, *335*, 120.

28. Kato, T., Saito, K., and Kurihara, N. *Appl. Radiat. Isotopes* **1994**, *45*, 693–698.
29. Hill, D. T., Sutton, B. M., Levinson, S. H., Meier, J., Villani, A. J., and Spainhour, C. B. *J. Labelled Compd. Radiopharm.* **1983**, *20*, 363–375.
30. Hadi, U. A. M., Malcolme-Lawes, D. J., and Oldham, G. *Int. J. Appl. Radiat. Isotopes* **1978**, *29*, 621–623.
31. Kaempfer, I., Heinicke, J., Sorger, D., Schulze, K., Schliebs, R., and Knapp, W. H. *J. Labelled Compd. Radiopharm.* **1996**, *38*, 1047–1052.
32. Galinier, E., Ombetta, J. E., Frangin, Y., Mertens, J., Besnard, J. C., and Guilloteau, D. *J. Labelled Compd. Radiopharm.* **1994**, *34*, 487–497.
33. He, X. S., Kiesewetter, D. O., Lee, K. S., Mattson, M. V., Weinberger, D. R., and de Costa, B. R. *J. Labelled Compd. Radiopharm.* **1993**, *33*, 573–581.
34. Rocca, P., Cochennec, C., Marsais, F., Thomas-dit-Dumont, L., Mallet, M., Godard, A., and Queguiner, G. *J. Org. Chem.* **1993**, *58*, 7832–7838.
35. Kabalka, G. W., Varma, R. S., Jinaraj, V. K., Huang, L., and Painter, S. K. *J. Labelled Compd. Radiopharm.* **1985**, *22*, 333–338.
36. Wilbur, D. S., Anderson, K. W., Stone, W. E., and O'Brien, H. A., Jr. *J. Labelled Compd. Radiopharm.* **1982**, *19*, 1171–1188.
37. Langone, J. J. *Methods Enzymol.* **1980**, *70(A)*, 221–247.
38. Bolton, A. E., Hunter, W. M. *Biochem. J.* **1973**, *133*, 529–538.
39. Bogdanove, E. M., and Strash, A. M. *Nature* **1975**, *257*, 426–427.
40. Ramsey, N. W., Dunn, M., Bhattacharyya, A. K., and Adam, T. *Br. J. Radiol.* **1980**, *53*, 357–358.
41. Vaidyanathan, G., and Zalutsky, M. R. *Bioconjugate Chem.* **1990**, *1*, 269–273.
42. Vaidyanathan, G., and Zalutsky, M. R. *Bioconjugate Chem.* **1990**, *1*, 387–393.
43. Filer, C. N., Fazio, R., and Ahern, D. G. *J. Org. Chem.* **1981**, *46*, 3344–3346.
44. Jones, A. N., Simpson, R. E., and Jenkins, H. J. *J. Labelled Compd. Radiopharm.* **1992**, *31*, 297–303.
45. Kudelin, B. K., Gavrilina, L. V., and Kaminskii, Y. L. *J. Labelled Compd. Radiopharm.* **1992**, *31*, 723–728.
46. Albarran, G., and Collins, C. H. *J. Chromatogr.* **1987**, *395*, 623–629.
47. Egli, P. *J. Labelled Compd. Radiopharm.* **1981**, *18*, 1677–1682.
48. Wolfgang, R. L., Anderson, R. C., and Dodson, R. W. *J. Chem. Phys.* **1956**, *24*, 16–23.
49. Aboagye, E. O., Price, P. M., and Jones, T. *Drug Discovery Today* **2001**, *6*, 293–302.
50. Hammond, L. A., Denis, L., Salman, U., Jerabek, P., Thomas, C. R., Jr., and Kuhn, J. G. *Invest. New Drugs* **2003**, *21*, 309–340.
51. Fowler, J. S., and Wolf, A. P. In *Positron Emission Tomography and Autoradiography: Principles and Applications for the Brain and Heart*, Phelps, M., Mazziotta, J., and Schelbert, H., Eds., Raven Press, New York, 1986, Chapter 9.
52. Brady, F., Luthra, S. K., and Gibson, A. M. *PCT Int. Appl.*, **2004**.
53. Snyder, S. E., Bryan, T., Gunupudi, N., Butch, E. R., and Kilbourn, M. R. *Physiol. Imaging Brain PET* **2001**, 105–108.
54. Cox, B., Denyer, J. C., Binnie, A., Donnelly, M. C., Evans, B., Green, D. V., Lewis, J. A., Mander, T. H., Merritt, A. T., Valler, M. J., and Watson, S. P. *Prog. Med. Chem.* **2000**, *37*, 83–133.

55. Dorman, G. *Top. Curr. Chem.* **2001**, *211*, 169–225.
56. Dorman, G., and Prestwich, G. D. *Trends Biotechnol.* **2000**, *18*, 64–77.
57. Hatanaka, Y., and Sadakane, Y. *Curr. Top. Med. Chem.* **2002**, *2*, 271–288.
58. Duncalfe, L. L., Carpenter, M. R., Smillie, L. B., Martin, I. L., and Dunn, S. M. J. *J. Biol. Chem.* **1996**, *271*, 9209–9214.
59. Bernstein, P. S., Choi, S.-Y., Ho, Y.-C., and Rando, R. R. *Proc. Nat. Acad. Sci. USA* **1995**, *92*, 654–658.
60. Chudziak, F., Schwanstecher, M., Laatsch, H., and Panten, U. *J. Labelled Compd. Radiopharm.* **1994**, *34*, 675–680.
61. Kim, C. S., and Roe, C. R. *Fundam. Appl. Toxicol.* **1992**, *19*, 222–227.
62. Garland, W. A., Miwa, B. J., Eliahou, C., and Dairman, W. *Synth. Appl. Isot. Labeled Compd. Proc. Int. Symp.*, 2nd. **1986**, 271–276.
63. Kaspersen, F. M., Vader, J. F., Funke, C. W., and Sperling, E. M. G. *Recueil Travaux Chim. Pays-Bas.* **1993**, *112*, 191–199.
64. Lappin, G., and Garner, R. C. *Anal. Bioanal. Chem.* **2004**, *378*, 356–364.
65. Zannikos, P., Hutchison, J. S., Williams, P. E., Weaner, L. E., Stubbs, J., Van Marle, S. P., and Garner, R. C. Presented to the American Society for Clinical Pharmacology and Therapeutics, March 24–27, **2002**, Atlanta, GA.
66. Combes, R. D., Berridge, T., Connelly, J., Eve, M. D., Garner, R. C., Toon, S., and Wilcox, P. *Eur. J. Pharm. Sci.* **2003**, *19*, 1–11.
67. Sandhu, P., Vogel, J. S., Rose, M. J., Ubick, E. A., Brunner, J. E., Wallace, M. A., Adelsberger, J. K., Baker, M. P., Henderson, P. T., Pearson, P. G., and Baillie, T. A. *Drug Metab. Dispos.* **2004**, *32*, 1254–1259.
68. Berenbom, M. *Cancer Res.* **1962**, *22*, 1343–1348.

12

SELECTION OF THE DRUG FORM IN EXPLORATORY DEVELOPMENT

George J. Quallich

Global Research and Development, Pharmaceutical Research and Development, Pfizer, Inc., Groton, CT 06340

INTRODUCTION

A critical element of drug discovery and the early stages of drug development is the identification of the appropriate drug form. Drug form refers to the nature of the solid drug entity, such as a free acid, free base, salt, nonionizable compound, and, additionally, to its solid state: amorphous, crystalline, polymorphic phase. The drug form impacts pharmacokinetics (PK), bioavailability, chemical stability of the drug, chemical stability in the dosing medium, yield/impurity profile obtained from the synthetic route, and other factors that affect the speed of drug candidate progression. Typically, numerous scientific disciplines such as chemistry, formulation, analytical, material science, and pharmacokinetics/dynamics/metabolism are involved in developing an understanding of the drug form characteristics. The resulting level of complexity in the drug form decision process, due to the drug form variables and disciplines involved, requires a clear approach in guiding these decisions at a time wherein candidate attrition is at its highest in the drug discovery and development process. Effective utilization of resources, therefore, amplifies the challenges in the selection process. There is no single model that applies to the entire pharmaceutical industry. This is due to the variability in the size of pharmaceutical companies, as well as the variability in size/expertise of sites within a particular company, and the nuances of each drug candidate. Therefore, this chapter is targeted toward (a) education of new

Fundamentals of Early Clinical Drug Development: From Synthesis Design to Formulation, edited by Ahmed F. Abdel-Magid and Stéphane Caron

medicinal and process chemists in describing the key fundamentals that are essential for any organization to be successful and (b) examples of potential pitfalls which are inherent in the drug form selection process.

IDENTIFICATION OF THE LEAD CANDIDATE

If success is to be garnered at the earliest stages of medicinal chemistry in identification of leads, the decisions require knowledge on the physical properties of drug candidates. The corollary is that the properties of the leads are inevitably found in the drug candidate progressing into development. There is a tendency to (a) select compounds with the highest binding and potency from initial high throughput screening and (b) ignore other attributes of the compounds under investigation. Guidance is provided in "The Rule of Five," which analyzed drug properties found in compounds that progressed into Phase II clinical studies.[1,2] If a compound complies with these rules ($\leq$5 hydrogen bond donors, $\leq$10 hydrogen bond acceptors, molecular weight <500, log $P < 5$), then typically acceptable oral bioavailability is achieved. These rules are predictive of drug candidates that will have poor solubility and permeability. However, once the leads are defined in the discovery process, there remains a temptation to focus solely on potency and selectivity in understanding structure activity relationships (SAR). Thus, it is imperative to stress the utilization of solubility and permeability data as pivotal in guiding the medicinal chemist in the identification of lead candidates. Significant advances in solubility and permeability methods have occurred which enable obtaining these data in a routine or high throughput manner.[3,4]

Solubility has the potential to increase exposure significantly more than permeability, by five orders of magnitude.[5] As a result, a general risk assessment can be made based on compound structure. Compounds **1**–**3** exemplify this point.

Compound **1** is nonionizable and therefore has no potential to form salts, which could increase solubility. The options for nonionizable compounds are to develop them as either crystalline or amorphous forms.[6] Herein, the associated risk with the use of crystalline compounds is "solubility-limited absorption," while chemical/physical instability is a risk associated with the use of amorphous compounds. Solubility-limited absorption is a nonlinear exposure as increasing quantities of active pharmaceutical ingredient (API) and/or drug product (DP) are dosed as depicted in Figure 12.1. This may not result in achieving sufficient drug exposure with crystalline material in initial toxicology studies, which require identification of target organ toxicity. Thus, a program could be delayed until a method of raising exposure is attained. By contrast, development of an amorphous form is clearly precedented, but maintaining the physical form is essential if consistent exposure is to be achieved. A contemporary example of not maintaining the physical drug form in DP is Norvir.[7,8] A new lower-energy and less soluble crystalline form emerged in manufacturing capsules that resulted in diminished bioavailability. Novir capsules could no longer be supplied to the market until reformulation of the new low-energy form gave acceptable human exposure. Amorphous compounds have the same

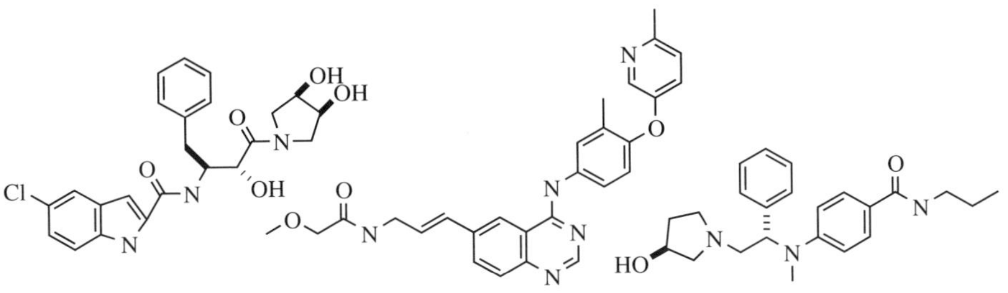

1 2 3

Figure 12.1.

potential issue if the drug form crystallizes. Therefore, nonionizable drug candidates represent (1) the highest risk level because no salt options exist and (2) potential to decrease exposure through either (a) transformation of a crystalline form to a lower-energy/less soluble one or (b) crystallization of amorphous material.

Compound 2 has weakly basic sites wherein few salt options exist, because acids of low pK_a would be necessary to enable ionization. Unfortunately, salts of weak acids or bases often dissociate in water. The dissociation may eliminate the perceived advantages in increasing solubility and exposure.

Compound 3 possesses a tertiary amine, enabling it to have numerous salt options with various counterions. Linear dose response, as depicted in Figure 12.2, is anticipated. The risk and development resources can be anticipated: Compound **1** has the

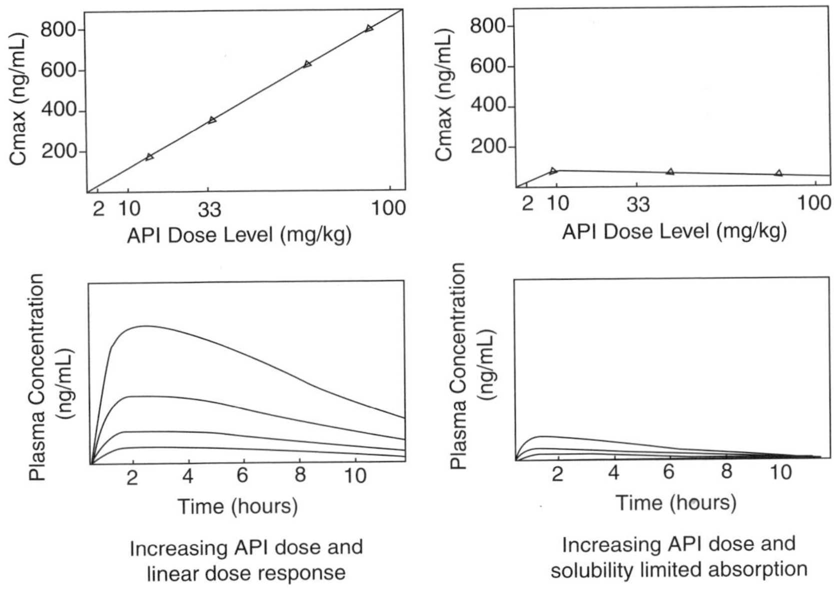

Figure 12.2. Linear and solubility-limited API dose response.

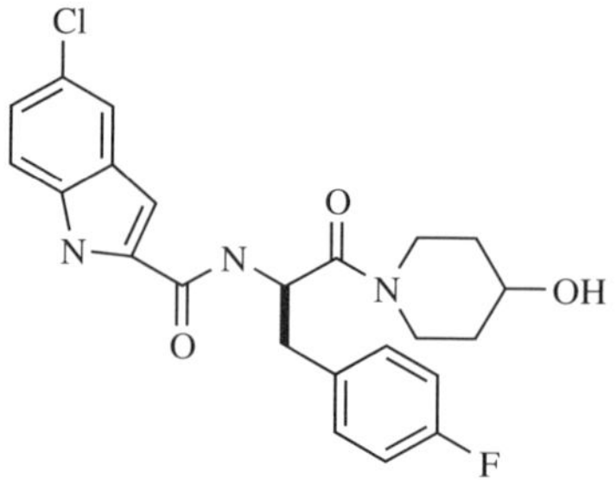

4

Figure 12.3.

highest risk due to few drug form options, while compound **3** with a multitude of options represents the lowest risk. This risk assessment does not preclude the development of compounds **1** or **2**, but does provide an understanding of the additional resources and time that may be required to define a drug form capable of being employed in exploratory development. The chemical structures of drug candidates provide useful insights in gaining drug form perspective and development complexity.

Understanding the drug form employed in aqueous solubility and PK/bioavailability studies is critical to prevent surprises during the exploratory development. Kinetic solubility is often conducted early as an indicator of a possible issue. The advantages are low compound requirements ≤ 1 mg, experimental time scale in minutes, and some correlation with *in vivo* animal SAR. However, thermodynamic solubility of a characterized drug candidate form is essential to prevent discontinuous development. Two examples to exemplify this potential impact are amorphous **4** (Figure 12.3) and the anhydrous benzoate salt of **3**. Acceptable bioavailability, 65% C_{max} 1.7 μg/ml, is attained with amorphous **4**. By contrast, the crystalline compound had nearly an order of magnitude less exposure, 8% bioavailability, and 0.3 μg/ml C_{max}. Compound **4** underscores the point made earlier—namely, that nonionizable compounds represent the highest degree of risk. If crystalline **4** was progressed, then much higher cost of goods would be attained because additional API would be required to achieve the desired bioavailability, if not precluded by solubility-limited absorption. The alternative, to develop the amorphous form of **4**, would take additional resources and time to ensure that no API crystallization occurs in DP. In the second example, anhydrous benzoate salt of **3** is highly crystalline and does not convert to another drug form in a kinetic moisture balance experiment.[9] The initial aqueous solubility is >150 mg/ml. However, under thermodynamic conditions the benzoate hydrate crystallizes and aqueous solubility of this entity drops precipitously to 7 mg/ml. Both **4** and the benzoate of **3** demonstrate that knowledge of all drug forms and aqueous solubility is key to selection in exploratory development. An additional utility of the thermodynamic aqueous solubility, along with the absorption rate constant, is to calculate the maximum absorbable dose of the drug and define particle size requirements.[10] Having outlined the rationale for understanding drug form in lead selection, the process for training new pharmaceutical chemists is outlined.

One of the surprises new graduates learn in becoming medicinal or process chemists in the pharmaceutical industry is the importance of crystallizing compounds. A half century ago, Woodward and other chemists used crystallization and melting points of strychnine's chemical degradation products to elucidate its structure.[11] Subsequently, melting point of crystalline material was one of the methods to confirm that Woodward's total synthesis of strychnine was achieved.[12] In more recent times, high-resolution exact mass spectroscopy and high-field NMR, with sophisticated techniques, have revolutionized the ability to analyze and characterize compounds with ever-decreasing quantities. The result of the advancing analytical techniques is a diminished academic focus on crystallization in synthetic organic chemistry over the last few decades. New chemists joining industry lack an understanding of (a) the crystallization consequences in drug development and (b) the fundamental aspects of crystallization techniques. A practical reality is a proactive education of new chemists into the importance of crystallization, and the attendant techniques pay significant dividends in lead development and at the discovery/development interface. The curriculum for this education entails a lecture on (1) understanding the impact of drug form on PK/bioavailibility, thereby eliminating development surprises; (2) concepts and theory in nucleation, supersaturation, crystallization, crystal packing, Ostwald ripening, powder X-ray diffraction (PXRD), and a brief exposure to recognizing when polymorphs may exist due to differences in melting points or PXRD; and (3) a rapid systematic laboratory approach to crystallization including solubility estimation and small-scale crystallization techniques (solvent evaporation, temperature-controlled heating/cooling, antisolvent addition, and vapor diffusion). The lecture is followed by the chemists practicing crystallization techniques with a variety of compounds and salt formation in a "hands-on laboratory experience." The lab experience reinforces lecture concepts of supersaturation, Ostwald ripening, crystallinity, PXRD, and polymorphism. This course, taught in a few days, arms process and medicinal chemists with a systematic approach to crystallization and a clear understanding of its importance in the pharmaceutical industry.

OVERVIEW OF THE EXPLORATORY DEVELOPMENT PROCESS

The drug candidate's physical properties play a key role in the investment strategy. The nuances of each drug candidate are derived from various factors including the projected human dose (a high dose translates to the properties in tablet formulation being dominated by the API such as flow during DP manufacture), method of administration (immediate release tablet or parenteral), a shorter than average development timeline (antibiotics require only a six-month toxicology study, versus two years for chronic therapy, and have sharper clinical endpoints, relative to depression, for example), and the stability of the API/DP (API hygroscopicity

TABLE 12.1. Overview on the Analysis and Decision Process of Solid-State Properties

Stage 1	**Stage 2**
Birefringence	Hydrate screening
PXRD	Low-energy form determination
Purity assessment	API accelerated stability
Thermal analysis	
Kinetic hygroscopicity	
Bioavailability[a]	
Stage 3	**Stage 4**
Polymorph screening	API and DP ICH[b] stability
DP stability projections	Manufacturing acceptance

[a]Based on solubility and biopharmaceutics classification.
[b]ICH: International Conference on Harmonization (see reference 25).

to the point of deliquescence). Thus choice of the exploratory development form, and the corresponding investment timing, is a case-by-case decision for every candidate. The rationale for changing the form of some drug candidates early is outlined in the following case studies that provide insight into the issues faced in exploratory development, along with the solution to these challenges.

Prior to describing case studies, an overview on the analysis and decision process of solid-state properties is depicted in Table 12.1. The intent is to highlight key elements in discriminating superior drug candidate forms, and how a thorough evaluation of the information is necessary for an accurate assessment. The first batteries of analytical measurements employed are Stage 1; birefringence → PXRD → purity → thermal analysis → kinetic hygroscopicity. Polarized Light Microscopy (PLM) is a fundamental technique allowing determination of compound habit and birefringence; splitting of a light wave into two unequally reflected waves by an optically anisotropic medium. Birefringence is often employed as the first tool to appraise crystallinity because of the speed in which the determination can be performed, and the low compound requirements, <0.001 mg.[13] Here both extinction and habit (shape) are indicators of a crystalline compound. Some crystalline compounds, however, do not demonstrate birefringence such as isotropic crystals, which have only one refractive index. Crystallinity is subsequently validated through PXRD, the gold standard of fingerprinting the uniqueness of a compound's form.[14,15] Note, however, that not all birefringent compounds are crystalline such as a meso phase (liquid crystal), they have a distinct habit and extinction, but are amorphous by PXRD. Combining birefringence and PXRD information enables an assessment of a compound's crystallinity, as exemplified in Table 12.2 and Figure 12.4.

Purity determination is critical to complete before conducting kinetic hygroscopicity, thermal measurements, and accelerated stability studies because it can

TABLE 12.2. Assessment of a Compound's Crystallinity

Classification	PXRD	PLM
Fully crystalline	Distinct peaks at 2θ to 40°. Exhibits flat baseline.	Birefringent demonstrates habit.
Crystalline with disorder	No peaks beyond 2θ 30°. Peak broadening.	Birefringent may not have distinct habit.
Crystalline with severe disorder	Reduced peak heights at 2θ >20°. "Amorphous hump" is apparent.	Birefringent may not have distinct habit.
Partially Crystalline (Mixture of crystalline and amorphous)	Reduced peak heights. "Amorphous hump" is apparent.	Mixture of isotropic and anisotropic solid.
Amorphous (isotropic)	No pattern. "Amorphous hump" is apparent.	No birefringence. No habit. Often appears as broken glass or solidified spheres.
Meso-phase (liquid crystal)	No pattern. "Amorphous hump" is apparent. May have one or two broad peaks.	Birefringent typically uniaxial. May have distinct habit at low 2θ.

adversely affect the outcome of these evaluations. Chemical purity, often performed by HPLC with UV detection, provides organic purity but is insufficient alone. Inorganics (ash) and counterion stoichiometry, not detected by HPLC, can dramatically impact the outcome of hygroscopicity and thermal analysis in a negative manner. For example, consider the dibasic compound **5** (Figure 12.5). If the compound had an inorganic impurity such as sodium chloride, it would appear to be hygroscopic (Figure 12.6). Similarly, if the mesylate salt of dibasic **5** had a stoichiometry of 1.5 moles of counterion, one would be evaluating a mixture of the mono- and di-salt properties of this compound and not a pure entity. Either the mono- or dimesylate may have excellent solid-state properties, but these could be obscured by the other component. An additional purity assessment such as combustion analysis, coupled with HPLC organic purity, defines whether ash or counterion stoichiometry was consistent with a single entity and also detects the existence of a hydrate or solvate. Finally, physical purity impacts stability. Amorphous material is less ordered and usually deemed to be less stable as a result. Low levels of amorphous material contained in the crystalline compound (partially crystalline) can be the cause of degradation by oxidation, hydrolysis, or other paths. Once the compound is determined to be crystalline by microscopy and PXRD, chemical and physical purity assessment is warranted.

Thermal analysis, after a purity perspective, provides further information on a compound's physical makeup. This includes hydration, solvation, and whether the sample is possibly a mixture of polymorphs.[17] Differential Scanning Calorimetry (DSC) measures heat flow of a sample relative to a reference pan and yields solid-to-liquid phase changes such as melt onset, melt peak, and heat of fusion with milligram quantities. Additionally, the phase changes are seen as endothermic or exothermic events in the thermagram. The highest melting form of a compound is often, but not necessarily, the

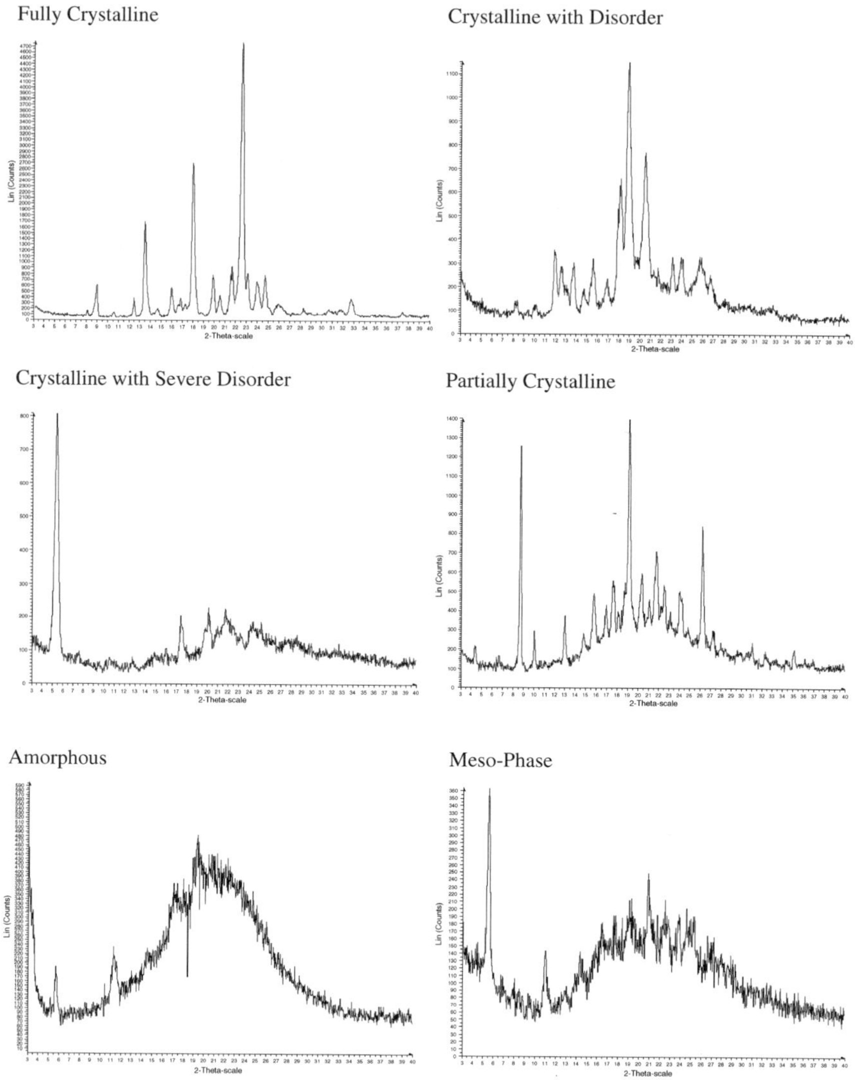

Figure 12.4. Depiction of PXRD classifications listed in Table 12.2.

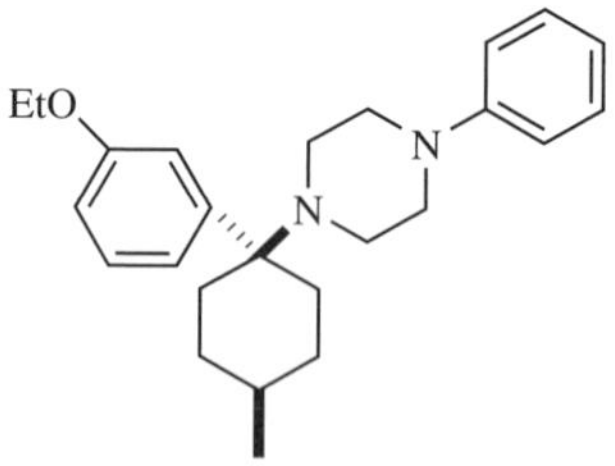

5

Figure 12.5. Dibasic compound.

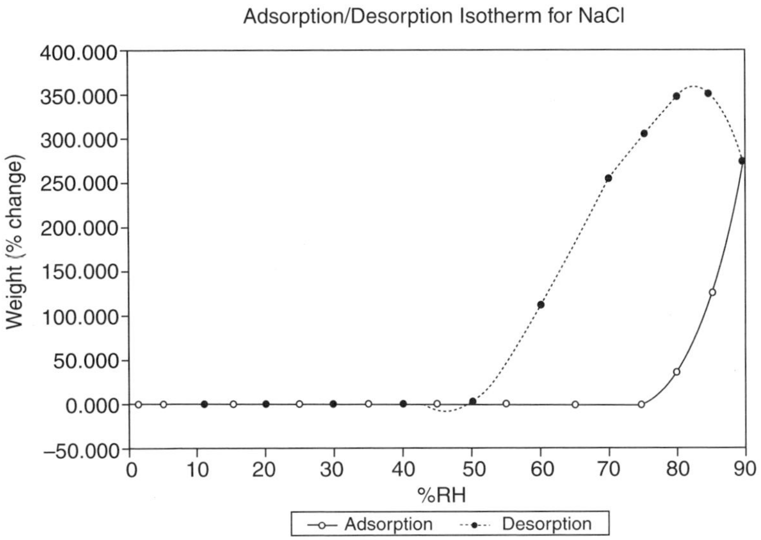

Figure 12.6. Moisture adsorption balance experiment with sodium chloride.

most stable. This is because anhydrous phases are either monotropically or enantiotropically related. Monotropic systems are those where the low-energy drug form is the most stable at ambient temperature and all temperatures below the melt. In enantiotropic systems, the most stable form is dependent on the temperature, and a crossover point in the stability of two forms can be identified. Note that in enantiotropic systems, multiple thermal events can be observed such as the melt of one form, recrystallization to a new form, followed by a melt of that new form (Figures 12.7, and 12.8). The heat of fusion rule states that the compound with the higher heat of fusion is the more stable, (exemplified later with compound **13**).[18] Exceptions where the rule is not applicable for anhydrous polymorphs are when conversion or decomposition occurs at the melting point. Hot-stage microscopy is utilized to validate the conversion(s) detected in the DSC such as: clean melt; water/solvent evolve; melting occurs with decomposition; melt and recrystallization into another form (using hot-stage temperature ramping up); multiple melts of different crystal habits; or melt crystallizes on cooling to ambient temperature.[19] Hot-stage microscopy provides essential information if the heat of fusion rule is applied to identify the more stable form. If the DSC and hot-stage microscopy suggest dehydration/desolvation, then thermogravimetric analysis (TGA) can determine the percent of volatiles. TGA can define the existence of a stoichiometric entity based on weight loss information. Solvate identification is determined with TGA coupled with FTIR or Raman spectroscopy. Birefringence and PXRD alone do not provide all the necessary evidence of whether a particular sample is a mixture of forms, although a mixture of crystal habits visualized by microscopy may be suggestive. Therefore, thermal methods provide insight on the composition of a compound and its form (phase) purity.

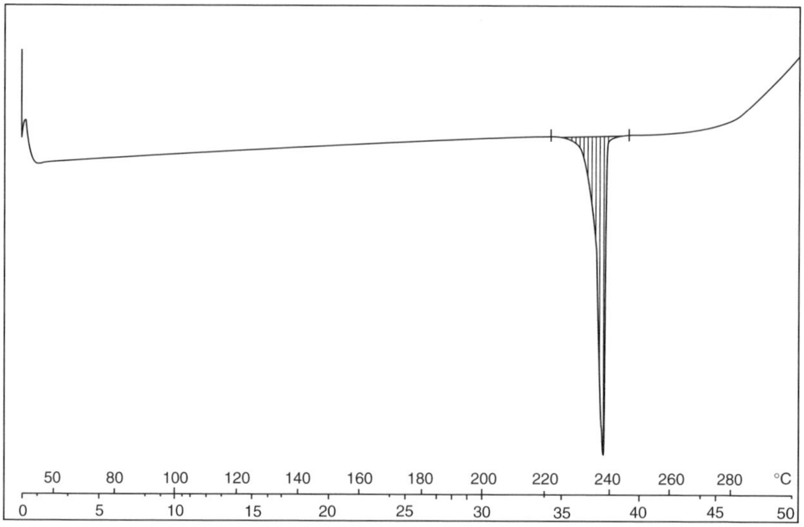

Figure 12.7. DSC of monotropic form of compound 1.

Hygroscopicity in exploratory development is initially conducted in a kinetic manner and is subsequently determined under thermodynamic conditions. The kinetic measurements are performed on moisture balances wherein the humidity is ramped from low to high relativity humidity (RH). Then ramping back from high to low RH completes the experiment. It should be recognized that hygroscopicity is a continuum and there is no specific development cutoff, although deliquescence represents an extreme endpoint, which is at best difficult to manage.

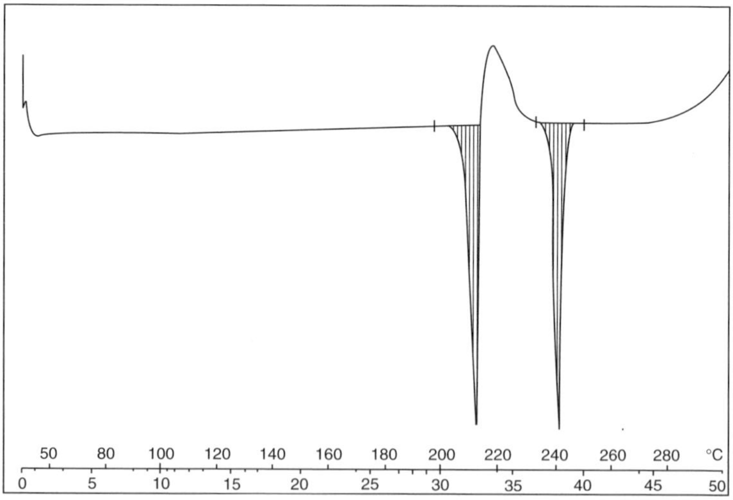

Figure 12.8. DSC of enantiotropic form of compound 1 wherein melt, recrystallization, and melt of the new crystalline form occurs.

The information for a series of drug forms, however, can provide a rank ordering of hygroscopicity. A useful guide is that if a compound has $\leq 2\%$ weight gain by 90% RH, then hygroscopicity should not be a developmental issue. The kinetic measurements conducted on anhydrous compounds should not be expected to detect hydrates. These are discovered through thermodynamic experiments detailed later. If hydrates are known, the humidity desorption ramping yields knowledge on hydrate robustness, as evident by how tightly water is bound in the crystal lattice.

The solid-state properties of drug candidates can have a significant impact on the exploratory development process. The preceding Stage 1 analytical assessments clarify whether the drug candidate properties could delay development timelines, or require more resources to deliver key milestones. Some rationale for each of the assessments as they relate to risk is now outlined. As stated earlier, understanding crystallinity has a direct impact in PK and bioavailability. Crystalline forms in general tend to be more chemically and physically stable, less hygroscopic, and more effective at purging impurities on isolation through crystallization relative to amorphous substances. Impurities or a mixture of entities due to counterion stoichiometry may provide misleading information on further characterization resulting in additional unnecessary drug form research. A low melting point can adversely affect particle size reduction through traditional milling operations by sticking as a result of the energy induced in processing. Similarly milling hydrates with poorly bound water, identified by thermal and/or kinetic hygroscopicity analysis, may result in loss of their hydration. This analysis also enhances the knowledge of alternative drug forms existence by thermal stressing. Molecular mobility is a cause for chemical instability; as a result, the more hygroscopic a compound, the greater the risk. The variability of water content due to hygroscopicity would be anticipated to change particle flow characteristics during DP manufacture, augment the challenge of obtaining the desired dose per tablet (because amount of drug per unit weight changes), and increase the difficulty in attaining content uniformity in drug product. The preceding issues highlight a number of risks that could arise for drug candidates in exploratory development progression. Having the information on solid-state drug properties thereby enables the strategic investment decisions at the drug Discovery/Development interface, where resource commitment should be appropriate for at this phase of high drug candidate attrition.

CASE STUDIES IN EXPLORATORY DEVELOPMENT

The following case studies are intended to expose issues that often need to be addressed in the exploratory development. Herein, the research tactics are defined and the resolution detailed in the examples.

Solubility-Limited Exposure: Compound 5

The initial form of compound **5** was the fumarate salt. Fumarate salts are usually isolated in excellent yields and are highly crystalline. The drawbacks of this counterion are often low aqueous solubility and the potential formation of Michael adducts with the drug candidate or excipients. In this case, solubility-limited exposure (Figure 12.2) was significant. The aqueous solubility of the fumarate salt was 50 μg/ml, preventing target organ toxicity even at 5000 mg/kg.[20]

Salt selection employing ionic bonds is a tactic wherein the physical properties, such as solubility of the covalently bonded drug structure, can be modified.[21] Pharmaceutically acceptable counterions employed in salt selection are "generally regarded as safe" because they have precedence in marketed pharmaceuticals, food, or are normal human metabolites that appear well-tolerated in relatively large amounts such as lactic acid. When considering precedence of counterions, it is important to consider the route of administration (parenteral, immediate release tablet, topical), targeted dose levels for toxicology studies and clinical trials, and duration of therapy (chronic or acute). In other words, counterions that have some precedence may still impact the outcome of toxicology and/or clinical studies. An example is bromide, which is present in marketed products. This counterion has sedating effects that are achieved with modest drug doses. Since one of the goals of toxicology and clinical studies is to understand the safety of the drug candidate, clouding the results with a pharmacologically active counterion is deleterious to continuous development. Furthermore, there have been publications on the overall precedence of counterions in marketed pharmaceuticals stating that chloride and sodium are the most commonly found anions and cations, respectively, >50% for all salts.[22] Their high precedence may not be due to their superiority in optimizing the physical properties in the solid state, but rather to being a default choice due to ready access, cost, and precedence. Hydrochloride and sodium salts are not a panacea as they frequently form hydrates, which add a further dimension to manage in the synthesis and formulation of these highly precedented salts.[23] Therefore, prior to commencing salt selection research, a thoughtful choice of the counterions employed considers the factors of precedence in the route of administration, doses levels, and duration of therapy.

Overcoming solubility-limited exposure of compound **5** commenced with attempts to crystallize numerous salts with precedent. The dibasic nature of **5** enables the consideration of hemi-, mono-, and di-salts, yet predicting crystallinity and solubility properties is beyond the capability of the present science. The initial synthesis phase is an empirical approach, but the organic solubility of free base **5** is an asset in determining appropriate organic solvents for salt selection. Additionally, if a crystalline salt is known, organic solubility of both free base and the salt provide a guide as the ionized compound often imparts different solubility in some solvents. Depicted in Table 12.3 is solubility of a free base and a corresponding salt wherein some of the empirical nature in salt selection is minimized. Acetone, acetonitrile, isopropanol, and tetrahydrofuran represent solvents having high solubility of the free base and low solubility of the ionized species. It is likely that supersaturation will be obtained in salt synthesis with these solvents. By contrast, 2-butanone and methanol appear to be poor solvent choices due to high solubility of both the free base and ionized species. Cyclohexane and hexane are antisolvents. Solvents not desirable in pharmaceutical manufacturing, such as chloroform, are included to

TABLE 12.3. Solubility of a Free Base and a Corresponding Salt

Organic Solvent	Free Base Solubility (mg/ml)	Salt Solubility (mg/ml)
Acetone	33–99	<1
Acetonitrile	20–34	2–4
2-Butanone	37–112	38–113
Chloroform	45–134	<1
Cyclohexane	<1	<1
Ethanol	31–94	>115
Ethyl acetate	6–13	<1
Isopropanol	31–92	6–11
Isopropyl ether	3–6	<1
Hexane	<1	<1
Methanol	>116	>124
Methylene chloride	34–102	<1
Tetrahydrofuran	>92	<1

have a broad range of polarity and solvent diversity to maximize identification of lead salts. A similar solubility screen with the free base and the final salt form can be performed subsequently to displace these undesirable solvents in commercial development (Stage 4). Having identified appropriate solvents for crystallization studies based on solubility of free base, or ideally solubility of the free base and a corresponding salt, a variety of crystallization techniques are viable.

While the intent of this chapter is not to describe in detail the numerous crystallization techniques, an overview is presented.[24] Straightforward crystallization techniques are facilitated by the aforementioned solubility determination. What is required for crystallization is to achieve supersaturation, and subsequently the nucleation process in a controlled fashion. The solubility determination allows the most direct and scalable crystallization process, specifically the addition of an acid to a solution of the free base. Because the salt has limited solubility relative to the free base, supersaturation will be obtained and nucleation may occur. If precipitation does not occur, evaporation, cooling, or heating followed by cooling can facilitate crystallization. Alternatively, an antisolvent can be added to achieve supersaturation. Vapor diffusion is a very slow addition of antisolvent to a solution of compound/salt to be crystallized (Figure 12.9). This technique, while not rapid is an excellent procedure to grow crystals for single crystals X-ray determination. If crystallization is unsuccessful, factors such as solvent(s) employed, experimental techniques, impurities, rate of achieving supersaturation, degree of supersaturation, and water content (if hydration is critical), are reexamined.

After completion of solubility and crystallization studies with compound **5**, a number of leads were secured. The systematic process for understanding the solid-state properties of the lead salts followed the Stage 1 outline described earlier: birefringence → PXRD → purity determination → thermal analysis → kinetic hygroscopicity. Those leads passing the criteria are then advanced to the significant issue

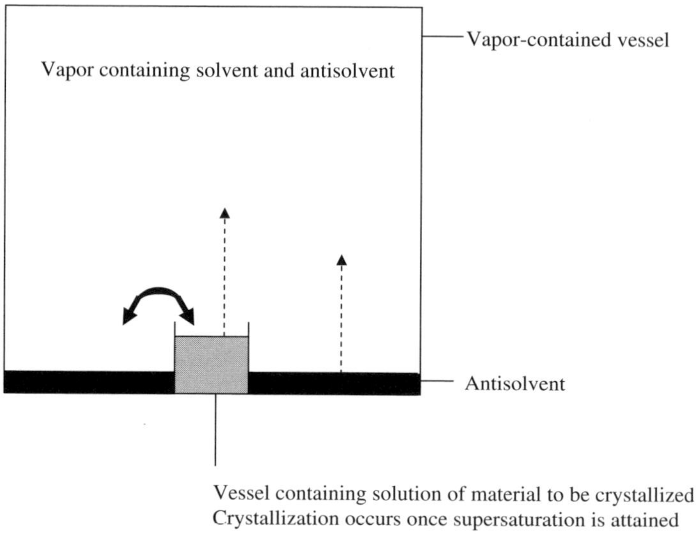

Figure 12.9. Crystallization by vapor phase diffusion.

that salt selection was targeted to overcome, namely, solubility-limited exposure for compound **5**. Solubility in water under thermodynamic conditions includes temperature cycling to 40°C, then 5°C prior to extended equilibration at ambient temperature. Aqueous solubility is then determined, with pH of the resulting aqueous solution measured, along with physical characterization of the undissolved solid. This later analysis is critical because conversion to a new polymorph, hydrate, or precipitation of the free base is vital in knowing the equilibrium solubility. Solubility (31 mg/ml), bioavailability, and accelerated chemical stability discovered that the dimesylate salt was superior to all other leads. Frequently, counterions with less rigidity increase solubility as exemplified in this case, fumarate versus dimesylate. The dimesylate allowed target organ toxicity to be obtained at 85 mg/kg, and thus it enabled candidate progression.

Crystallization of an Amorphous Drug Candidate: Compound 4

As stated earlier, nonionizable compounds represent the highest level of risk for emerging drug candidates because ionized salt options do not exist. Solubility screening described earlier enables a thoughtful approach to crystallization. However, compound **4** was amorphous and had not yielded to crystallization after dedicated attempts. In examples like this one, increasing the purity level often leads to success as impurities retard or prevent nucleation. Additionally, hydrates and solvates can be obtained in initial crystallization, which on desolvation may produce the anhydrous or nonsolvated crystalline compound. Even if the development of an amorphous form is planned, evaluation of the solid-state properties of the crystalline form is necessary to underwrite development robustness.

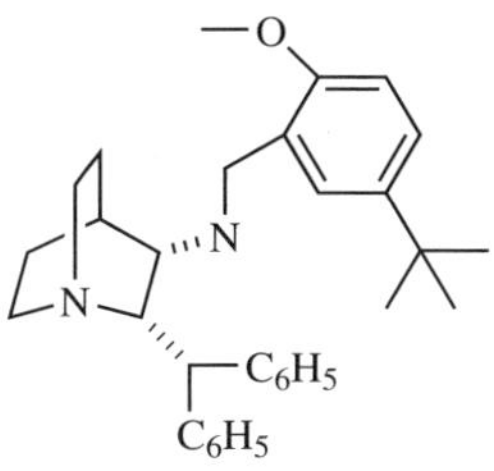

6

Figure 12.10.

API Stability: Compound 6

Understanding API stability is a tremendous exploratory development challenge. The short-term aim is to have rapid assessments that allow predictive rank ordering of various drug forms. There is also the need for a commercial perspective, wherein a viable API form can possess suitable stability to be advanced to the market in a DP formulation. Compound **6** (Figure 12.10) had an immediate concern because the benzoate salt was unstable at even 5°C due to extensive oxidative degradation. Alternative crystalline salts failed to improve the persistent oxidative degradation pathway. These situations are especially resource-intensive in that synthesis and characterization of a myriad of lead salts precedes accelerated stability studies. If the characterization step is skipped, then the outcome of the stability studies is questionable due to homogeneity and purity of the compounds evaluated. Careful assessment of stability studies is warranted because they may not be predictive of ICH conditions due to their accelerated nature.[25] In this example, salt selection accelerated stability conditions were 70°C/75%RH (relative humidity) for three weeks, targeting growth of a single impurity to $\leq$0.1%. The research hypothesis was that a salt containing an antioxidant would shut down the degradation path. This tenet proved to be valid. The citrate monohydrate emerging from salt selection research met the accelerated stability criteria and adequate DP stability was also attained. The reasons for the significant stability increase over previous salts could be twofold. Citric acid is known to chelate metals.[26] During crystallization, chelation of multivalent metals in the API solution that catalyze free radical oxidation, such as iron, may occur. Second, the original tenet of employing an antioxidant counterion to prevent degradation may also be operative in the solid state.[27] The key learning is to identify the degradation mechanism and thereby minimize empirical approaches.

API Stability: Compound 7

The free acid of compound **7** (Figure 12.11) reinforces the importance of comprehending the API degradation mechanism. In this case, degradation occurred more rapidly at temperatures above ambient conditions, (40°C/75%RH), and the key observation was that degradation leveled off after 2–3% of a new impurity formed. The pseudo dimer, **8**, was identified as the new impurity. Salt selection could be a

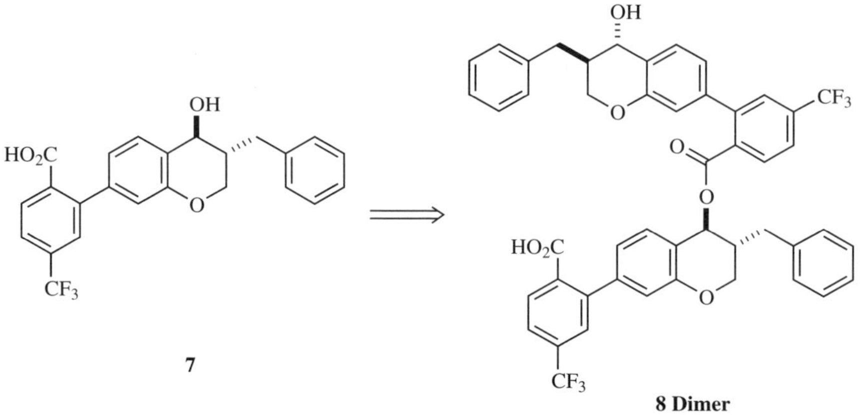

Figure 12.11.

tactic to eliminate formation of the ester impurity **8**. In due course, it became evident that low levels of amorphous material in the crystalline free acid were responsible for the ester formation. PXRD is not necessarily the best method to detect low levels of amorphous content. Microscopy, thermal methods, and solid-state NMR are useful in providing alternative analytical assessments. A point not mentioned earlier in salt selection is that free acids and free bases should be routinely explored as the API and not excluded by salts. Most marketed drug forms between 1983 and 1996 were not salts.[28] Due to their ionized nature, salts have the potential to be more hygroscopic, form hydrates (particularly hydrochloride and sodium salts), and possess different stabilities versus free acids and bases. Furthermore, salt formation adds another API manufacturing step, lowering the overall yield and thereby raising the cost of goods for the commercial product. The partially crystalline nature, crystalline free acid **7** containing some amorphous component, could have led to a significant commitment of salt selection research. Instead, controlling nucleation and the crystallization process delivered fully crystalline **7**. Herein, identification of the degradation path and rate facilitated the best API outcome.

Drug Product Stability and the Thermodynamics of Hydration: Compound 9

The tactics employed to resolve API instability has been described with compounds **3** and **4**. Yet stability of the API in DP is an equally momentous hurdle, although there have been a few instances wherein remarkable stability has been achieved.[29] The monosuccinate salt of compound **9** (Figure 12.12) met all Stage 1 and 2 API criteria including 70°C/75%RH accelerated stability for three weeks with $\leq$0.1% growth of any single impurity. Moreover, formation of the succinate salt increased the chemical purity of the free base from 95% to >99%. Unfortunately, Stage 3 tablet development uncovered instability in drug product. The new impurity was identified as amide **10** resulted from condensation of succinic acid with the free base **9**. Additional formulations were unsuccessful in eliminating this degradation path. Salt selection produced

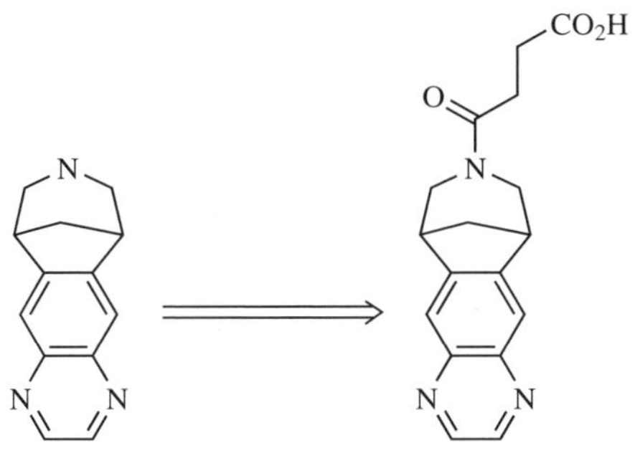

Figure 12.12.

numerous crystalline leads, but unacceptable hygroscopicity at times to the point of deliquescence became a routine obstacle. Ultimately, the monocitrate and mono-L-tartrate were the best salts discovered. In characterizing the two salts, each possessed a different hydration propensity. The citrate appeared to be a channel hydrate, which was confirmed by single crystal X ray. As a result, it adsorbed water based on the relative humidity present, from ~0% to 3.3% weight gain. Three forms of the L-tartrate were discovered in hydrate and polymorph screening (Stages 2 and 3): two anhydrous forms and a hydrate. The low-energy form of the two anhydrous tartrates was defined, and understanding the phase boundary between the hydrate and the anhydrous forms was at this point essential. A 50/50 mixture of hydrate and anhydrous forms was made and analyzed by near-IR and PXRD. The mixture was then stored at a variety of relative humidities (RH): 0%, 23%, 33%, 43%, 59%, 75%, 85%, 94%, 100%. Periodically, the samples were examined by near-IR to assess if the percentage of hydrate or anhydrous forms at the various RH conditions was increasing. Near-IR is extremely effective in rapid determination of these RH conversions, which are accelerated by the presence of both anhydrous and hydrate seeds.[30] At the end of the study, PXRD validated the near-IR determinations. The anhydrous form was found to be stable at 85%RH and below. The choice between the citrate channel hydrate and the anhydrous L-tartrate was difficult because both possessed hydration properties. Stability in drug product was slightly better with the L-tartrate, which became the decision tipping point. Thus the tartrate, a different 1,4-dicarboxylic acid, became the development form displacing the nonhygrospcopic succinate salt. This case (free base reaction with counterion in DP, hygroscopicity, hydration behavior) underscores the complexity in decision making when selection of "ideal" solid-state properties is precluded. It also exemplifies the extensive research commitment required to resolve instability in DP when the API is stable.

Multiple Drug Candidate Issues—"A Complex Salt" Compound 2

Manipulation of one physical property in a drug candidate often affects other solid-state properties as detailed with compound **5**. This cause-and-effect state of affairs

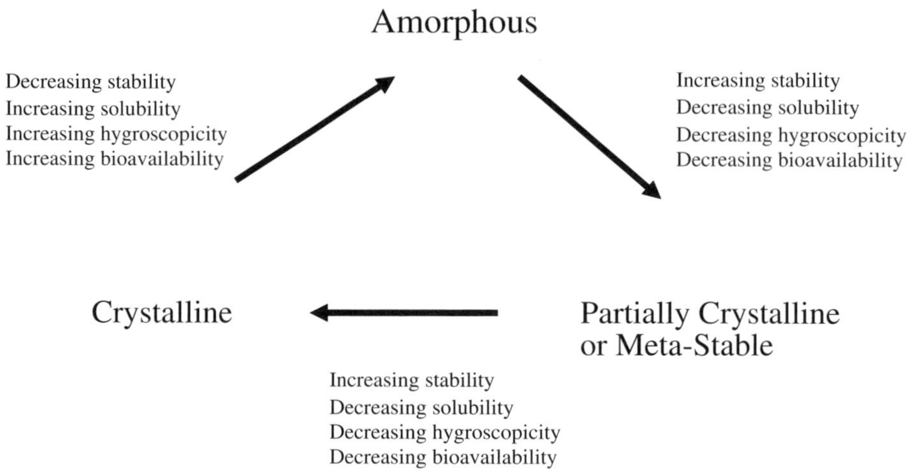

Figure 12.13. Balancing of solid-state properties to achieve an optimum result.

(Figure 12.13) requires a balancing of properties with the aim of achieving the optimum result. In cases where more that one physical property necessitates improvement, the attainment of the requisite delicate harmony is not only a challenge, but forces what may seem to be unconventional approaches. This challenge is not due solely to the drug candidate's inherent structure, but is instead the result of the extenuating factors such as the present example, wherein a high drug dose and exposure level was sought due to the therapeutic area.

The immediate issue with the mesylate salt of compound **2** was hygroscopicity to the point of rapid deliquescence under ambient conditions. Stated in another manner, it was difficult to weigh the mesylate salt with any accuracy because it picked up water quickly; however, it did provide acceptable bioavailability. A concern was that decreasing hygroscopicity often results in the lowering of solubility and bioavailability. As stated previously, weak carboxylic acids and bases have a moderate level of risk as few ionized salt options exist. Compound **2** is a weak base with pK_a values of 4.6 and 5.1. Therefore it is anticipated that only strong acids such as hydrochloric acid would be sufficiently acidic to protonate the molecule. The challenge was to maintain acceptable drug exposure, reduce markedly the hygroscopicity, and achieve the goal with apparently few salt options available.

Ionization of an acid and base is represented by a continuum of proton transfer. This extends from a completely ionized species to one that is a solely hydrogen-bonded complex, absent of ionization in the solid state. The concept of hydrogen-bonded complexes is not new, because the literature is replete with examples over the decades.[31,32] Yet, it is not within the present art to predict their existence for any drug candidate.[33,34] Methods to prepare complexes often entail grinding the organic compound of interest with another organic component(s).[35] Thus a contemporary concern is the robustness of hydrogen-bonded complexes in DP formulation (necessitating wet or dry granulation with excipients and the subsequent high compression forces utilized in tableting), because examples have not been described. The lack of formulation knowledge at present augments the resource requirements and risk of failure for API hydrogen-bonded complexes.

Drug form finalists for compound **2** became the hydrochloride, dimaleate, dimalonate, and sesquisuccinate. All of these met the Stage 1 physical criteria including hygroscopicity (outlined earlier), and each API candidate was identified as a single phase. Based on the pK_a of the counterions, it is not surprising that these four entities spanned the range from full ionized species to hydrogen-bonded complexes in the following order: hydrochloride, dimaleate, dimalonate, sesquisuccinate.[36] The optimum balance of these non-hygroscopic API options, led to selection of the sesquisuccinate that provided superiority in bioavailablity, DP formulation, API, and DP stability over the alternatives. Balancing physical properties to achieve the optimal result requires a predetermination of the key requirements in advance, such that decisions are made in an objective manner.

Discovery of Hydrates: Compound 11

The development of a drug candidate as a hydrated free acid, free base, nonionizable compound, or salt form represents an increase in developmental research. Physical stability issues were described in the L-tartrate salt of compound **5**. Other concerns that arise in the development of hydrates are solubility, bioavailability, the ability to dry and mill the API to particle size specifications on commercial scale without dehydration, and tableting behavior in DP formulation.[37] In addition, the discovery of hydrates during development is not a straightforward task. Moisture balances utilized to measure kinetic hygroscopicity should not be relied on to identify hydrates, although in some instances they may be formed. Compound **11** (Figure 12.14) passed kinetic hygroscopicity under typical moisture balance conditions, picking up just 0.03% water weight at 90%RH during the Stage 1 physical profiling. Three conditions useful in forming hydrates from anhydrous compounds include accelerated stability conditions (70°C/75%RH), thermodynamic solubility determination in water, and water activity studies.[38,39] While the first two of these approaches demonstrate conversions to hydrates at times, water activity studies

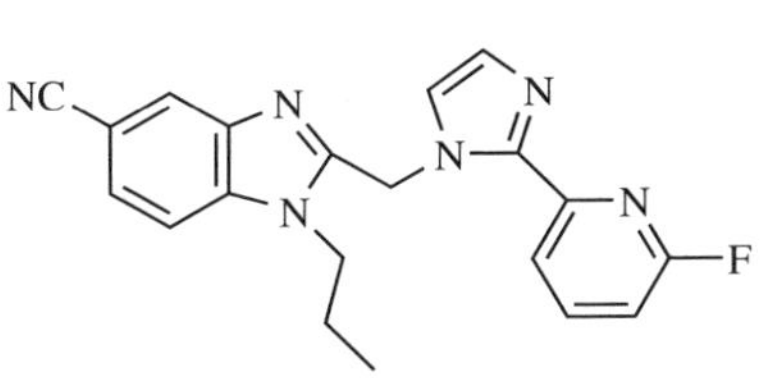

11

Figure 12.14.

TABLE 12.4. Water Activity Studies on Compound 11[a]

Mole Fraction of Water in Isopropanol	Water Activity	Isolated Solid
0.03	0.1	Anhydrous & hydrate
0.11	0.3	Anhydrous & hydrate
0.20	0.5	Hydrate
0.33	0.7	Hydrate
0.45	0.8	Hydrate
0.90	0.9	Hydrate

[a]All experiments employed the anhydrous free base at ambient temperature over two weeks.

have a significantly higher incidence of hydrate formation. This is because the compound of interest often has higher solubility in organic/water media versus water, and thus it attains equilibrium in a shorter time frame.[40] By contrast, stability conditions and equilibrium water solubility possess much slower kinetics in the transformation, especially where the compound has low aqueous solubility. The hydrate of compound **11** was first discovered in water activity studies, and an abbreviated version is depicted in Table 12.4. Having observed the anhydrous to the hydrate transformation in water activity studies, validation of the facile conversion was evidenced by hydrate PXRD peaks beginning to form in anhydrous free base under stability challenge, 12 weeks at 40°C/75%RH. An additional value of water activity studies is that it provides an initial perspective on the phase boundary between the anhydrous and hydrate forms.[41] Determination of the phase boundary is essential to define which API form, anhydrous or hydrate, should be developed. In the present case, the hydrate was advanced over the anhydrous form. The decision was driven mainly by projecting acceptable physical stability and by the ability to achieve particle size specifications on scale. Recall that the opposite was true for compound **9** wherein the anhydrous form was nominated over the hydrate. In some instances, advancement of either a particular anhydrous or hydrate form represents nonviable choices. In these cases, a new research effort is initiated to identify a viable drug form to progress.

The Additional Challenges of Parenteral Development: Compound 12

Parenteral formulations have increased requirements beyond the solid-state properties presented in the previous case studies. Achieving the highest aqueous solubility possible, understanding the impact of additives such as buffers, and preventing particulates/haze/color in solution are formidable constraints for this dosage formulation. Salt selection should occur for every parenteral formulation, and one should have knowledge of the thermodynamic solubility at various pH levels and temperatures. This research underwrites formulation development below supersaturation, thereby ensuring continuous development. During salt selection, integration of the additives and buffers that are potentially utilized in formulation research is essential to prevent surprises. Compound **12** (Figure 12.15) is illustrative of this

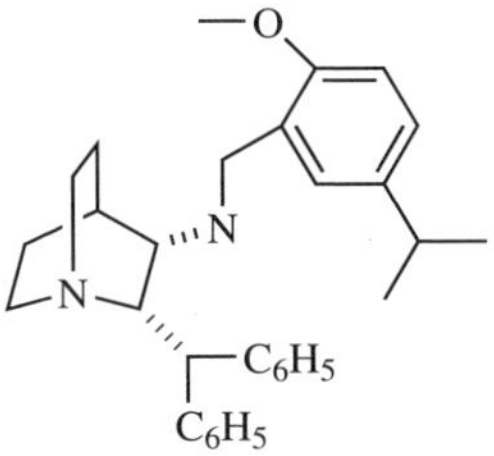

12

Figure 12.15.

point. The dihydrochloride dihydrate low-energy form of **12** was advanced as a result of its excellent solid-state properties and highest aqueous solubility of any salt form (75 mg/ml, pH 2.2). Adjustment of the pH with buffers to acceptable levels for use in parenterals provided adequate solubility (>27 mg/ml, pH 4.3). Citric acid along with other counterions and additives were an element of the salt selection research. The solubility of the crystalline citrate monohydrate salt (4.3 mg/ml, pH 3.56) is significantly lower than the dihydrochloride dihydrate salt. Hence if citric acid was to be used as a buffer in the parenteral formulation of the dihydrochloride dihydrate salt at concentrations $\geq$4 mg/ml, then precipitation of the citrate salt would occur. Finally, securing a parenteral formulation without detectable particulates, haze, and/or color represents a requirement distinctive to this dosage form. The presence of particulates, haze, and/or color could be due to one or more of the following: API degradation; insolubility of API impurities; new API impurity profile resulting from a change in crystallization conditions or new synthetic route; DP manufacture; and leaching of septum components. The ability of the human eye to detect extremely low levels of particulates, haze, and color adds to the complexity of this dosage form and therefore increases the development resources. While the exceptional power of the eye to detect these changes could be viewed as setting an extraordinarily high hurdle, it is essential to recognize that the route of administration is intravenous, and ensuring quality at the maximum level is obligatory. Parenteral formulations are significantly more demanding to develop than traditional dosage forms such as immediate release tablets because of a high aqueous solubility requirement, the impact of additives in the dosage form, and the nature of the DP, which enables very low levels of change to be observed.

ANALYTICAL TOOLS

The analytical techniques discussed previously are routinely employed in characterizing all exploratory development candidates. Many additional tools are extremely valuable in providing critical information to guide candidate progression. Two techniques that provide fundamental insight into the drug form are single-crystal X-ray and solid-state NMR.

Single Crystal X Ray

The most basic knowledge of the solid state is obtained from single-crystal structure determination.[42,43] Medicinal and process chemists are able to obtain absolute and relative stereochemistry of a compound to guide SAR and facilitate the development of new synthetic routes. The calculated powder pattern, derived from the single-crystal structure, provides all peaks that can exist in the experimentally derived PXRD. The calculated pattern in comparison with the experimental PXRD of a crystalline solid can define if the crystalline solid is a single phase or a mixture. In other words, if the experimental PXRD possessed more diffraction lines than the calculated PXRD obtained from the single-crystal X ray, the experimental PXRD is a mixture of phases or components. The opposite can also occur; there are fewer diffraction lines in the experimental PXRD than in the calculated PXRD, and/or the peak intensities vary significantly. This phenomenon is known as preferred orientation. Comprehension of the Bragg equation is necessary to clarify the point further. Powder X-ray spectra are defined by the Bragg equation, $2d\sin\theta = n\lambda$, where d is the distance of between crystal planes, θ is the half-angle of diffraction, and λ is the X-ray wavelength (Figure 12.16). Preferred orientation relates to the tendency of particular crystalline habits to stack or lay in a non-random manner due to physical structure (tablets, laths, Figure 12.17). Imagine tablets poured on a flat surface. They will tend to orient in a preferred fashion, laying flat instead of on each end. As a result of preferred orientation, some experimental PXRD diffraction lines will be diminished or absent in intensity because all the crystal planes are not statistically represented. If PXRD spectra from a number of pure phase samples are examined, then there is the potential that some diffraction lines will be missing as a result of preferred orientation. In the absence of the calculated PXRD, this would suggest a mixture of phases in some experimental PXRD, which possess apparently extra diffraction lines. Therefore, having single-crystal data is particularly important in resolving issues of preferred orientation. Without single-crystal data though, proving that preferred orientation exists can be achieved by a number of techniques. Thermal methods, discussed previously, are capable of determining phase purity. Another technique is to gently grind

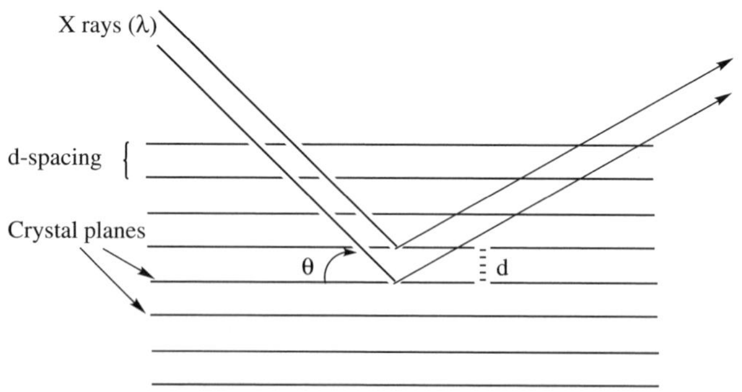

Figure 12.16. Depiction of X-ray diffraction and the terms employed in the Bragg equation.

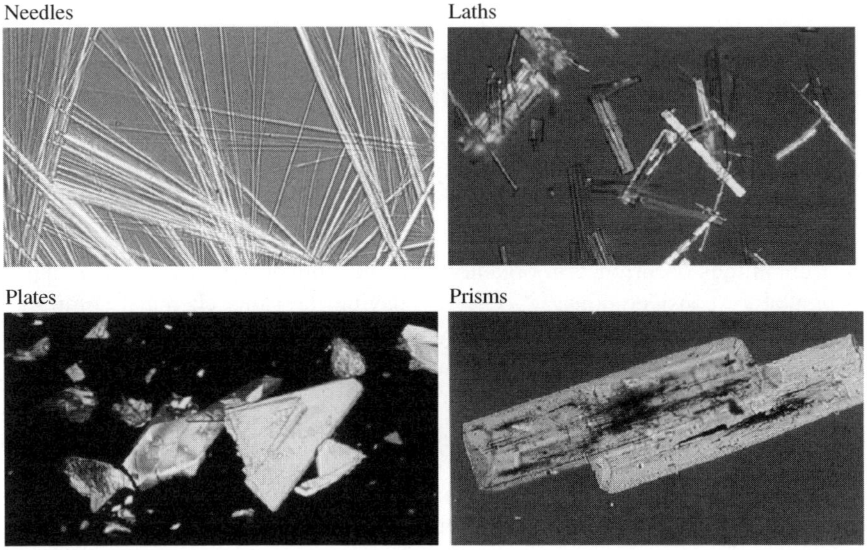

Figure 12.17. Crystal habits that may demonstrate preferred orientation in PXRD spectra.

samples with a mortal and pestle to reduce particle size/shape, thereby minimizing or eliminating preferred orientation in the PXRD. The caveats are that grinding may convert one phase to another and/or generate amorphous material.[44] Finally, solid-state NMR is an ideal tool to assess preferred orientation phenomenon. The value of securing single-crystal X ray is broad in proving the covalent bond structure and the phase purity, and it assists in determining the low-energy polymorph, which is discussed later.

Solid-State NMR (SS NMR)

A complementary approach to single-crystal X-ray concerning phase purity is solid-state NMR.[45] Advantages include (a) the nondestructive nature of the experiments and (b) the ability to use a variety of nuclei such as ^{13}carbon, ^{19}fluorine, ^{15}nitrogen, and ^{31}phosphorus. In the crystalline solid state, each atom's distinctive position relates to its chemical shift.[46] As a result of the different crystal packing in polymorphs, each form has distinct spectra. The utility of the technique enables the determination of phase purity, level of amorphous content, phase conversions, and the number of molecules in the asymmetric unit cell.[47] SS NMR was employed with compound **2**. While it was clear from the pK_a values of compound **2** that carboxylic acids such as succinic acid could not protonate the weak bases, it was initially ambiguous whether a complex had formed or the product was a mixture of free base **2** and succinic acid. Because of the short timeframe of the SS NMR experiment relative to growing single crystals for the series, SS NMR substantiated all four salts or complexes were pure phases prior to obtaining single-crystal X-ray data. While this is a relatively new analytical tool, it has such broad application that it will continue to revolutionize our knowledge in characterizing and understanding the solid state.

POLYMORPH SCREENING AND DETERMINATION OF THE LOW-ENERGY DRUG FORM

A through-screening program to determine if polymorphs exist is imperative because of the ubiquitous occurrence in organic molecules, impact in API/DP development, and regulatory considerations.[48,49] Crystals having the same chemical composition but different molecular packing are polymorphs. Anhydrous crystals are therefore distinct entities from solvates and hydrates, which are referred to as pseudopolymorphs, because their chemical composition is not identical. Hydrates are more precedented than solvates because toxicity of the latter limit their utility in toxicology studies and pharmaceuticals. Knowledge of solvates is important, however, because crystallization in the appropriate media is required to generate the desired drug form.[50] The following section on polymorphs will not specifically cover the selection of hydrates or solvates versus anhydrous forms, because with compounds **9** and **11** elucidated key tactics and criteria. However, polymorph screening approaches will encompass the synthesis and characterization of hydrates and solvates.

The basic polymorph screening methods are (1) evaporation of API solutions to dryness, (2) antisolvent addition to API solutions, (3) cooling API solutions to induce precipitation, and (4) reslurrying solid API in solvents. A polymorph screening experimental design includes the use of a variety of solvents, solvent polarities, temperatures, and compound solubilities with these four methods. The comprehensiveness of the screen is targeted to cover as much experimental space as possible. All studies begin with pure compound because impurities can inhibit transformations and/or impact the polymorphic preference by their presence. The four methods mentioned above can be structured to increase the likelihood of generating the lowest-energy form or a high-energy state of the compound. For example, isolating a compound by fast evaporation of a solution to dryness may effect a spontaneous precipitation. Immediate crystallization increases the possibility of a high-energy drug form. Similarly, rapid antisolvent addition or swift cooling achieves supersaturation quickly and amplifies the possibility of generating of high-energy drug form.[51] Ostwald's rule of stages and Ostwald ripening are the basis for understanding these observations and the subsequent formation of the low-energy form.[52] Crystallization occurs in stages, from a high-energy solution state to a solid form closest to it in energy. If the medium is appropriate, this solid-state form will continue to cascade down in stages to the lowest-energy form. If rapid precipitation occurs or the compound is isolated without a sufficient equilibration time, the low-energy form of the drug candidate may not be isolated. Slurrying the initial precipitate in proper solvent(s) for a prolonged period of time, a day or longer, allows the ripening dynamics of the precipitate dissolving and recrystallizing as long as the drug form has some solubility in the solvent(s). This enhances the potential cascade to lower-energy forms and improves the crystallinity of the isolated compound. Chemists observe the process at times on large scale when an initial thick slurry resulting from precipitation thins out and subsequently thickens. A corollary is that low compound solubility in the media, as a result of antisolvent addition or cooling, may retard or preclude Ostwald ripening. While the first three methods described commence with a solution, the fourth technique, reslurrying compound in solvents, has more than one starting point. Reslurrying crystalline material is usually pursued; however, commencing with a higher-energy form such as an amorphous solid, it allows the beginning of the energy cascade at an earlier point. This approach is capable of identifying higher-energy forms than the crystalline one in hand. Once polymorph screening is complete, experiments to elucidate the relative stability of polymorphs secured commences.

The lowest-energy polymorph is the slowest to nucleate (Ostwald's rules), has the lowest solubility and bioavailablity, is not necessarily the highest melting (enantiotropic/monotropic relationship), and has the highest crystal density (exceptions are due to hydrogen bonding in the crystal lattice). Characterization of the new forms follows Stage 1 analysis; birefringence → PXRD → purity determination → thermal analysis → kinetic hygroscopicity. During characterization, some new forms could emerge from thermal analysis such as a dehydrated/desolvated form. Completion of Stage 1 analysis will identify the number of new forms as well as mixtures. Then studies are conducted by reslurrying a 50/50 ratio of two polymorphs in solvents of varying polarities and drug solubility to "bridge" to a single form. Thus bridging employs Ostwald's rules of stages to assess the transformation to the lower-energy form. Bridging results in the lowest-solubility component as long as sufficient solubility allows the conversion to transpire. A number of solvents are utilized to eliminate the potential of solvent effects that could predispose the outcome to a particular form. Once bridging studies are concluded, a rank ordering of polymorphs may be emerging.

Forward steps in identification of the low-energy polymorph are case-by-case dependent. Plausible scenarios at this stage are as follows: (1) No new forms are discovered, (2) new forms are discovered and bridging clearly defines the rank ordering, and (3) new forms are discovered, but their relative stability remains uncertain. During the bridging studies or subsequently, measurements to support relative polymorph stability ordering are essential. Determination of the solubility of each polymorph enhances the findings of the bridging experiments. Care in analyzing the solubility results requires determining if a form conversion happens during the experiment. A form transformation would result in the solubility determination of the new component's solubility, not the original entity. Similarly, solution calorimetry (heat of solution) results can be compromised by form conversion taking place during the analysis. Physical stability of polymorphs is a continuum from extremely metastable to those capable of being isolated, characterized, and stable under ambient conditions. Some of the most metastable polymorphs preclude characterization due to transformation to a new phase. Another class can be fully characterized, but disappear when the more stable form is isolated.[53] Concomitant polymorphs relate to mixtures that are routinely obtained.[54] The metastable end of the spectrum is contrasted with apparently stable polymorphs that coexist and do not readily interconvert. A prime example is carbon, wherein graphite is the

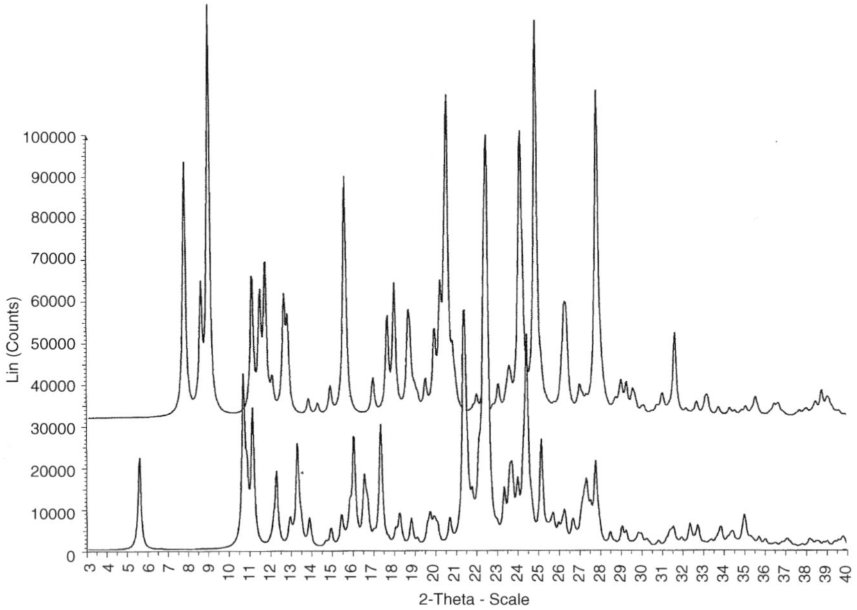

Figure 12.18. PXRD of polymorphs Form A (top spectra) and Form B (bottom) of compound 13.

most thermodynamically stable phase, but diamonds are clearly long-lived. The industrial importance of diamond underscores not only that polymorphs coexist, but also that the varying properties can be unexpected and possess enormously different utilities. A specific example of the challenges to identify the most stable polymorph is detailed below to illustrate the utility of the analytical tools in providing key insights into decision process.

Upon polymorph screening the anhydrous mesylate salt of compound **13** (Figure 12.19) (form A), another anhydrous form (B) was discovered as evidenced by PXRD (Figure 12.18). Bridging studies suggested that the new form B was the thermodynamically more stable form; however, the data were not conclusive. Additional assessment was essential to support that form B was the lowest-energy anhydrous form. Some of the characterization data for both forms is depicted in Table 12.5. The heat-of-fusion rule and the single-crystal density support the

TABLE 12.5. Characterization Data for Forms A and B of Anhydrous Mesylate Salt of Compound 13

	Form A	Form B
Melting point	206°C	203°C
Heat of fusion (joules/g)	103.6	114.7
Single crystal density (g/cm^3)	1.338	1.372

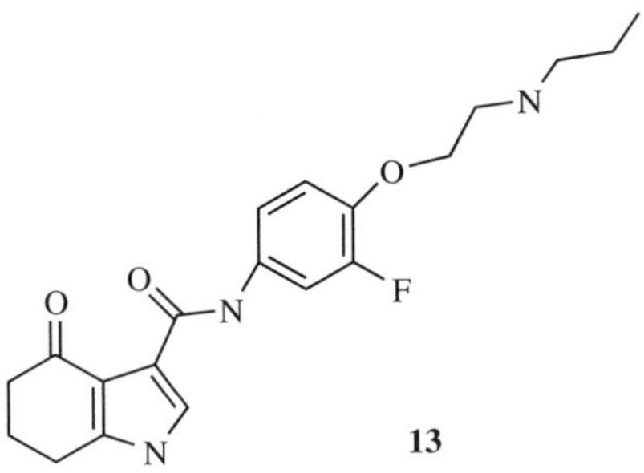

13

Figure 12.19.

bridging studies that form B is the lowest-energy form. Note that B is not the highest melting and is therefore enantiotropially related to form A. Solubility and other characterization reinforced the relative stability of form B. This example was provided to demonstrate that the information necessary to determine the lowest-energy polymorph is complex.

This chapter has focused on the selection of the drug candidate form in exploratory development. At some point in development it is necessary to define the polymorphs, hydrates, and solvates of the chemical intermediates in the synthetic process. The research rationale is to build in the robustness necessary for a commercial manufacturing process. If this is not established and a new form of an intermediate arises, then the following potentially occur; intermediates fail to dissolve in the solvents for the chemical reaction, crystallization fails to purge impurities to targeted levels, physical properties preclude material movement in the process, filtration times become excessively long, and the process falls out of regulatory filing specifications. Geodon/Ziprasidone **14** (Figure 12.20), an important pharmaceutical for treatment of schizophrenia, illustrates an example of multiple compound forms. The free base is the immediate precursor to the commercial hydrochloride salt. A new form of the free base (B) was identified during development as evidenced by PXRD. Characterization determined one form to be anhydrous (A depicted in top spectra) and the other the monohydrate (B bottom) (Figure 12.21). Thus the importance of understanding physical form encompasses more than drug candidates and commercial products.

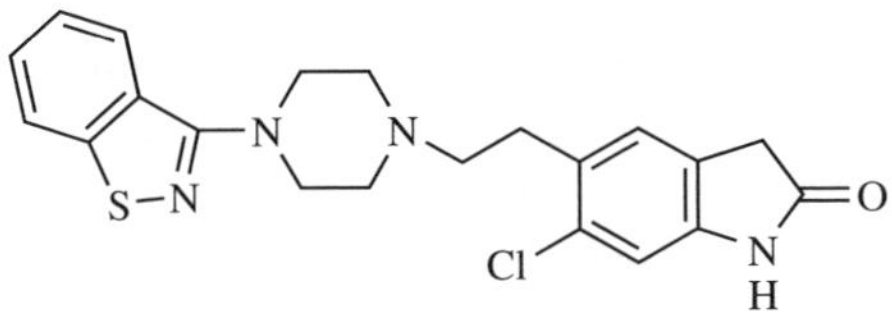

14

Figure 12.20.

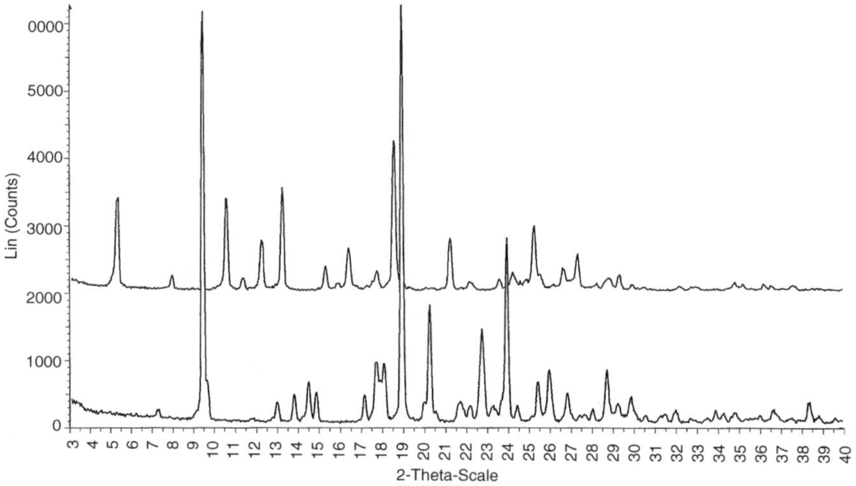

Figure 12.21. PXRD of the anhydrous form (top spectra) and monohydrate form (bottom) of compound **14**.

HIGH-THROUGHPUT SCREENING

Automation has advanced in capability and utility to facilitate the identification of the API form.[55] The scope includes salt selection, hydrate, solvate, and polymorph screening. Use of 96 well-plate concepts developed in other applications has evolved to be the platform of choice. The value of the technology is to cover more experimental phase space without an increase in resources. There is exceedingly high value in having the capability as an experimental tool. In covering the phase space, experiments still require a fundamental understanding of solubility. High-throughput automation of solubility is routine, but can't be ignored or the experimental space researched is compromised by inadequate design. Birefringence, PXRD, and Raman spectroscopy are utilized to analyze and "bin" high-throughput experiment hits, thereby saving significant resources in the Stage 1 effort. However, the major automation limitations are twofold. In cases where crystallization readily occurs, the initial automated Stage 1 analysis effort of birefringence, PXRD, and Raman spectroscopy does not reduce the number of samples for downstream evaluation of purity, counterion stoichiometry, and so on. The second constraint is due to the very nature of high-throughput screening. Small samples (1–5 mg) with high surface area may not be physically stable to dehydration or desolvation either changing into an amorphous solid due to crystal lattice collapse or appearing hygroscopic due to desolvation.[56] This could result in not detecting a fleeting hydrate/solvate, which under other techniques could be coaxed into a stable anhydrous form. High-throughput screening is a significant advance in understanding and identifying options to solid-state issues, but it should be recognized as an additional research tool and not a universal solution.

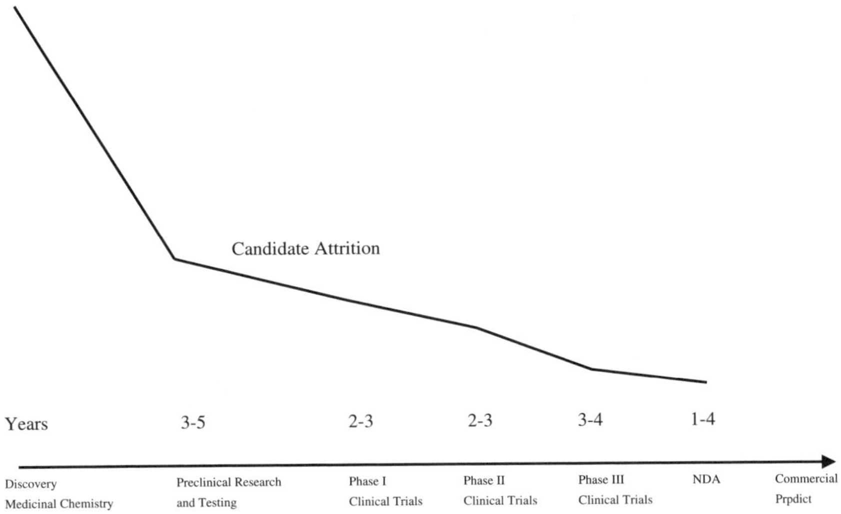

Figure 12.22. Development timelines and drug candidate attrition.

SUMMARY

Identification of the drug form for exploratory development requires a coordinated team effort of numerous scientific disciplines to achieve a successful outcome. Education of new discovery and process chemists on the impact of drug form and crystallization techniques pays high dividends in an exploratory development setting. Balancing PK, bioavailability, amorphous/crystallinity form, hygroscopicity, API chemical and physical stability, stability and performance in DP, and the effect of other physical characteristics is a daunting challenge. This becomes a strategic case-by-case decision on the level of resources to be deployed at a time of high drug candidate attrition. The attrition results from toxicology studies and other information learned in Phase I clinical trials (Figure 20.21). A systematically staged approach therefore provides clarity throughout an organization wherein all drug candidates are not equal, but defining each development paradigm is essential. Within a systematic development strategy, the level of solid-state science underwriting the level of API understanding requires an uncompromised process. A four-stage process was defined and the underlying rationale was described. The eight case studies and other examples illustrated critical elements of both a practical and an efficient approach to selection of the exploratory development drug form. Success in selection of the exploratory development form requires a focused team, structured scientific approach, and predetermined API/DP criteria.

ACKNOWLEDGMENTS

I am indebted to D. J. M. Allen, Joseph Krzyzaniak, and Z. Jane Li for the development of the crystallization course curriculum employed for discovery and process chemists, as well as for their keen solid state insight and scientific dialogue.

REFERENCES AND NOTES

1. Lipinski, L. A. Experimental and Computational Approaches to Estimate Solubility and Permeability in Drug Discovery and Development Settings. *Adv. Drug Delivery Rev.* **1997**, *23*, 3–25.
2. Lipinski, L. A. Compound Properties and Drug Quality. *Prac. Med. Chem.* **2003**, 341–349.
3. Kerns, E. H., and Di, L. Physicochemical Profiling: Overview of the Screens. *Drug Discov. Today: Technologies* **2004**, *1*, 343–348.
4. Saunders, K. C. Automation and Robotics in ADME Screening. *Drug Discov. Today: Technologies* **2004**, *1*, 373–380.
5. Curatolo, W. Physical Chemical Properties of Oral Drug Candidates in the Discovery and Exploratory Development Settings. *Pharm. Sci. Technol.* **1998**, *9*, 387–393.
6. Desuiraju, G. R. In Search of Clarity. *Nature* **2003**, *423*, 485.
7. Chemburkar, S. R., Bauer, J., Deming, K. Spiwek, H., Patel, K., Morris J., Henry R., Spanton S., Dziki W., Porter, W., Quick, J., Baauer P., Donaubauer J., Narayanan B. A., Soldani M., Riley D., and McFarland, K. Dealing with the Impact of Ritonavir Polymorphs on the Late Stages of Bulk Drug Process Development. *Org. Process Res. Dev.* **2000**, *4*, 413–417.
8. Bauer, J., Spanton, S., Henry, R., Quick, J., Dziki, W., Porter, W., and Morris, J. Ritonavir: An Extraordinary Example of Conformational Polymorphism. *Pharm. Res.* **2001**, *18*, 859–866.
9. Moisture balance instrument manufactures: Surface Measurement Systems, 24040 Camino del Avion, Suite A250, Monarch Beach, CA 92629 and VTI Corporation, 7650 W. 26th Ave., Hialeah, FL 33016.
10. Johnson, K. C., Swindell, A. C. Guidance in the Setting of Drug Particle Size Specification to Minimize Variability in Absorption. *Pharm. Res.* **1996**, *13*, 1795–1798.
11. Woodward, R. B., and Brehm, W. J. The Structure of Strychinine. Formation of the Neo Bases. *J. Am. Chem. Soc.* **1948**, *70*, 2107–2115.
12. Woodward, R. B., Cava, M. P., Ollis, W. D., Dafniker, H. U., and Schenker, K. The Total Synthesis of Strychnine. *Tetrahedron* **1963**, *19*, 247–288.
13. McCrone, W. C., McCrone, L. B., Delly, J. G. *Polarized Light Microscopy.* McCrone Research Institute, 2003.
14. Jenkins, R., and Snyder, R. L. *Chemical Analysis Introduction to X-Ray Powder Diffractometry.* John Wiley & Sons, New York, 1996.
15. Isomorphs are an exception as they have the same crystal packing with different solvents in the crystal lattice. Stephenson, G. A., Stowell, J. G., Toma, P. H., Dorman, D. E., Greene, J. R., and Byrn, S. R. Solid-State Analysis of Polymorphic, Isomorphic, and Solvated Forms of Dirithromycin. *J. Am. Chem. Soc.* **1994**, *116*, 5766–5773.
16. Amidon, G. L., Lennernas, H., Shah, V. P., and Crison, J. R. A Theoretical Basis for a Biopharmaceutic Drug Classification: The Correlation of *in Vitro* Drug Product Dissolution and *in Vivo* Bioavailablity. *Pharm. Res.* **1995**, *12*, 413–420.

17. Threlfall, T. L. Analysis of Organic Polymorphs. *Analyst* **1995**, *120*, 2435–2460.
18. Berger, A. *Acta Pharm. Technol.* **1982**, *28*, 1–20.
19. McCrone, W. C. *Fusion Methods in Chemical Microscopy.* Interscience Publishers, New York, 1957.
20. Jacobson-Kram, D. *Toxicology Testing Handbook.* Marcel Dekker, New York, 2001, pp. 55–60.
21. Stahl, P. H., and Wermuth. *Handbook of Pharmaceutical Salts.* Wiley-VCH, New York, 2002.
22. Berge, S. M., Bighley, L. D., Monkhouse, and D. C. Pharmaceutical Salts. *J. Pharm. Sci.* **1977**, *66*, 1–19.
23. Haynes, D. A., Jones, W., and Motherwell, W. D. S. Occurrence of Pharmaceutically Acceptable Anions and Cations in the Cambridge Structural Database. *J. Pharm. Sci.* **2005**, *94*, 2111–2120.
24. Mullin, J. W. *Crystallization.* Butterworth Heinemann, 1997.
25. International Conference on Harmonization guidance for API and DP can be found at the following site. www.fda.gov/cder/guidance/index.htm.
26. Kirk-Othmer. Citric Acid. *Encyclopedia Chem. Technol.* **1993**, *6*, 359.
27. Akers, M. J. Antioxidants in Pharmaceutical Products. *J. Parenter. Sci. Technol.* **1982**, *36*, 222–228.
28. Ando, H. Y., and Radebaugh G. W. *Remington: The Science and Practice of Pharmacy.* Lippincott Williams & Wilkins, New York, 2000, p. 704.
29. Cohen, L. P. Safe and Effective: Many Medicines Prove Potent for Years Past Their Expiration Dates. *Wall Street Journal* March 28, 2000, 1,16.
30. Norris, T., and Santafianos, D. Discovery of a new stable polymorph of 4-(3-ethynylphenylamino)-6,7-bis(2-methoxyethoxy)quinazolinium methanesulfonate using near-infrared spectroscopy to monitor form change kinetics. *J. Chem. Soc. Perkin Trans.* **2000**, *2*, 2498–2505.
31. Etter, M. C., and Frankenbach. Hydrogen-Bond Directed Cocrystallization as a Tool for Designing Accentric Organic Solids. *Chem. Mater.* **1989**, *1*, 10–12.
32. Pedireddi, V. R., Jones, W., Chorlton, A. P., and Docherty, R. Creation of crystalline supramolecular arrays: a comparison of co-crystal formation from solution and by solid state grinding. *Chem. Commun.* **1996**, 987–988.
33. Fleishchman, S. G., Kuduva, S. S., McMahon, B. M., Walsh, R. D. B., Rodriguez-Hornedo, N., and Zaworotko, M. J. Crystal Engineering of the Composition of Pharmaceutical Phases: Multi-Component Crystalline Solids Involving Carbamazepine. *Crystal Growth Des.* **2003**, *3*, 909–919.
34. Childs, S. L., Chyall, L. J., Dunlap, J. T., Smolenskaya, V. N., Stahley, B. C., and Stahley, P. G. Crystal Engineering Approach to Forming Cocrystals of Amine Hydrochlorides with Organic Acids. Molecular Complexes of Fluoxetine Hydrochloride with Benzoic, Succinic, and Fumaric Acids. *J. Am. Chem. Soc.* **2004**, *126*, 13335–13342.
35. Trask, A. V., Motherwell, and W. D. S., and Jones, W. Solvent-Drop Grinding: Green Polymorph Control of Cocrystallization. *Chem. Commun.* **2004**, 890–891.
36. Li, Z. J., Ambramov, Y., Bordner, J., Leonard, J., Medek, A., and Trask, A. V. Interactions in Complexes of Heterocyclic Bases and Dicarboxylic Acids: Crystallography, Hydrogen Bond Analysis, and ^{15}N Spectroscopy. In press., *J. Am. Chem. Soc.*

37. Khankari, R. K., and Grant, D. J. W. Pharmaceutical Hydrates. *Thermochim. Acta* **1995**, *248*, 61–79.
38. Zhu, H., and Grant. D. J. W. Influence of Water Activity in Organic Solvent + Water Mixtures on the Nature of the Crystallizing Drug Phase. 2. Ampicillin. *Int. J. Pharm.* **1996**, *139*, 33–43.
39. Zhu, H., Yuen, C., and Grant, D. J. W. Influence of Water Activity in Organic Solvent + Water Mixtures on the Nature of the Crystallizing Drug Phase. 1. Theophylline. *Int. J. Pharm.* **1996**, *135*, 151–160.
40. Gu, C. H., Young, V., and Grant, D. J. W. Polymorph Screening: Influence of Solvents on the Rate of Solvent-Mediated Polymorphic Transformations. *J. Pharm. Sci.* **2001**, *90*, 1878–1890.
41. Ticehurst, M. D., Storey, R. A., and Watt, C. Application of Slurry Bridging Experiments at Controlled Water Activities to Predict the Solid-State Conversion Between Anhydrous and Hydrated Forms Using Theophylline as a Model Drug. *Int. J. Pharm.* **2002**, *247*, 1–10.
42. Clegg, W. *Crystal Structure Determination*. Oxford University Press, New York, 1998.
43. Inkpen, M., Palmer, R. A., Ladd, M. F. C. *Structure Determination by X-Ray Crystallography*. Springer, New York, 2003.
44. Ojala, W. H., and Etter, M. C. Polymorphism in Anthranilic Acid: A Reexamination of the Phase Transisitons. *J. Am. Chem. Soc.* **1992**, *114*, 10288–10293.
45. Offerdahl, T. J., and Munson, E. J. Solid-State NMR Spectroscopy of Pharmaceutical Materials. *Am. Pharm. Rev.* **2004**, *7*, 109–112.
46. Dong, Z., Padden, B. E., Salsbury, J. S., Munson, E. J., Schroeder, S. A., Prakash, I., and Grant, D. J. W. Neotame Anhydrate Polymorphs I: Preparation and Characterization. *Pharm. Res.* **2002**, *19*, 330–336.
47. Dong, Z., Padden, B. E., Salsbury, J. S., Zhou, D., Munson. E. J., Schroeder, S. A., Prakash, I., Vyazovkin, S., Wight, C.A., and Grant, D. J. W. Dehydration Kinetics of Neotame Monohydrate. *J. Pharma. Sci.* **2002**, *91*, 1423–1431.
48. Haleblain, J., McCrone, W. Pharmaceutical Applications of Polymorphism. *J. Pharma. Sci.* **1969**, *58*, 911–929.
49. Byrn, S., Pfeiffer, R., Ganey, M., Hoiberg, C., and Poochikian, G. Pharmaceutical Solids: A Strategic Approach to Regulatory Considerations. *Pharm. Res.* **1995**, *12*, 945–954.
50. Mersmann, A. *Crystallization Technology Handbook*. Marcel Dekker, New York, 2001.
51. Threlfall, T. Crystallization of Polymorphs: Thermodynamic Insight into the Role of Solvent. *Org. Process Res. Dev.* **2000**, *4*, 384–390.
52. Ostwald, W. *Z. Phys. Chem.* **1897**, *22*, 289–330.
53. Dunitz, J. D., and Bernstein, J. Dissappearing Polymorphs. *Acc. Chem. Res.* **1995**, *28*, 193–200.
54. Bernstein, J., Davey, R. J., and Henck, J. Concomitant Polymorphs. *Angew. Chem. Int. Ed.* **1999**, *38*, 3440–3461.
55. Avantium, Solvias, and Symyx are some of the companies utilizing or selling automation for API high throughput screening.
56. Stephenson, G. A., Groleau, E. G., Kleemann, R. L., Xu, W., and Rigsbee, D. R. Formation of Isomorphic Desolvates: Creating a Molecular Vacuum. *J. Pharm. Sci.* **1998**, *87*, 536–542.

13

STRATEGIES TO ACHIEVE PARTICLE SIZE OF ACTIVE PHARMACEUTICAL INGREDIENTS

David J. am Ende and Peter R. Rose
Chemical Research & Development, Pfizer Global R&D, Groton, CT 06340

In general the control of particle size of active pharmaceutical ingredients (APIs) is critical to the successful manufacture of most drug product formulations. The particle size of the API can impact various attributes of the formulation. There are various methods used in the industry to achieve the desired particle size distribution, including the use of crystallization, wet milling, or dry mechanical milling approaches. This chapter will review the rationale for particle size selection for a specific formulation application, the common methods to achieve the desired particle size distribution, and strategies for selection of particle size control methods for API candidates.

IMPORTANCE OF PARTICLE SIZE OF API

As a development scientist working on the synthesis and scale-up of a new drug candidate, it is important to understand the rationale for the particle size requirements for the new API and the selection process used to determine that range. The following section describes the importance of particle size in a dosage form and approaches to selecting the particle size.

Particle size control is important to the development of dosage forms that require solid API. For the discussion in this chapter, we will focus primarily on those dosage forms using oral administration. However, similar requirements are needed

Fundamentals of Early Clinical Drug Development: From Synthesis Design to Formulation, edited by Ahmed F. Abdel-Magid and Stéphane Caron

for other formulation types such as for inhalation or suspensions. In order for an API to be effective after oral administration, it needs to be dissolved in the gastrointestinal tract and then diffuse through the intestinal wall and absorbed into the body.[1] The first step in that process is the disintegration of the dosage form followed by dissolution of the active ingredient. For relatively insoluble compounds the rate-limiting step in the overall absorption process is generally the rate of dissolution. The impact of particle size on crystalline API dissolution rates and their corresponding absorption has been well-described in the literature. Examples of compounds with differences in dissolution and absorption due to particle size include nifedipene, griseofulvin, indomethacin, and spironolactone.[2–6]

Particle size reduction increases the specific surface area of the powders, resulting in modifications of dissolution rate. Noyes and Whitney described the relationship of dissolution rate to surface area of the particle in their 1897 paper "The Rate of Solution of Solid Substances in Their Own Solution."[7] They describe the dissolution rate of the solid being controlled by a stagnant layer of saturated solution that forms around a solid particle. The dissolution rate described by Noyes and Whitney can be expressed by the Equation (13.1):

$$\frac{dC}{dt} = \frac{DA}{Vh}[Cs - C(t)] \tag{13.1}$$

where dC/dt is the dissolution rate, D is the diffusivity of the solute, A is the surface area of the particle, h is the stagnant layer film thickness (diffusion layer), V is the volume of solution, Cs is the solubility of the solid, and $C(t)$ is the concentration of the solute at time t.

Based on Equation (13.1), higher surface area particles result in increased dissolution rate. As the particle size of a compound is decreased, the specific surface area increases. Therefore, smaller particles will tend to have increased dissolution rates compared to larger particles. It is also evident by Equation (13.1) that the rate of dissolution is directly proportional to its solubility value taken at the pH conditions of the dissolving surface.[1]

In addition to dissolution and absorption, the control of particle size is generally required to ensure that uniform distribution of particles is obtained in a solid dosage formulation often referred to as content uniformity. Tablets and capsules are produced in equipment that controls the mass of the API and excipients and ultimately blend composition by volumetric filling. Solid materials with different particle size and shape have different flow and packing properties, giving rise to potential segregation within the formulation blends. These differences can impact the specific bulk volumes of the powder during tablet compression and encapsulation, resulting in the potentially inconsistent fill volumes and subsequently inconsistent content uniformity of the dosage form. Content uniformity is usually achieved through careful selection of the particle size distribution especially important for highly potent APIs where, for example, precise milligram quantities are required in each tablet during manufacturing.

SELECTION OF PARTICLE SIZE

Several approaches have been used by formulation scientists to select the correct particle size of the API used for dosage formulation of a new drug candidate. Current approaches typically include empirical methods combined with mathematical predictions. Trial and error or empirical methods usually involve the production of various particle size ranges of the API followed by formulation into the proposed dosage form. These dosage forms can then be characterized for dissolution, content uniformity, and any other required testing. This approach is often used when significant quantities of API are available, the physical characteristics of the API are not fully known, and the dose is not well-established. While this method can be used successfully, it is not preferred in today's environment where bulk-sparing strategies are employed and compressed development times are required. The trial-and-error method tends to be material-, equipment-, and personnel-intensive.

More recently, mathematical methods for prediction of formulation mixing have been used to determine the dosage requirements and the particle size needs to achieve successful formulation.[8] One such approach developed by Johnson and co-workers proposed a model to describe the theoretical dissolution rate of a polydisperse powder on its weight-averaged particle size distribution.[9] The method can be used to establish the required particle size distribution needed to achieve the desired oral dissolution rate. This approach includes mathematical modeling of the dissolution, absorption, and pharmacokinetics of API to aid the selection of particle size.[10,11] The predictive models show, for example, that high-potency APIs with low doses require smaller particle size distributions. Thus, increasing the number of particles of the API and decreasing the particle size will increase the probability of achieving the needed content uniformity in the formulation.[12] This approach has been described in the literature for a variety of pharmaceutical products, including nifedipine, digoxin, and nitrofurantion.

APPROACHES TO ACHIEVING PARTICLE SIZE

Once the target particle size of the API has been chosen, the method of achieving particle size has to be determined. There are a variety of approaches that can be used to achieve the desired particle size range for an API. These approaches include (1) methods to directly achieve the desired particle size through crystallization or precipitation, (2) the reduction of particle size using wet milling, and (3) particle size reduction by dry milling methods via jet mills or mechanical mills. This section will review several approaches used in the pharmaceutical industry and will describe the benefits and disadvantages of each as it relates to new API candidates. The approaches will be reviewed in order of their typical unit operations. Specifically, crystallization, wet milling, and dry milling will be described.

CRYSTALLIZATION APPROACHES TO ACHIEVING PARTICLE SIZE

Crystallization from solution is the most common separation and purification process used in the pharmaceutical industry.[13] Besides purity control, crystallization processes impact various other physical attributes of the API molecule being crystallized, including the final solid form (e.g., hydrate, solvate, polymorph), the crystal size distribution, and crystal shape. Modification and control of the crystallization method of the API can have a profound impact on these physical characteristics.

Development of any crystallization process requires a good understanding of the solubility and metastable zone.[14] A full analysis of the relationship of solubility to the methods of generating supersaturation, such as temperature, solvent composition, or pH, are required to develop a controllable and robust crystallization process. Crystallization to achieve particle size directly from solution is very feasible, but the development time, material required, and information needed to ensure scalability of the process make this unfavorable for most early (i.e., Phase I) drug candidate programs. This approach is more favorable in mid to late development (typically for Phase II and Phase III programs) when more synthetic experience and understanding of the impurity effects and process changes common in the development of the commercial process are known. Crystallization for particle size control can be an excellent way to avoid downstream processing steps, including milling or other methods of particle size modification, and should be evaluated as the candidate progresses. An excellent review of design approaches for crystallization control was recently written.[15] Although the focus of this chapter is on the API, the application of crystallization control is also needed during processing of pharmaceutical intermediates as it relates to improved robustness of filtration and control of impurities.

The four main methods to generate supersaturation to induce crystallization are as follows[16]:

1. Temperature change—that is, typically cooling crystallization
2. Evaporation
3. Chemical reaction
4. Changing solvent composition—that is, typically via drowning out with an antisolvent

The following section describes the common modes of crystallization and the potential opportunities and limitations of each technique.

Cooling Crystallizations

Cooling of a saturated solution to generate supersaturation is probably the most common method of crystallization. The manipulated variable is the temperature of the cooling medium, which determines the magnitude of the cooling rate as a function of time. Concentration- or time-dependent cooling rates can be used to design a

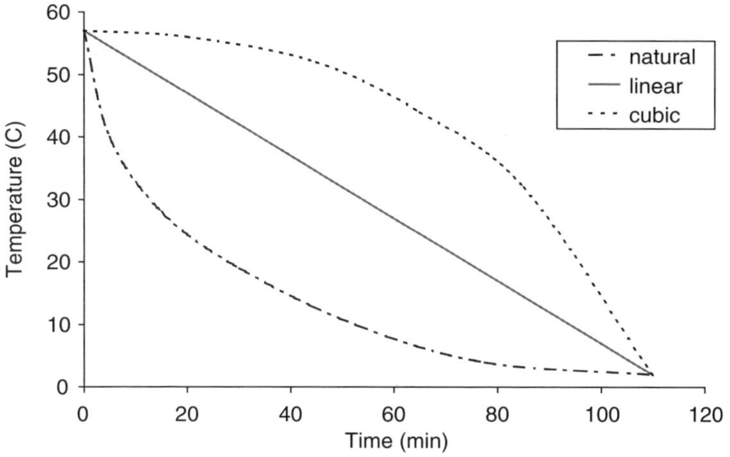

Figure 13.1. Typical cooling profiles for cooling crystallizations where the left axis is the temperature of the solution versus time. *From top to bottom*: Parabolic cooling profile provides low supersaturation to induce nucleation followed by faster cooling during crystal growth. Linear cooling profile refers to the direct proportionality of temperature with time. Natural cooling profile refers to a sudden exposure of a warm saturated solution to ambient temperature in which a rapid temperature increase is followed by gradual cooling until equilibrium is reached at the ambient temperature.

particular cooling profile. The three main types of cooling profiles to achieve supersaturation are natural, linear, and parabolic as shown in Figure 13.1. The natural cooling rate simulates the temperature profile when a reactor is suddenly exposed to significantly constant lower temperature without any temperature control. For example, a round-bottomed flask that is being heated by a heating mantle is suddenly removed from the heat and allowed to cool "naturally" to room temperature. Such natural cooling profiles are characterized by an initial steep cooling rate, followed by a much slower cooling rate in the latter part of the cooling period. These types of cooling profiles result in high supersaturation at the beginning of the crystallization and low supersaturation at the end of the crystallization.

Linear cooling profiles can be used very efficiently by adjusting the rate of cooling to suit the purposes of the crystallization. For example, a steep linear cooling can be used to generate small particles where primary nucleation is dominated by the rapid formation of supersaturation, but this method may not be as efficient as a purification approach if impurity purging is needed. The main issue with linear cooling profiles is the variable levels of supersaturation generated which tend to result in a polydispersed crystal size distribution. The main advantage to this type of cooling profile is that it can be easily transferred to a manufacturing facility with simple temperature control systems. A summary of advantages and disadvantages of the various cooling profiles is given in Table 13.1.

To maximize particle growth and to ensure maximum purification, a parabolic (or cubic) cooling profile should be used. During a parabolic cooling profile, an

TABLE 13.1. Characteristics of Various Cooling Profiles

Natural Cooling	Linear Cooling	Cubic (Parabolic) Cooling
Particle Size Obtained	**Particle Size Obtained**	**Particle Size Obtained**
Highly variable, but often generates smaller particles	Large and small particle sizes can be obtained but typically with wide distribution	Typically large particles with narrow distribution of size and shape
Advantages	**Advantages**	**Advantages**
General high initial supersaturation	Ease of programming	Consistent supersaturation possible
	High initial supersaturations using fast cooling profiles	More efficient than slow linear cooling profiles
Disadvantages	**Disadvantages**	**Disadvantages**
Inconsistent supersaturation	Inconsistent supersaturation at end of crystallization	Multivariable programming required
Difficult to reproduce on scale-up		
Multivariable programming required		

initial slow period of cooling is followed by a steep cooling period. Parabolic cooling profiles minimize the rapid formation of supersaturation and typically result in larger crystal size. The use of parabolic cooling profiles provides the potential to control supersaturation levels better than linear cooling profiles.[17] In addition, parabolic cooling profiles are more efficient than slow linear cooling rates. The efficiency is gained by starting with slow cooling rates at the beginning of the crystallization to allow growth followed by increasing the rate of cooling as the systems becomes desaturated. Braatz[18] and Roberts[19] have shown that control loop approaches to achieve consistent particle size generate parabolic cooling profiles for batch crystallization processes. In addition to controlled cooling methodologies, seeding strategies should also be considered for materials that are slow to nucleate or when large crystals are desired.

Understanding the solubility as a function of temperature is of fundamental importance to designing cooling crystallizations. Classical thermodynamics can be used to initially estimate the solubility of organic solute via the van't Hoff equation.[14]

$$X_{\text{solute}} = \frac{1}{\gamma_{\text{solute}}} \exp\left[\frac{\Delta H_m}{R}\left(\frac{1}{T_m} - \frac{1}{T}\right)\right] \tag{13.2}$$

where X_{solute} is the mole fraction of the solute in the solution at a temperature T (in K), γ is the activity coefficient of the solute in the solvent, ΔH_m (in J/mol) and T_m (in K) are the heat of fusion and melting point, respectively, and R is the gas constant (8.314 J/mol·K). Equation (13.2) allows the simple calculation of ideal solubility ($\gamma = 1$) to estimate differences in solubility of structurally similar analogues.[16] By using heat of fusion and melting point data obtained from a simple Differential Scanning Calorimetry (DSC) experiment, ideal solubility can be calculated. Equation (13.2) further illustrates that the solute solubility increases as the temperature increases, or as the heat of fusion increases, or as the melting temperature decreases. By definition, the ideal solubility ($\gamma = 1$) is independent of the solvent. Solubility in specific solvents can be estimated by a variety of advanced methods such as group contribution methods or it can be measured.

Antisolvent and Reactive Crystallizations

Antisolvent crystallizations (also known as drowning-out crystallizations) result when an antisolvent is added to a solution of the product in order to increase supersaturation and induce crystallization. These types of crystallizations are typified by dramatic changes in supersaturation with the addition of the antisolvent.

Reactive crystallizations, sometimes referred to as precipitations, are induced by a chemical reaction and often result in the creation of large supersaturations. In some pharmaceutically relevant systems, the desired particle size range can be achieved directly through careful control of mixing of the antisolvent and drug solutions.

These approaches can be very useful in producing small crystal size, but care must be taken to analyze the crystalline phase because high supersaturation can result in the formation of metastable crystalline phases. For fast nucleating systems with time scales on the same order as the times scale for mixing in agitated vessels, the scale-up parameters will influence the level of supersaturation created and ultimately particle size distribution. The scale-up parameters include addition rates, feed location, order of addition, agitation speed, and so on. Approaches to control particle size for fast nucleating APIs typically involve high-intensity mixing devices. These include twin impinging jets, ultrasonic mixers, high-shear rotor–stator mixers, and static mixers.

In the case of impinging jets (IJ), two liquid streams pumped at velocities typically in the range of 5–15 m/sec are contacted at a point of impingement, creating a high-intensity zone of mixing and affording a uniform and consistently high supersaturation condition for precipitating small particles. Using solvent and antisolvent liquid streams in an impinging jet device has been exemplified with lovastatin, simvastatin, and omeprazole.[20,21] Using a reactive intermediate (API free base) solution in one stream and a reactive dilute HCl solution in the other liquid stream, impinging jets have been exemplified for the reactive crystallization of ziprasidone HCl monohydrate salt, with an average of 15-μm particle size. Similarly, the particle size control of voriconazole was afforded via impinging jet technology through the intense mixing of an ethanol–water solution of the reactive camphorsulfonate salt of voriconazole with a second reactive stream of dilute sodium acetate solution at similar flow rates.[22,23]

Similar approaches using a mixing-T or other high-energy mixing device have shown improved scalability as the mixing zone is minimized and controlled through continuous feed at constant velocity. Rivera and Randolph described a simple continuous precipitation method using a mixing T for crystalline pentaerythritol tetranitrate (PETN). PETN was dissolved in acetone and precipitated with water in a mixing T followed by a static mixer. Narrow particle size distributions, controlled from 12 μm up to 60 μm, were demonstrated through control of the flow ratio of acetone solution to water.[24] Scale-up of impinging jet crystallization is relatively straightforward with knowledge of the jet velocity and jet diameter. To maintain similar mixing times and energy dissipations used in the lab, for example, the following scale-up criterion should be used as reported recently by Johnson:

$$\varepsilon = \frac{u^3}{d} \tag{13.3}$$

where ε is the turbulent energy dissipation, *u* is the jet velocity, and *d* is the inside jet diameter.[25] Thus scale-up from the lab is accomplished by manipulating the volumetric flow rate in the jets of a given diameter to maintain a similar energy dissipation. For example, to scale-up from 0.02-in. ID jet at 5 m/sec in the lab to 1/8-in. ID jet in the plant requires a jet velocity of approximately 9 m/sec based on Equation (13.3) to maintain similar energy dissipation. It should be emphasized that the use of high-intensity mixers (such as impinging jets) to affect particle size is limited to cases when the time scale for nucleation kinetics are similar to time scales of mixing. In this way, control of mixing can be used to ensure uniform saturation for nucleation and growth. In cases where nucleation and growth kinetics are fairly slow relative to mixing, control techniques other than mixing must be considered.

Evaporative Crystallization

Evaporative crystallization occurs when the concentration of the product in solution is increased by distilling off the solvent until nucleation occurs. Alternatively, evaporative crystallization can be completed by removing a dissolving component by azeotropic distillation. Batch evaporative crystallization processes are often difficult to scale-up with results similar to lab experience due to significant equipment differences at the differing scales. In addition, the rate of evaporation that can be obtained in batch evaporative crystallizations is generally slow, resulting in low levels of supersaturation. Therefore, small particle size is often not obtainable using batchwise evaporative crystallization. However, the use of spray crystallizers as fast evaporators has shown success to control particle size.[26] Recently, supercritical fluid crystallization using solvents such as carbon dioxide has also been used to produce small-size crystalline API through fast evaporative methods. Micron and submicron crystals have been obtained using this technology.[27,28] General applicability is limited in supercritical fluid media by limitations of solubility of the drug in supercritical CO_2.

MILLING APPROACHES TO ACHIEVING PARTICLE SIZE

Wet Milling High-Shear (Rotor–Stator) Mixers

High-shear mixers are employed in the pharmaceutical industry for various mixing operations in drug product formulations and are becoming more prevalent for particle size reduction of APIs.[29,30]

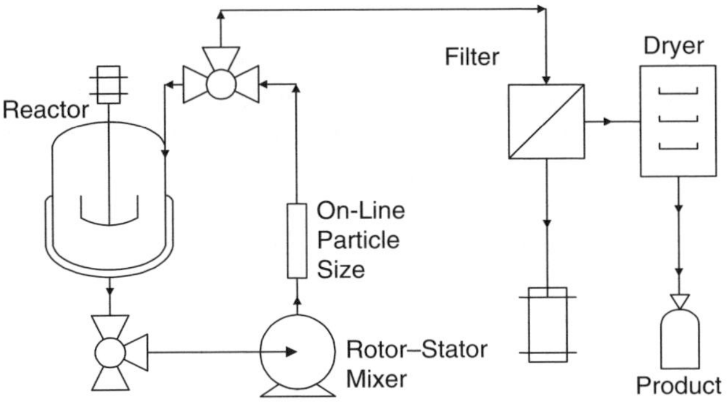

Figure 13.2. Flow diagram of a typical configuration of a rotor stator (wet mill) with on-line particle size analytics.

For reducing particle size of APIs, high-shear mixers can be integrated into the final crystallization step and can provide a convenient alternative to dry milling for the manufacture of API lots, especially during early process development. High-shear wet milling involves the use of high-speed rotor blades within a work-head to exert suction and draw liquids and solids into the rotor–stator assembly. Centrifugal force drives the materials into the work-head where milling occurs in the area between the rotor and stator.[31] Evaluation of particle size reduction for a particular API can be completed readily in the laboratory using commercially available bench scale rotor–stator mixers. At kilo-lab or pilot plant scale, the high shear-milling device is typically configured in a recirculation line and operated in a continuous fashion until the particles are reduced to a suitable size. A typical configuration is shown in Figure 13.2. The monitoring of particle size distribution on-line can provide additional process understanding and robustness.

Potential Advantages of High-Shear Wet Milling. High-shear wet milling can offer a number of advantages over dry milling operations. One major advantage of high-shear wet milling is reduced potential occupational exposure through minimizing the number of handling operations as compared with typical dry milling operations. The high-shear wet milling unit operation is performed on the reactor slurry typically through a recirculation loop. A single isolation by filtration of the solid product is completed instead of multiple isolations and product handling typical of dry milling operations. Reducing dust generation typical of dry milling operations also enhances safety. High-shear wet milling may be the preferred method of particle size reduction for compounds of high potency or with high dust explosivity ratings.

The capital equipment costs are generally one order of magnitude less than dry powder mills and do not require additional dedicated containment facilities. A high-shear mixer can be retrofitted easily into existing plant facilities. Compounds with low melting points ($<100^\circ C$) may also benefit from high-shear wet milling. Typically, hammer and other impact mills generate significant temperature at the mill head that can make low-melting APIs difficult to process. High-shear wet milling provides superior temperature control due to the presence of solvent and the cooling capacity of the process equipment. Materials that are susceptible to loss of crystallinity during dry milling may benefit from the increased heat dissipation capabilities of wet milling.

Another advantage is the reduced unit operations required when using high-shear wet milling. Eliminating conventional dry milling operations increases productivity and reduces overall yield loss. Typical yield losses due to dry milling on commercial scale range from 1% to 5% and can be potentially eliminated with high-shear wet milling. In early development, where batch sizes are small, yield losses can be significantly larger. In addition, elimination of the dry milling step may reduce the overall processing cycle time if the wet milling operation occurs as part of the final API crystallization step.

Limitations of High-Shear Wet Milling. A practical lower limit on particle size reduction exists with conventional rotor–stator mixer technology. For instance, high-shear mixers typically produce particles with a size between 20 and 50 μm. Smaller particle size requirements (<20 μm) would likely require additional or alternate methods of particle size reduction or crystallization control. The size achieved is in part due to the energy and rotor–stator selection as well as physical properties of the slurry and the individual particles. The ability of the particles to break is attributed to various material properties, including brittleness, shape, plasticity, and availability of fracture planes.

Although one normally expects the filtration times to be excessive for high-shear milled API, in actuality we find that filtration and drying times are often only slightly longer and in many cases do not pose a significant issue. Some crystalline materials will form crystalline bridges across the crystals during tray drying and may require a delumping operation. It is recommended, therefore, that filtration and drying rates and post-drying flow properties be evaluated in the laboratory prior to scale-up.

Slurry Flow Recycle Loops. The preferred equipment set up for a high-shear wet mill is using a recycle loop. The design of high-shear mixers is a high-speed rotor surrounded by a stator screen generating shear. The high-shear mixer provides some pumping capacity but is limited. It requires positive flooding through gravity or a supplemental pump. The high-shear mixer can deliver moderate pressures (5–20 psi) and flow rates. Experience to date in 100- to 200-liter vessels has shown that the high-shear mixer is sufficient to ensure continuous slurry flow and recycle using a Silverson Model 200UHLS mixer. Larger-scale operations will likely require supplemental pumps to move the product slurry through the recycle loop and back into the reactor.

Scale-Up of High-Shear Rotor–Stators

Starbuck and co-workers have proposed the shear frequency as a scale-up parameter for rotor–stators and have shown that the rate of particle size reduction is highly dependent on the shear frequency parameter.[30] It was further shown that lab scale and pilot plant scale wet milling resulted in similar particle size distributions when the shear frequency was maintained constant.

The shear frequency is proportional to rotation rate of the rotor, the number of slots on rotor, and the number of slots on the stator according to the equation

$$\gamma = RPM^{*}N^{*}M \tag{13.4}$$

where γ is the shear frequency, RPM is the rotation rate, N is the number of slots on the rotor, and M is the number of slots on the stator as reported by Starbuck and co-workers.[30]

Small-scale lab pilots using a suspension containing between 1 and 10 g of API are typically used to evaluate the feasibility. A suspension of the compound of interest is suspended in a suitable solvent system. A small lab-style high-shear mixer is inserted into the reactor, and mixing is commenced until the particle size reduction is determined to be complete by off-line or on-line analysis. The choice of solvent for high-shear wet milling is typically one where the compound has very low solubility; more preferably, the choice is the isolation solvent system used in the final crystallization. Laboratory evaluations can be performed using either in-line or in-reactor configurations based on the batch size.

Recent work performed in our laboratories employed the use of a high-shear rotor–stator to reduce particle size in a highly agglomerating system.[32] Varenicline tartrate, a drug candidate for treatment of smoking cessation, is prepared in the last step of the synthesis by treating varenicline free base with tartaric acid in methanol. When nucleation occurs the crystals tend to agglomerate as shown in Figure 13.3 (left). The API was wet-milled using a high-shear rotor–stator mixer resulting in efficient deagglomeration as shown in Figure 13.3 (right). The particle size of the agglomerates was on the order of 200 μm. The particle size of the deagglomerated API was on the order of 25 μm.

Figure 13.3. (*Left*): Agglomerated varenicline tartrate obtained during the API crystallization step of the synthesis. (*Right*): Deagglomerated varenicline tartrate after wet milling via rotor–stator.

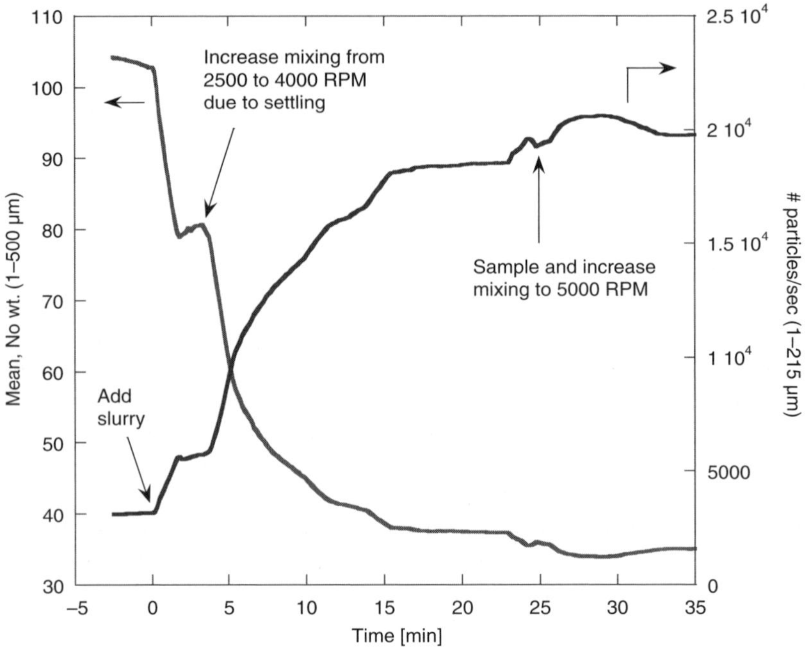

Figure 13.4. Particle size reduction of agglomerated varenicline tartrate during high-shear wet milling in a laboratory-scale pump-around loop, recorded via *in situ* laser light scattering probe as a function of time. Left axis is the number-weighted average of large particles with chord lengths in the range of 1–500 μm. Right axis is the number-weighted average of small particles with chord lengths in the range of 1–215 μm. As shown, the rotational rate of the rotor was increased at particular stage of the experiment to ensure adequate suspension of the solids in the slurry.

On-line analytics can provide additional process understanding for high-shear milling operations. An example of using on-line analytics using a focused beam reflectance measurement (FBRM) probe is shown in Figure 13.4.[33] In this example, the initial slurry of large agglomerates undergoes a rapid reduction in total particle count on high-shear mixing (as shown by the reduction in the mean size between 1 and 500 μm). The formation of smaller particles (between 1 and 215 μm) shows a significant increase during the first 15 minutes of high-shear milling. On continued milling, the number of small particles and the mean size level-out to a quasi-steady state. This quasi-steady state defines the endpoint for high-shear wet milling.

Dry Milling to Achieve Particle Size

The primary goal of dry milling is to fracture particles to a smaller size. Size reduction by dry milling is typically accomplished by forming cracks that propagate through the deformed particle to form fractures. The resistance to particle breakage depends on various attributes of the material being milled, including the tensile strength, ductility, and brittleness, as well as sensitivity to heat such as softening or melting. The type of mill useful for a particular particle size reduction is dependent on the material properties and the required particle size and distribution. In addition, the ingoing particle size and the powder flow characteristics of the ingoing material will also be important considerations for efficiency and robustness of milling operations.

In general the energy required to mill to a particular size is related to the particle size of the ingoing material. Several theories have been developed to relate the energy input to particle size reduction, although none have been able to predict the particle breakage versus energy input.[34] Kick's theory states that the energy used for size reduction is directly related to the reduction ratio X_F/X_P, where X_F is the feed size and X_P is the product size. Rittinger's law states that the energy is proportional to the new surface produced. Bond's law relates the energy required for size reduction as inversely proportional to the square root of the diameter of the milling material. The overriding principle from these theories is that higher-energy impaction is required to achieve smaller particle size.

There are four typical mechanisms of size reduction: impaction, compression, shear, and attrition. Impaction occurs by a particle hitting a rigid object or other particle at a high differential velocity. Compression occurs when a particle is compressed between two rigid surfaces at a force great enough to cause particle fracture. Shear occurs by a fluid or particle–particle interaction and results in cleavage of the particle into multiple parts. Attrition occurs by particles scraping against other particles or other surfaces, resulting in particle reduction.

A variety of milling types are typically used in the pharmaceutical industry based on these milling mechanisms. The FDA's SUPAC Guidance for Immediate Release Solid Dosage Forms (1999)[35] describes the common classes of mills used in the pharmaceutical industry. These include fluid energy mills, impact mills, cutting mills, compression mills, and screening mills. In addition, other milling approaches have been used to make micron and submicron sizes. These techniques include wet ball milling, Dyno® milling, and cavitation milling. In this section, we will briefly review these types of mills and their operational parameters.

All milling operations have practical limits of particle size reduction that may be obtained. These particle size limits are related to the specific milling mechanism employed (related roughly to energy input), the operating conditions, the milling time, and the properties of the material to be milled. As an example, milling time affects the particle size produced. When a milling operation is short, the stress effects on the particles going through the mill are relatively constant. However, as the residence time increases, the milling environment may not be constant. As a particle reduces in size, the stress required to cause fracture increases due to depletion in cracks present in the particle. The result is increased energy input needed to reduce the size further.[34] Table 13.2 lists common examples of the types of mills and the expected lower limits for particle size for each mill type.

TABLE 13.2. Reference of SUPAC Milling Categories for Various Fine Grinding Machines Along with Nominal Particle Size Range Achievable

Class	Typical Achievable Particle Size Range	Subclass
Ball mills	Submicron	Wet ball milling
Fluid energy mills	2–10 μm	Tangential jet
		Loop/oval
		Opposed jet
		Opposed Jet with dynamic classifier
		Fluidized bed
		Fixed target
		Moving target
Impact mills	10–50 μm	Hammer air swept
		Hammer conventional
		Pin/disc
		Cage mills
Cutting mills	Various sizes	
Screening mills	50–150 μm	Rotation impeller
		Rotating screen
		Oscillating
		Hammer conventional

Fluidized Energy Mills[36–38]

Fluidized energy mills (also commonly known as micronizers or jet mills) operate by particle impaction and attrition. A typical fluidized milling operation involves sending a fluid, typically compressed air or nitrogen, at high pressure through a jet nozzle into a mill chamber. The high-velocity nitrogen creates turbulence within the milling chamber, where solid particles are fed into the mill. The high-energy mixing and turbulence inside the mill chamber allow significant particle-on-particle collisions where size reduction occurs. Typically, a particle size classifier is used to remove the smaller particles and retain the larger particles. The main variables used in fluidized energy mills are volumetric flow rate, the particle feed rate, and the diameter of the mill chamber. In addition, the design and placement of the jet nozzles and the shape of the milling chamber (e.g., spiral, loop) are variable, depending on the manufacturer (Figure 13.5). Typical particle size ranges obtained from fluidized energy mills are between about 2 and 10 μm. Advantages to fluidized energy mills such as jet mills (e.g., spiral or loop) include: no moving parts, reducing the likelihood of contamination due to wear of parts; the ability to easily sterilize the equipment; the small particle size obtainable; and the cooling effect of the expansion of the compressed gas that allows heat-sensitive compounds to be processed with little degradation. Disadvantages include the propensity to form aggregates or agglomerates post milling, the generation of amorphous content due to high-energy impacts, and the potential to form ultra-fine particles.

Figure 13.5. Spiral jet mill. Diagram courtesy of Hosokawa Micron Powder Systems, Summit, NJ.

Impact Mills[39]

Impact mills operate by reducing the particle size using a high-speed mechanical device that can impact with the particles or allow particle-to-particle impact. There are several types of impact mills used in the pharmaceutical industry, depending on the goal of the particle size reduction. Hammer mills and pin mills are two common types of impact mills, differing in the grinding head design used to achieve particle reduction.

Hammer Mills[39]

Hammer mills operate by rotating a series of hammers (typically four or more) hinged on a central rotating shaft enclosed in a cylindrical housing. The particle size reduction is related to the rotational speed, the size of the screens used for discharge, and the size of the particles undergoing impact. The screen size contributes to the residence time of the particles within the mill head. The smaller the screen mesh, the longer the particles will tend to stay within the high-intensity region of the mill. Hammer mills generally produce relatively narrow size distributions, with a minimum of fines due to the self-classification of the screens and the reduced fracture potential as the particles reduce in size. As the particles reduce in size, the inertia of particles impacting with the hammers decreases and the subsequent particle size reduction is decreased. Hammer mills are not suitable for low-temperature melting, sticky, or plastic-like materials, because mill fouling is common due to heat generation in the mill head. Variables that affect the particle size reduction operations include the feed rate to the mill head, the screen size and thickness, and the speed of the rotating hammers in the mill. The ingoing particle size and particle characteristics will also influence the milling capability. Hammer mills typically give particle sizes ranging from 15 to 50 μm.

Pin Mills

Pin mills use a series of pin breakers attached to disks instead of hammers in the rotating grinding head to achieve high-energy impacts between the mill and the particles (Figure 13.6). The configuration may consist of a rotor–stator design with one

Figure 13.6. Universal Pin Mill (UPZ). Photo courtesy of Hosokawa Micron Powder Systems, Summit, NJ.

moving set of pins and another stationary set or may have two opposing rotors with pins that further increase the differential speed of the pins. Pin milling is effective in reducing particle size typically down to 10 μm. Cooling of the rotating disks is useful in removing the heat generated in the mill. Advantages of pin mills include (a) the ability to achieve smaller particle size than hammer mills and (b) recent advances of using cryogenic conditions to improve heat removal. Disadvantages include a tendency to form wide particle size distributions, cleaning after milling, and mill fouling.

Air-Classifier Mills

Air-classifier mills (ACM) are impact mills that utilize an internal circulation system to produce fine grinds. Most ACM mills have variable speed controls that allow a range of particle sizes to be obtained. The air classification system has a classifier wheel to control the final particle size. The speed of the classifier and the airflow are adjustable to allow for changing particle size. The classifier allows only particles below a certain particle size to exit the mill. All other particles remain in the mill head to be further reduced. ACM mills can also be fitted with cooling devices to control temperature within the mill. Advantages of air classier mills include the ability to obtain narrower particle size distributions than standard impact mills. However, this technology is difficult to evaluate on small scale due to engineering issues associated with predictive scale-up.

Screening Mills

Screening mills reduce particles through the use of mechanical force to break particles when going through a screen. Screening mills are typically used for deagglomeration and deaggregation operations. This is useful when minimal primary particle breakage is required, but large chunks of particles formed during earlier processing (filtration and drying) need to be eliminated to ensure consistency and uniformity of the API feedstock.

Cutting Mills

Rotary cutting mills consist of a horizontal rotor with multiple knives spaced around the outside of the mill, turning at moderate to high speed in a housing fitted with additional stationary knives. The bottom of the mill has a screen attachment to control the residence time of the particles inside the mill head. Cutting mills are used when impact-, attrition-, or compression-type milling are not effective, especially for fibrous or resilient compounds. Cutting mills tend to give wide particle sizes ($\sim$150 μm to 5 μm) and are not suitable for most pharmaceutical applications.

Micron and Submicron Milling

A significant challenge for the development of many new drug candidates is low water solubility of the active pharmaceutical ingredients. One approach to increase bioavailability is to reduce the particle size of the API to the submicron size range. The milling techniques described above generally result in particle sizes much greater than the range necessary. In the past several years, a variety of methods have been developed to produce submicron particles. In the past few years, a variety of methods have been developed to produce submicron particles. Wet ball (or bead) milling (e.g., Dyno®), high-pressure homogenization, and cavitation milling have been developed by companies such as Baxter (Nanoedge®), Skyepharma, and Elan (Nanocrystal®).[40] One common issue with submicron particle size is a tendency of the particles to aggregate and agglomerate as the size decreases. This has been mitigated by the use of surfactants or polymers as stabilizers to prevent reaggregation of particles.

Wet ball milling (such as Dyno® milling) involves milling in the presence of grinding beads. The grinding stress causes a dispersion and/or break-up of the solids. The product to be ground is pumped via an inlet nozzle through the grinding chamber and exposed to the stress of the moving grinding beads. The transmitted energy is partially converted into heat that is carried off by the cooling liquid. The ground slurry of particles is separated out through an outlet.

Homogenization or cavitation milling involves sending a slurry of API in suspension flow through a narrow gap at very high velocity. The particles break due to the cavitation forces as the bubbles break. Issues with cavitation milling include the low solid loading and the need for pre-milling of particles by jet milling prior to homogenization milling.

Mill Selection. With an understanding of the most common mill types used in the pharmaceutical industry and the targeted particle size range, we can begin selecting the appropriate mill for the job. Two common approaches used to determine the type of mill for particle size reduction are empirical testing on lab or pilot milling equipment and predictive testing using single crystals (nanoindentation). Empirical testing on lab- or pilot-scale equipment provides an evaluation on miniaturized equipment using the same operating principle as the large-scale mills. Empirical testing typically uses rules of thumb for particle size ranges that may be obtained from the various mills. It is generally performed on samples ranging between 10 g and 1 kg. Dry milling operations are generally scalable using parameters (such as rotor tip speed, classifier speed, feed rate, or screen size) that are dependent on the features of the dry mill tested. The main advantage of test milling is the ability to use the material produced for formulation trials.

More recently, nanoindentation testing on single crystals of API have been shown to predict the type of mill suitable for particle size reduction.[41,42] Nanoindentation involves indenting single crystals of a compound to form cracks in the crystal face. The hardness (H) of the compound is determined by the depth of the indentation. The lengths of cracks formed during indentation are measured and used to determine the fracture toughness (K_c) of the compound. Use of these measures generates the brittleness index (BI):

$$\mathrm{BI} = \frac{H}{K_c} \tag{13.5}$$

Measurement of the BI provides a guide for the selection of mill choice. Materials with higher BI can use lower-energy mills such as hammer mills, while materials with low BI require higher-energy mills such as jet mills or air-classifier mills. The main advantage to nanoindentation testing is that only small quantity of API is required. This is especially useful for early drug candidates that have yet to be scaled-up to multi-kilogram quantities.

STRATEGY FOR SELECTION OF PARTICLE SIZE REDUCTION METHOD

The manufacture of bulk API is often on critical path to initiate drug safety studies. It is at this stage that selection of the particle size reduction method be considered to ensure the delivery of consistent API to efficiently advance the program to clinical study. Typical dry milling operations can be material-intensive, with significant yield losses anticipated (typically greater than 5%) associated with dusts and equipment coating. Yield losses in the final step of an early compound can be cost prohibitive.

For a standard oral tablet development program with a typical dose profile, high-shear wet milling is an attractive default technique for particle size reduction. Advantages include the following: Yield losses associated with dry milling are eliminated, the obtainable particle sizes do not require specialized filtration equipment, and the ease of lab scale wet-milling evaluation.

Where very small particle size range is required, as in formulation of suspensions, inhalations, low-solubility or high-potency compounds, micronization is

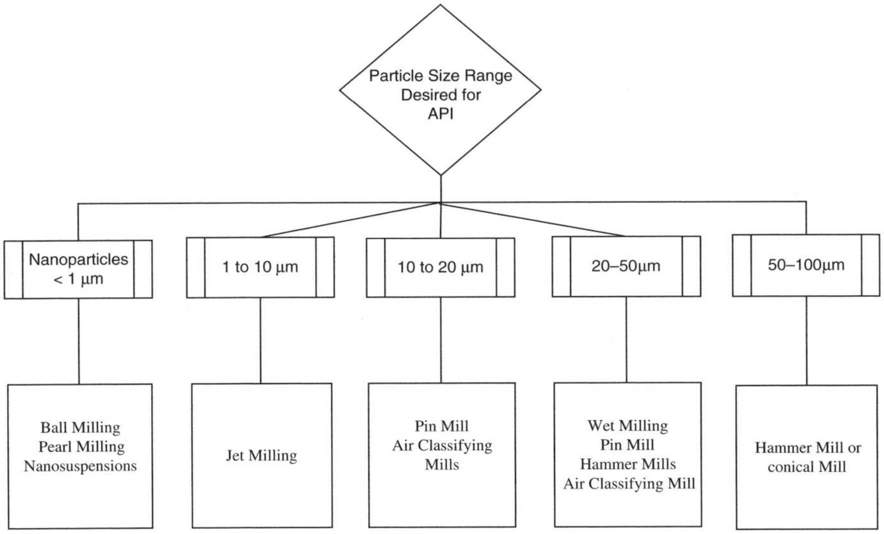

Figure 13.7. General overview of mill options based on particle size required.

often the easiest and fastest approach: Typical sizes are in the 2- to 10-μm range and proof of concept can be evaluated. If further understanding is needed, more comprehensive analysis and evaluation can be completed. As with any of the small particle size range milling methods, the input particle size impacts the final milling. Even if the particle size needed ultimately requires jet milling, the use of high-shear wet milling in the final isolation step can provide a more uniform and consistent particle size into the jet mill to ensure more consistent and more efficient micronization. Figure 13.7 provides mill selections based on particle size needs.

Later in the development timeline, designed crystallization to achieve targeted size is generally best suited as more material is typically available for development studies. It is important to remember that the cost of developing crystallization for particle size will likely be higher than for a milling operation. One may ask whether it is worth designing the crystallization for particle size at all. The long-term benefits (cost savings, less unit operations, etc.) make particle engineering worth the effort. Milling is a cost-intensive unit operation that can be potentially eliminated through careful control of the API crystallization step. In addition, many compounds undergo undesired changes during the milling operation such as form changes—including loss of hydrate or formation of amorphous content, agglomeration due to static charge, and so on—that direct crystallization to desired size could eliminate.

In addition, crystallization can sometimes achieve particle sizes that typical milling cannot readily generate. Supercritical fluid crystallization may be able to obtain micron and submicron crystals where typical micronization cannot. Direct crystallization methods to produce small particle sizes can sometimes result in more challenging downstream processing (filtration/drying) which needs to be evaluated on a case-by-case basis.

CONCLUSIONS

The particle size required for formulation will generally dictate what milling options are feasible. Early in process development, milling will typically be the most rapid method for achieving particle size. Wet milling is recommended over dry milling whenever possible due to the potential safety and cost benefits. Later in process development, particle-engineering approaches should be considered in an effort to obtain the desired particle size directly during the crystallization step. The closer the direct crystallization approach can be designed to obtain the desired particle size, the less the energy that will be required for subsequent particle size reduction, thus saving time and ultimately increasing productivity.

REFERENCES AND NOTES

1. Lieberman, H. A., Lachman, L., and Schwartz, J. B. *Pharmaceutical Dosage Forms: Tablets*, Vol. 1, 2nd edition, Marcel Dekker, New York, 1989.
2. Hegasy, A., and Ramsch, K. D. US Patent Number 5,264,446.
3. Paul, H. E., Hayes, K. J., Paul, M. F., and Borgmann, A. R. *J. Pharm. Sci* **1967**, *56*, 882–885.
4. Ridolfo, A. S., Thompkins, L., Bechtol, L. D., and Carmichael, R. H. *J. Pharm Sci.* **1979**, *68*, 850–852.
5. Jounela, A. J., Pentikainen, P. J., and Sothmann, A. *Eur. J. Clin. Pharmacol.* **1979**, *8*, 365–370.
6. Kabasakalian, P., Katz, M., Rosendrantz, B., and Townley, E. *J. Pharm. Sci.* **1970**, *59*, 595–600.
7. Noyes, A. A., and Whitney W. R. *J. Am. Chem. Soc.* **1897**, *19*, 930–934.
8. Aulton, M. *Pharmaceutics—The Science of Dosage Form Design*, Churchill Livingstone, New York, 2002.
9. Hintz, R. J., and Johnson, K. C. *Int. J. Pharm.* **1989**, *51*, 9–17.
10. Johnson, K. *Pharm. Rev.* **1996**, *13*, *12*, 1795–1798.
11. Johnson, K. *Drug Dev. Ind. Pharmacy* (29), 8 **2003** 833–842.
12. Kaye, B. H. *Powder Mixing*, Chapman and Hall, London, 1997.
13. Myerson, A. S. *Handbook of Industrial Crystallization*, Butterworth-Heinemann, Boston, 1993.
14. Mullin, J. W. *Crystallization*, Butterworth, Boston, 3rd edition, 1997.
15. Fujiwara, M., Nagy, Z. K., Chew, J. W., and Braatz, R. D. *J. Proc. Control* **2005**, *15*, 493–504.
16. *Handbook of Industrial Crystallization*, Myerson, A. S., Ed., Butterworth, Boston, 2002, Chapter 1.
17. Braatz, R. D, *Experimental Design and Control Strategies for Batch Crystallization*, Association for Crystallization Technology Conference, Mystic, CT, 2003.
18. Fujiwara, M., Chow, P. S., Ma, D. L., and Braatz, R. D. *Crystal Growth Des.* **2002**, *2*, 363–370.

19. Gron, H., Borissova, A., and Roberts, K. J. *Ind. Eng. Chem. Res.* **2003**, *42*, 198–206.
20. Midler, M., Paul, E. L., Whittington, E. F., Futran, M., Liu, P. D., Hsu, J., and Pan, S. H. US Patent 5,314,506 (1994).
21. Mahajan, A. J., and Kirwan, D. J., *AIChE J.* **1996**, *42*(7), 1801–1814.
22. am Ende, D. J., Crawford, T. C., and Weston, N. P. US Patent, 6,558,435 (2003).
23. am Ende, D. J., and Brenek, S. J. *Am. Pharm. Rev.* **2004** *7*(3).
24. Rivera, T., and Randolph, A. D. *Ind. Eng. Chem. Process Des. Dev.* **1978**, *17*(2), 182–188.
25. Johnson, B. K., and Prud'homme, R. K. *AIChE J.* **2003**, *49*(9), 2264.
26. Bennett, R. C. Crystallizer Selection and Design. *Handbook of Industrial Crystallization*, Butterworth-Heinemann, **1993**, p. 119.
27. Jung, J., and Perrut, M. *Supercritical Fluids* **2001**, *20*(3), 179–219.
28. York, P. *Pharm. Sci. Technol. Today* **1999**, *2*(11), 430–440.
29. *IKA Process Equipment—IKA Works High Shear Inline Mixers Catalogue*, IKA Works, Inc.
30. Lee, I., Variankaval, N., Lindermann, C., and Starbuck, C. *Am. Pharm. Rev.* **2004**, *7*(5), 120–123.
31. *Silverson Laboratory Mixers—A Closer Look*, Silverson Inc., Weblink:www.silverson.com.
32. Rose, P. R., Brenek, S. J., and am Ende, D. J. Association for CrystallizationTechnology, North Park, IL, October 2004.
33. Lasentec, Focused Beam Reflectance Probe, Real Time Analytics, Mettler Toledo AutoChem, Millersville, MD.
34. Parrott, E. L. *J. Pharm. Sci.* **1974**, *63*, 813–829.
35. Guidance for Industry—SUPAC-IR/MR Immediate Release and Modified Release Solid Oral Dosage Forms: Manufacturing Equipment Addendum, US Department of Health and Human Services, Food and Drug Administration, Center for Drug Evaluation and Research (CDER), January 1999 (CMC 9, Revision 1).
36. *Perry's Chemical Engineers' Handbook*, 7th edition, Green, D. W., Editor, McGraw-Hill, New York, 1997, Chapter 20.
37. Midoux, N., Hosek, P., Pailleres, L., and Authelin, J. R. *Powder Technol.* **1999**, *104*, 113–120.
38. Clement, S., and Purutyan, H. *CEP Mag.* **2002**, 50–54.
39. Hajratwala, B. R. *J. Pharm. Sci.* **1982**, *71*, 188–190.
40. Hite, M., Turner, S., and Federici, C. *Pharm. Manuf. Packing Sourcer* **2003**, *2*, 000–000.
41. Taylor, L. J., Papadopoulos, D. G., Dunn, P. J., Bentham, A. C., Mitchell, J. C., and Snowden, M. J. *Powder Technol.* **2004**, 143–144, 179–185.
42. Taylor, L. J., Papadopoulos, D. G., Dunn, P. J., Bentham, A. C., Dawson, N. J., and Mitchell, J. C., Snowden, M. J. *Org. Process. Res. Dev.* **2004**, *8*, 674–679.

14

CHALLENGES IN EARLY FORMULATION: TURNING DRUG SUBSTANCE INTO DRUG PRODUCT

MARK H. KLEINMAN AND BEEAH LEE
Pharmaceutical Development, GlaxoSmithKline, King of Prussia, PA 19406

INTRODUCTION

Okay. So you have now synthesized your drug substance, obtained the most suitable salt version, and chosen the right polymorphic form. How do you get the drug into the patient in the timelines presented by the project manager? For the chemist, the answer is usually to head over to Pharmaceutical Development and hand over the material. The goal of this chapter is to remove the black box that surrounds preformulation and formulation and open up the subject for chemists to afford easier and more fruitful discussions with formulator colleagues. As such, the aim is to provide a concise summary of choices and decisions involved in early development formulation. The primary focus is on physicochemical characterization, the ADME (absorption, distribution, metabolism, and excretion) model, and the rationale for preformulation and formulation for first dosing in humans.

Why Formulate?

Giving a patient the active pharmaceutical ingredient as a powder in a bottle is the simplest formulation. It provides a quick route to the clinic, but does not help

Fundamentals of Early Clinical Drug Development: From Synthesis Design to Formulation, edited by Ahmed F. Abdel-Magid and Stéphane Caron

overcome any known roadblocks that limit the utility of the drug. Formulation is one of the most important enabling tools in the pharmaceutical industry. Formulation can help enhance the stability, increase the solubility of the drug in the gut, increase permeability, change the rate of dissolution, alter the half-life of the drug *in vivo,* reduce potential toxicity, and yield greater patient adherence to drug therapy.

As the drug candidate moves closer to candidate selection—that is, the time point at which a company decides to move a drug from discovery into development—the preformulation scientists carry out the physicochemical characterization of the drug substance. While the scientists assay the solubility in a variety of solvents and aqueous buffer systems, the solution and solid stability, the hygroscopicity, and other solid-state parameters such as crystallinity and crystal habit, they are creating a reference database of knowledge for the formulator or formulation team. As more information about a drug candidate is discovered, the type of drug that is evolving becomes clearer. For example, the preformulators may discover that a synthesized drug candidate with an amine moiety (free base) may be quite lipophilic ($\log P = 5$), have a $pK_a = 6$, and be poorly soluble in aqueous solutions where $pH > pK_a$. Looking at just this small subset of information, a knowledgeable formulator may immediately infer that the drug in question may be potentially soluble in the stomach, be aware of the possibility that the drug may precipitate out in intestinal fluid, and be highly permeable through cell membranes once dissolved. This is just a small piece of the puzzle, but with the additional information presented in the subsequent sections discussed later, the formulator can choose the proper excipients and the correct drug delivery tool for the job at hand.

The goal of formulation in early development is to obtain a formulation that is simple, flexible, and fit for its intended purpose. As many candidates coming from discovery don't make it through to Phase II, let alone to the market, it is vital for the formulation to be easily developed, easily sourced (for the excipients), and inexpensive. Furthermore, the formulation should be flexible. This criterion is important because timelines for clinical studies are shrinking and the formulation created for Phase I may well be used as a starting point for Phase II or later clinical trials. During Phase I, pharmacokinetic measurements are carried out, so different doses may be used in the clinic. As a consequence, the formulation should be adaptable to these different amounts of API in the drug product. The developed drug product should also be created with its purpose in mind. For instance, if the drug is targeted for local action in the lungs, then it would likely be appropriate to select an inhaled dosage form so that the onset of action and the efficacy of the drug can be maximized.

Drug Substance to Drug Product

The key question that needs to be addressed before starting to think about the complexities of formulation is, Why is a drug substance not typically dosed on its own? In this example, we will assume that a solid oral dosage form is required since patients typically request this form of drug, and patient adherence to dosing regimens is strongest with oral dosage forms. In order to gain a fuller understanding of these characteristics, one needs to understand how an API in a solid oral dosage form goes from being swallowed to acting at the target site of action. We do not aim to produce the detailed picture of the gastrointestinal tract, but rather summarize the basics. A fuller presentation is considered in other sources.[1] Once swallowed, the "pill" (capsule, tablet, or other solid oral dosage form) passes from the mouth and goes down the esophagus and into the stomach. During this part of the journey, it encounters a change of pH ranging from 6 (in the mouth) to 1.5 (in the stomach).[2,3] Typically, the drug dissolves in the acidic environment before being passed into the small intestine, where the majority of drug absorption takes place, and then into the large intestine. By the time the API gets to the colon, the pH may already be >7. From the perspective of the example of dosed solid active pharmaceutical ingredient powder alone, the API dissolves based largely on its solubility. The speed of dissolution is also important becasue there are specific residence times for all parts of the GI tract.[4] Even if the drug substance powder by itself had an appropriate dissolution rate and completely solubilized in the requisite section of the GI tract, in order for it to be the perfect drug it would need to have the appropriate permeability, so that it would be able to pass through the cells that line the small intestine. This is done either in a transcellular manner, paracellularly by passive diffusion, transcellularly via carrier-mediated transport, or by endocytosis and then must find its way into the bloodstream on the other side of the cell. Once in the bloodstream, the drug then needs to be specific for its target, so that it can act at the expected site of action. Throughout its entire journey, the drug should also remain stable and not degrade. While this view may simplistic, it captures the importance of solubility, dissolution, permeability, stability, and transport. It is only the exception to the rule that drug substance powder offers the perfect characteristics to be delivered as a drug product on its own. As a consequence, formulation is essential to convert drug substance into a dosable drug product.

Formulating for the Patient

The solid-state properties also interrelate with basic pharmacokinetic properties. When pharmacokineticists look at the results from an early clinical study, they are looking for key parameters such as bioavailability, $C_{\max}$, $t_{\max}$, and $t_{1/2}$ (not to mention others such as clearance, protein binding, and volume of distribution). While all of the parameters are important to the formulator, we would like to pay special attention to the former parameters. Bioavailability (F) is defined as the fraction of an administered drug that reaches the systemic circulation intact. Absolute bioavailability is the bioavailability obtained using a dose that enters the bloodstream directly—for example, intravenous (i.v.). Relative bioavailability is a key parameter for the formulators because it compares the amount of drug that reaches the systemic circulation intact by two different routes of administration formulations, where one is typically i.v. and assumed to be 100% absorbed and the other, for example, is oral. Generally, high relative bioavailabilities are sought, so the number of molecules that reach the site of action can be maximized. This increases efficacy and keeps sourcing costs low. However, because there is usually an issue with the drug substance—such as it being poorly soluble, dissolving slowly, unstable metabolically, or being poorly permeable—the absolute bioavailability of an oral dose is usually lower than that obtained by i.v. administration. (This is indicated by a relative bioavailability $<100\%$, where the value is the bioavailability of the oral dose relative to the bioavailability of the i.v. dose multiplied by 100%.) Adding excipients to the mix, the formulator can address one or more of the issues mentioned above, thereby improving the relative bioavailability. The proper excipients can also alter the maximum concentration of API in plasma ($C_{\max}$), the time at which the maximum concentration of API in plasma is reached ($t_{\max}$), and the elimination half-life ($t_{1/2}$). In some cases, this is essential because if the release of a drug is too fast, a high, toxic concentration of the drug may be released, which is obviously undesirable. Excipients not only can alter the physical and chemical properties of the drug substance, but, by design, may also alter the release rate of the API in modified release formulations. In the case where slower release is desired, a pulsed or sustained release formulation may be developed. In some cases, the pharmacokinetics is the underpinning of the required pharmacodynamic response. For example, if a narrow spike in hormone levels is desired, the drug is dosed so that $t_{\max}$ would occur at a specific interval with a narrow time window. Because drug substance by itself dissolves at its intrinsic rate in a specified environment, excipients may be used so that dissolution and solubilization are modified to occur in the desired narrow timeframe. In this way, the drug reaches its target in a set, timely fashion.

The rationale for creating a formulation instead of dosing pure drug substance should also include understanding the requirements of the patient. The dosing administration should be convenient so that patients adhere to their prescribed dosing regimen. The patient should get the required dose at the specified time. In early clinical trials, the formulator should also acknowledge the intent of the clinical trial and understand the pharmacological response required by the clinician in order to properly choose the type of formulation and excipients. For example, will the clinical study be based on a range of doses (flexible formulation) or will it be using only a single dose? Furthermore, how often will the doses be given? Will the patient adhere to this type of regimen with the selected dosage form? Because some excipients cause toxicity on their own, it is vital for a formulator to understand not only how much of the API will be dosed, but also how much of each excipients can be administered.

Since there are many parameters and variables that may require alteration, the formulator is challenged to understand all of them as well as the science and art in their craft. Choosing the proper excipients to obtain the proper clinical response, the underlying pharmacokinetics, and the physicochemical parameters of the formulation is a challenge, but one that formulators are ready for. In the next section, we will discuss the relevant physicochemical characteristics required to understand the production of a successful formulation.

PHYSICAL AND CHEMICAL PROPERTIES OF FORMULATIONS

Physicochemical characterization of drug candidates is a task usually undertaken before formulation to aid in the formulation design. Typically, all drug substance batches for candidate drugs are subjected to a series of tests in order to determine if the drug is worth developing further. The physicochemical characteristics of the drugs typically determine how difficult it will be to develop the drug into a drug product. Though efficacy and toxicity are the major determinants of a drug being selected for development, the physicochemical properties usually determine the degree of difficulty for developing a successful formulation.

Oral dosage forms are typically the first intent and the most common formulations marketed today. The investigation of the solid-state properties of candidates is thus very important. The physical and chemical properties exhibited by the drug substance in the solid state are vital to understand because they play a large role in the production process of tablets or capsules (or other solid oral dosage forms). Powder properties such as cohesion, powder flow, micromeritics (the science of small particles in which particles are defined as any unit of matter having defined physical dimensions), yield strength, hygroscopicity, and surface area are some of the physicochemical properties that are studied while developing a formulation for a solid dosage form. Physical stability is a factor that needs to be taken into consideration because changes in the physical form of the drug substance can lead to differences in physical properties.[5] While the physical properties are important, it is also vital to consider the solid-state chemistry of the API in the context of the formulation. Though the reactivity and rates of reactions in the solid state are typically much slower than in solution or gas phases, chemical transformations need to be monitored carefully. The drug substance should be chemically stable since instability may cause safety and toxicological problems and may even factor into the amount of drug that may be administered to the patient. The presence of chemical stability also ensures that no toxic degradation product(s) are produced that may harm the patient.

Crystalline and Amorphous States

Solid-state pharmaceuticals can exist in the crystalline or amorphous forms. Crystalline forms are characteristically more stable than amorphous forms. Amorphous forms have the risk of changing in both a physical and chemical sense. As a consequence, it is imperative that pharmacokinetic (PK) testing is conducted on different forms and that comparisons between forms (different crystalline and/or amorphous) take place. In the case where amorphous material is used in this PK testing, conclusions should be qualified based on the solid form used for testing because this difference may affect the results.

The Crystalline State. Most drugs are present in crystalline forms. These forms are held together by molecular forces, and the molecules are arranged in an ordered manner. Crystals could be polymorphs or pseudo-polymorphs (solvates). Polymorphs are crystals that have the same chemical composition but different internal structures resulting in differences in properties such as melting points, spectral properties (typically in the infrared), X-ray diffraction or more formulation relevant properties such as solubility, density, hardness, and crystal shape. Pseudo-polymorphs (solvates) are crystals that have different chemical composition whereby solvent molecules occupy regular stoichiometric positions in the crystal structure.

The Amorphous State. In the case of amorphous forms, there is no long-range order and the disordered region constitutes the entire solid. Amorphous forms are thermodynamically metastable with respect to the crystalline form, and thus there is always the potential for an amorphous form to transform into the more stable crystalline form. Above the glass transition temperature, the amorphous solid is in the rubbery state; below the glass transition temperature, the amorphous solid is in the glassy state. The difference between these two states is that in the glassy state the molecules are configurationally frozen but in the rubbery state they are not and thus they will flow.

Physical Properties of Solids as Applicable to Formulation. At the molecular level, some of the key properties of a compound that depend on structural differences include density, hardness, cleavage, solubility, hygroscopicity, optical, electrical and thermoanalytical properties, solid-state reactivity, and physical and chemical stability. In terms of oral solid dosage formulation, properties at the macroscopic level may be more important to consider since virtually all of the formulations are manufactured from powders (API and excipients). Size and shape distribution control as well as solid-state form have the greatest influence on formulation related activities such as yield, filtration, washing, drying, milling, mixing, tableting, flowability, dissolution, suspension formulation, and lyophilization.

The properties described above need to be taken into account, but not all of them need to be measured and characterized *a priori* for formulation. Most tests except for the few basic ones (such as tests for crystallinity, stability, and hygroscopicity) can be performed as needed or required during the formulation design and development stage once the properties of the candidate molecule or the requirements for the dosage form are better defined.

For a minimum set of characterization experiments to help with formulation development in early clinical drug development, it is usually necessary to perform the physicochemical characterization of the compound and establish the drug substance stability. Once the baseline characteristics are determined for the drug substance, further studies investigating both the compatibility with excipients and the compaction profile are needed. In this manner, one can start with the most flexible, affordable, and convenient dosage form—tablet or capsule. Furthermore, after the tablet or capsule is made, it is typically necessary to carry out the characterization of the drug product in order to determine if it meets the criteria of an acceptable formulation for the purpose of being developed. This includes tests regarding the integrity, content uniformity, dissolution profile, and the stability of the drug product.

Alternate dosage forms such as parenterals, inhaled, and topicals require more in-depth studies or characterization of specific physicochemical properties. For example, for parenteral dosage forms the physicochemical properties of the solution state may be more relevant than the solid state as they are formulated in solutions or nano-emulsions. For inhaled dosage forms there is more emphasis on particle size reduction of powders because there is an ideal range to maximize lung deposition. For topical delivery the rheological properties of the dosage form are very important, because it will determine the diffusion properties of the molecule as well as the cosmetic or appearance of the gel or cream to be applied.

FROM PHYSICOCHEMICAL PROPERTIES TO *IN VIVO* BIOAVAILABILITY

Once the preformulation scientists have investigated the physical and chemical properties of the new molecular entity, the next step is to prepare a formulation for animal model and human studies. While physical parameters of the solid state allow us to understand the solubility of the API in different environments and how fast it will dissolve, the eventual fate of the API *in vivo* is still unknown at this point in development. Thus, obtaining an understanding of ADME (absorption, distribution, metabolism, and excretion) principles is the next step toward a successful formulation. In this section, we describe the fundamentals of ADME and explain how the rationale for formulation and choice of formulation are linked to *in vivo* pharmokinetics. Because formulation is an enabling activity, knowledge of the mechanism that you are trying to bypass or alter is an essential piece of the puzzle and frequently determines the course of action. Drugs may not be completely bioavailable due to a lack of absorption, poor solubility, or lack of stability. Below, the example of a tablet formulation is presented to demonstrate the principles involved in absorption.

Absorption

Absorption is the process by which a drug proceeds from the site of administration to the systemic circulation. From a formulation and solid oral dosage form perspective, this mechanism will involve the disintegration of the tablet with concomitant liberation (dissolution) and solubilization of the API followed by the passage of the API from the gastrointestinal system into the systemic circulation. As a consequence, from a pharmaceutical development view, absorption consists of three main mechanistic parameters: dissolution, solubility, and permeability. While this is a simplification, it does serve as a good summary model for formulation in early clinical development. The major site of absorption is the small intestine. Therefore, anything that affects stomach emptying may influence the rate of absorption.

Dissolution. The first step after ingestion of a solid oral dosage form such as a tablet is the disintegration of the tablet. As the tablet disintegrates, the API is liberated and then dissolved in the surrounding solution. The kinetic aspect that corresponds to the rate of solubilization is termed *dissolution* and can be written using modified Diffusion Layer Theory as

$$\frac{dC}{dt} = k\frac{DS}{Vh}(Cs - Ct)$$

where dC/dt is the dissolution rate, k is the intrinsic dissolution rate constant, D is the diffusion coefficient, V is the volume of dissolution medium, h is the thickness of film layer, S is the surface area, Cs is the maximum solubility, and Ct is the concentration at time t.

Dissolution is a vital parameter to control when formulating and is even one of basic quality control tests to study batch-to-batch reproducibility after tablet processing. The underlying importance of dissolution in terms of the formulation is that it may be the kinetic rate-limiting step that defines the entire absorption process:

$$\text{Dissolution} \rightarrow \text{Solubility} \rightarrow \text{Permeability} \rightarrow \text{Absorption}$$

Consequently, if the rate of dissolution is slow, absorption and relative bioavailability may be limited. Therefore, if poor relative bioavailability is seen in preliminary PK studies, dissolution of the formulation in species-specific relevant solutions should be closely investigated.

Dissolution—Potential Solutions for Solid Oral Dosage Forms. If poor dissolution is causing low bioavailability, the normal implication is that the drug is not dissolving fast enough. Understanding the relevant parameters in the dissolution equation will help the formulator recognize the key parameters that influence dissolution such as wettability (the surface tension at the interface and the contact angle), pH-solubility, particle size, and stability of the crystalline state. The following all reduce dissolution rates: poor wettability (high surface tension or large contact angle), low solubility, large particle size, and very stable crystalline solids with high melting points. Understanding these effects is a key step toward developing an enabling formulation.

Once the mechanism that is decreasing the rate of dissolution is understood, development of formulations becomes a more facile process. A polymorph screen may yield a different form with a more acceptable dissolution profile. This new polymorph may be a solvate such as a hydrate. In many cases, dissolution rates of the solvate is faster than for the anhydrous material.[6] However, in reality, the formulator will likely find it difficult to get the project team to change polymorphs simply due to poor dissolution because new chemical manufacturing processes will be required for the conversion of polymorphs. As a consequence, evaluating the particle size and its effects on dissolution is also common and may provide a path forward for the polymorph under current consideration. Based upon the dissolution equation presented above, one can see that a reduction in particle size (an increase in overall surface area) should enhance the dissolution rate. While there are some limitations to reducing particle size (such as agglomeration or entrapped air), micronization is a key tool for the formulator. There are several gram scale micronization approaches available to today's formulator, including but not limited to mills (bead, jet, or ball) and microfluidizers. In the ball mills, small stainless steel (or other unreactive balls) are placed in a heavy-duty (usually stainless steel) enclosure and are shaken (side-to-side or in a centrifuge). Grinding of the particles to the appropriate size occurs over time via physical crushing by the balls. Care needs to be taken to ensure a homogeneous particle size reduction and a suitable endpoint. For the microfluidizer, a stream of air, nitrogen, or other (preferably) unreactive gas is used at high velocity to collide with the particles. In all of the processes detailed here, heat is an important ingredient to control as friction is created between particles and during the flow of the materials. The additional heat may change the milled drug substance either chemically (degradation) or physically (different polymorph); consequently, these potential changes should be investigated.

Issues derived from local pH, local solubility, and wettability may also be handled by the formulator. Wettability is a surface parameter that describes the affinity of a drug particle for the dissolving medium. As mentioned earlier, wettability depends upon both surface tension and contact angle (measured as the angle formed between the edge of a liquid droplet and the surface). Surfactants or surface active agents are substances that absorb to surfaces or interfaces to reduce surface or interfacial tension. They may be used as wetting agents, detergents, or emulsifying agents. Different types of surfactants such as sodium dodecyl sulfate (sodium lauryl sulfate), water-soluble lipophilic polymers [including polysorbates and polyoxyethylene sorbitan monostearates (TWEENs)], sorbitan monopalmitate, benzalkonium chloride, nonoxynol-10 (*p*-nonylphenyl polyoxyethylene ether), and cholate and taurocholate bile salts all solve wetting problems by decreasing contact angles; however, the extent of the interaction depends upon the drug being solubilized.[6–8]

Dissolution rates may also be affected by the local solubility. This occurs because as the drug particle dissolves, there is a local increase in the concentration of the drug. As a consequence, if the dissolution rate is increased, local solubility may begin to be the limiting factor due to thermodynamic equilibrium limits. The surrounding and local (micro-) pH values are also important factors to consider. In the case of weakly basic drugs, dissolution rates are enhanced in the stomach (low pH). While little can be done about the surrounding medium in the GI tract, the local solubility and pH may be modified by the excipient or choice of salt counterion. As a case in point, a weak base with high lipophilicity may be rendered more soluble with a more hydrophilic counterion[9] or excipient. Thus a change in salt version or excipients can help locally solubilize the drug. Excipients with ionizable groups play a role in altering the local pH. These types of excipients are called buffering agents and include compounds such as phosphate salts (K_2HPO_4, KH_2PO_4), sodium acetate, and sodium citrate. These materials are used to resist changes in pH upon dilution or addition of acid or alkali and may even be incorporated into a solid oral dosage form to help minimize acid- or base-catalyzed reactions.[7,8] Solubility is an issue for both dissolution as well as bioavailability of the drug and is further discussed in the following section.

Solubility. Solubility is defined as the equilibrium point where undissolved solid drug substance is in equilibrium with the drug substance in solution. Solubility in the stomach is vital for many drugs since the next section of the GI tract for the API to traverse is the small intestine. It is in the small intestine where most of the absorption into the systemic circulation typically takes place. As a consequence, the key to having a tablet formulation perform well *in vivo* is to ensure that the drug is solubilized readily in the stomach and is still soluble enough at the pH levels seen in the small intestine so that the API is readily permeable. If the API is not soluble in the small intestine, much of the drug may not be absorbed and instead will just pass through the entire GI tract and be excreted. This may result in low bioavailability because of the missed window of absorption in the high surface area of the small intestine. Though there is still some absorption in the large intestine, the drug will in large extent be eliminated from the body. Furthermore, from a clinical or pharmacodynamic perspective, if the drug is slightly soluble, the drug may elicit a weak response over a broad time period, rather than a desired sharp effect. Even if the tablet disintegrates fast and dissolution of the drug occurs readily, the rate-limiting step of absorption could be solubility. In this case, the bioavailability suffers due to the precipitation of the API in the gut caused by an initial high concentration of dissolved drug in the stomach.

Solubility—Potential Solutions for Solid Oral Dosage Forms. After conducting *in vitro* solubility assays and determining that the likely cause for poor bioavailability is low solubility, there are numerous options available to the formulator. The solubility may be increased in the appropriate GI compartment (we will use the stomach for simplicity) by modifying either the drug substance or the formulation. Modifications to the drug substance include a change of salt version or polymorph, whereas changes to the formulation may include altering the pH of the formulation or altering the lipophilicity (creating a local hydrophobic neighborhood).

Changing polymorph or salt version is a possible way to alter the solubility. Solubility is a thermodynamic parameter that can be related to the energy required to break apart the crystal lattice. Generally, as the stability of the solid increases (higher melting point), the solubility decreases. So, even though the polymorph or salt version contains the exact same API, different solubilities are possible. In the case of poor solubility described above, a less stable polymorph (but still stable enough to manufacture) or a more soluble salt may be investigated. In some cases it is advantageous to have salts that are less soluble. In these cases, more stable salts could be chosen with the objective to mask taste, slow down dissolution, or increase the chemical stability.

While there are phase differences between solid oral dosage forms and other dosage forms such as parenteral liquid formulations, the approaches taken to resolve any solubility issues are similar. One could choose from pH manipulation, surfactants, complexation agents, co-solvents, and nonaqueous vehicles.[9,10] These approaches may also be undertaken in oral dosage forms such as capsules and softgels.

Co-solvents may be added to the formulation to help solubilize the API. Common co-solvents in capsules and softgel formulations include glycerine, ethanol, polyethylene glycol, and *N,N*-dimethylacetamide.[11–14] In addition, semisolids and liquids may be formulated in capsules or softgels with the addition of complexation agents or other vehicles. For example, cyclodextrins (CD), cyclic oligosaccharides in the shape of a bucket with holes at both ends, come in a variety of sizes (6 (α), 7 (β), or 8 (γ) sugars). A number of references have been published describing the use of cyclodextrins in pharmaceutical products and thus will not be detailed here.[15,16] The unmodified cyclodextrin possesses hydroxyl groups at the top end of the bucket and hence is soluble in aqueous media. The internal surface of the cyclodextrin is hydrophobic and hence size-permitting; a thermodynamically stable equilibrium complex may be formed with the lipophilic API inside the CD. So, essentially, small drug molecules are partially or fully encapsulated within the cyclodextrin bucket. Note that cyclodextrin technology has advanced and sulfonated, and hydroxypropylated versions are currently marketed for use with pharmaceuticals.[17,18]

Permeability. Permeability is a measure of the rate of drug transport from one side of the membrane to the other and is typically measured in cm/sec. Membranes within the GI tract, as well as elsewhere in the body, act as barriers to most drugs. The membranes, which are constructed of hydrophilic head groups and hydrophobic tails, permit the absorption of lipid soluble substances by passive diffusion. If a drug is lipid-insoluble, it may have significant difficulty crossing the membrane. The permeability (P) is related to the diffusion coefficient within the membrane (D_{membrane}, cm^2/sec), the membrane thickness (Δx), and the partition coefficient (K) as shown by the following passive transport equation:

$$P = D_{\text{membrane}} \times \frac{K}{\Delta x}$$

The partition coefficient is an equilibrium constant based upon the equilibrium concentrations found on the aqueous side and within the membrane. For lipophilic compounds (higher log P), this value will be larger and a greater concentration of the API will be found within the membrane.

Based on the simplified equation above, one notes that the membrane distance, Δx, cannot be modified by excipients. At this stage of development, the formulator cannot typically just change a functional group on the API to get increased hydrophilicity, because this means introducing a new chemical entity that may have different activity or potency and may introduce intellectual property issues, and this also means the restart and repeat of the development activities. A formulator may, however, recommend these changes to the discovery chemists to consider in designing a next generation or a backup drug in the series. Thus for a given API, the equilibrium constant K cannot be varied unless a strong complexation agent is added. Since these strong complexation agents increase the mass size, permeability is often decreased.[19] Thus, most excipients geared toward improving permeability attempt to change the API diffusion coefficient within the membrane or alter active transport mechanisms (and thus bypass the passive transport described above).

Permeability—Potential Solutions for Solid Oral Dosage Forms. Excipients that enhance permeability are called *bioenhancers*. Because overall permeability is a function of passive diffusive transport, as well as active and passive/facilitated transporters, it is important to understand the effects of excipients on each mechanism. For passive diffusion, use of bioenhancing excipients in solid oral dosage forms is usually required when the bioavailability is low (possibly) due to poor intestinal permeability. Examples of these excipients in solid oral dosage forms include conventional excipients such as Tween-80 (Polysorbate-80) and sodium lauryl sulfate (sodium dodecyl sulfate) or newer excipients that are typically mixtures Labrafil (PEGylated oils) or Labrasol (PEGylated caprylic/capric glycerides). The mechanism of action of these different bioenhancers are somewhat different, but the mechanism are always based upon their chemical structure and noncovalent interactions. For example, nonionic surfactants or bile salts can disrupt the membrane by forming mixed micelles with membrane lipids or even alter the composition of the membrane. Fatty acids and their derivatives increase the fluidity of the membrane, thereby making permeation much easier and faster for the drug. In addition, other cationic excipients can be added to interact with the negatively charged lipids at the surface of the membrane to afford a more hydrophobic environment for the drug as it enters the membrane. A nonexhaustive listing is shown in Table 14.1.[20,21]

TABLE 14.1. Examples of Bioenhancing Excipients for Oral Dosage Forms (Solids, Semisolids, and Filled Capsules)

Excipient Class	Examples of Excipient	Mechanism of Action
Anionic surfactants	Sodium lauryl sulfate	
Cationic	Quaternary ammonium salts (e.g., cetylpyridinium chloride)	
Nonionic	Poloxamer, Brij, Span, Myrj, Tween	Perturbation of intercellular lipids, protein domain integrity
Bile salts	Sodium glycodeoxycholate, sodium glycocholate, sodium taurodeoxycholate, sodium taurocholate, sodium deoxycholate	
Fatty acids and derivatives	Oleic acid, caprylic acid, octanoic acid, decanoic acid, palmitoyl-DL-carnitine, *N*-myristoyl-L-propyl-L-propyl-glycinate	Increase fluidity of phospholipid Domains
Cyclodextrins	α-, β-, γ-Cyclodextrin; methylated β-cyclodextrins	Inclusion of membrane compounds
Cationic compounds	Chitosan, trimethyl chitosan, Poly-L-arginine, L-lysine	Ionic interaction with negative charge on the membrane surface

Source: Adapted from references 20 and 21.

Permeability is often evaluated by *in vitro* Caco-2 cell screens. Caco-2 cell monolayer have an epithelial membrane barrier function which is similar to that found in the human colon,[22] and it has carrier-mediated systems similar to the small intestine (such as bile acid transporter, glucose carriers, vitamin B_{12}, and the dipeptide carrier).[23] Using this type of method, the permeability enhancement caused by some common excipients has been investigated on a number of low permeability drugs. No enhancement was seen with lactose, hydroxypropylmethylcellulose (HPMC), EDTA, propylene glycol or polyethylene glycol 400 (PEG-400). In contrast, sodium lauryl sulfate (SLS) and Tween-80, which are both surfactants, were able to enhance the permeability of the drugs studied.[24]

Excipients have been shown to alter both passive and active transporters of drugs. There are varieties of active and passive transporters (including monocarboxylate, organic cation, nucleoside, and the dipeptide transporter).[25] As an example of the effect of excipients on active transporters, we highlight *p*-glycoproteins (p-gp). P-gp is a transport membrane protein that exists in many tissues of the human body. It actively transports a variety of structurally unrelated molecules out of cells.[26] *p*-Glycoproteins represent a mechanism of drug efflux which involves the active transport of drugs from the enterocyte back into the gut lumen and thus represents a mode of drug excretion that could lead to poor absorption and bioavailability. One of the most common p-gp-inhibiting agents is grapefruit juice.[27]

As an example, using both *in vitro* (Caco-2 cell) and *in vivo* (human patients) conditions, the effects of D-α-tocopheryl polyethylene glycol 1000 succinate (TPGS) and Poloxamer 188 (a block copolymer with oxirane and methyloxirane monomers) on *p*-glycoprotein were studied.[28] It was determined that TPGS, but not Poloxamer 188, inhibited the p-gp-mediated talinolol transport in Caco-2 cells. Another recent study[29] on a number of PEG-related excipients indicated that some excipients used for topical formulations (PEG-20 glyceryl laurate, PEG-20 glyceryl oleate, and PEG-40 sorbitan lanolate) largely inhibit p-gp transporters relative to saline. In addition, Gelucire 44:14 (a mixture of PEG-32 lauryl glycerides), a novel excipient from Gattefossé, demonstrated large inhibition of p-gp transporters. Furthermore, in another study[30] other excipients were tested. The results using digoxin showed that (among other excipients) Labrasol, Imwitor 742 [a blend of mono-, di-, and triglycerides, mainly composed of caprylic ($C_8H_{16}O_2$) and capric acid ($C_{10}H_{20}O_2$)], Cremophor EL (a mixture of hydrophobic and hydrophilic parts, including glycerol–polyethylene glycol ricinoleate and fatty acid esters of polyethyleneglycol as well as the hydrophilic part consisting of polyethylene glycols and ethoxylated glycerol), and Miglyol 840 (propylene glycol dicaprylate/dicaprate) were greater inhibitors of p-gp than Polysorbate 20 (Polyoxyethylene sorbitan monolaurate), TPGS, and Polysorbate 80 (Polyoxyethylene (20) sorbitan monooleate). As one can see from the studies described above, there are many issues that should be considered when choosing excipients, such as the chemical structure of the API, site of absorption, and solubility, because these will all affect permeability in some manner. As such, there does not seem to be good correlation with any one specific parameter at the present time.

A proposal has recently been made to incorporate predictions of the effect of transporters based on Biopharmaceutical classes.[31] The authors contend that based on the Biopharmaceutical Classification System (BCS) classifications, one can predict whether transporter effects may be significant. The original BCS model was developed to identify rationales for circumstances where additional clinical studies do not need to be performed based on available information (*in vivo–in vitro* correlation studies). It can be used to classify drugs based on their solubility and permeability characteristics. Class 1 (drugs that are highly soluble and highly permeable) are expected to have minimal transporter effects because the high concentrations of the API in the gut will saturate any transporters (efflux and absorptive). For class 2 (drugs with high permeable, but low solubility), it is suggested that the efflux transporters may predominate. Likewise, for class 3 (drugs that have low permeability and high solubility) drugs, absorptive transporters are predicted to have larger effects. Finally, class 4 compounds, which are both poorly soluble and have poor permeability, would be expected to demonstrate larger effects due to both absorptive and efflux transporters. These predictions/classifications will certainly help guide the formulator going forward in terms of addressing permeability issues. While permeability may be still be a somewhat complex issue for the formulator, there are certainly some available solutions so that drug candidates may be enabled and progressed to first time in human (FTIH).

Distribution

When formulating, it is also important to note the distribution of the drug. In terms of distribution, the body is a container in which a drug is distributed throughout by the systemic circulation. The apparent volume of distribution depends upon key items of interest to the formulator such as protein binding and tissue sequestration. If a drug is trapped within tissue, is amply lipophilic such that it enters a cell membrane lining the intestine and does not re-emerge in the bloodstream, or if it interacts with blood proteins so that the API never reaches the site of action, the bioavailability and efficacy of the drug will be reduced. This is an obvious issue for formulators. However, altering the distribution presently remains a challenge using standard excipients.

Metabolism and Excretion

The latter parts of the ADME acronym are *metabolism* and *excretion*. In terms of excretion, there are many routes whereby a drug is cleared from the body. The major organs capable of clearing a drug are the kidneys and the liver. It is difficult for the formulator to change the rate of excretion while not altering the gross metabolism of the organism. As such, this section deals primarily with metabolism.

Metabolism involves mostly biotransformations, which are processes by which xenobiotic (molecules foreign to the body) or endobiotic (molecules formed in the body) compounds are typically transformed via enzymatic processes into more polar, more water soluble, and, hence, more excretable metabolites. Biotransformations typically take place in the liver, but may also take place in the GI or in other metabolic organs such as the skin, kidney, and lungs. There are two categories of metabolic reactions, termed Phase I and Phase II. Phase I transformations group together simple reactions such as oxidation, reduction, hydrolysis, hydration, and isomerism. Phase II transformations include reactions with other small endogenous molecules such as glucuronidation (coupling with glucuronic acid), sulfate, acetate, glutathione, and amino acids. Among the key enzymes that may perform oxidations are cytochrome P450 enzymes. These groups of heme proteins contain iron and are embedded in the phospholipid bilayer of the smooth endoplasmic reticulum with a portion exposed to the cytosol. The CYP450 enzymes are primarily located in the liver, but are also located in the small intestine, kidneys, lungs, and brain. CYP450 enzymes are the major pathways for oxidative and reductive drug metabolism in humans.

In terms of altering metabolism, the formulator usually thinks in terms of bypassing metabolic routes rather than changing the metabolic path. The rationale for this approach is that changing the metabolism for the drug can, in many cases, have adverse effects on the ability to metabolize other foreign compounds, whether the compounds are endogenous or ingested. Knowledge of likely biotransformations, inherent stability, and chemical structure of the parent drug help the formulator to understand where and how the drug is being metabolically cleared. For example, if the chemical structure of the drug has a preponderance of aliphatic nitrogen or sulfur atoms, *N*-oxidation or *S*-oxidation may be expected.

One way to reduce the extent of metabolism (and therefore a way to increase the bioavailability of the drug) is to reduce the likelihood of metabolism in the liver. All drugs absorbed in the GI tract will have to pass through the liver and be eliminated before they can reach the target organ or receptors throughout the body. If the drug gets metabolized at this point, it is termed first-pass metabolism. Since all orally dosed drugs have first-pass metabolism to contend with, alternate delivery methods that bypass the first-pass metabolism—such as mucosal modes, which include sublingual (absorbed under the tongue), buccal (absorbed through the inside of the cheek), intranasal (through the nose) or inhaled (breathed in through the mouth) delivery—should then be considered.

Modified-Release Dosage Forms

In contrast to conventional (immediate release) forms, modified-release dosage forms provide either delayed or extended (sustained) release of drug. For solid oral dosage forms, the way the drug dissolves in the stomach is modified by the addition of excipients that slow the release of the drug. The two main types of modified-released dosage forms are delayed and extended release forms. Extended-release products are typically designed to release their medication in a controlled manner at a predetermined rate. For example, extended release tablets have an increased $t_{1/2}$, and the duration and location of drug release is modified to achieve and maintain optimum therapeutic blood levels of drug. Modified-release formulations that provide delayed release are exceptionally useful when combating acid instability. For example, enteric coated tablets or capsules designed to pass through the stomach unaltered and release their medication within the intestinal tract. This altered release may be used to bypass gastric toxicity, have a longer-acting drug product, help with patient adherence to therapy, and also play a role in product line extensions. For a more in-depth discussion of modified release dosage formulations, refer to the cited books.[7,8]

DESCRIPTIONS OF COMMON FORMULATIONS

In the following section we will describe the most common formulations used in first time in human (FTIH) clinical trials. Drug substances are usually not administered alone but in combination with inactive ingredients (pharmaceutical excipients) that serve varied and specialized functions. These pharmaceutical excipients may solubilize, suspend, thicken, dilute, emulsify, stabilize, preserve, color, flavor, lubricate, and/or add bulk to the final dosage form. The drug product to be dosed in humans is manufactured under strict manufacturing conditions and appropriately tested for stability, efficacy, ease of administration, and safety. The field of study concerned with the formulation, manufacture, stability, and effectiveness of pharmaceutical dosage forms is termed *pharmaceutics*.

The objective of the formulation strategy for FTIH is to develop a simple but flexible formulation that can accommodate later changes to the dose and dosing regimen. The formulation should be robust and scalable. The first choice in the majority of cases is an immediate-release tablet or capsule. Dry powder blends or granules are usually used to make the tablets or capsules. Currently, a modified release option is frequently being considered early in development to circumvent issues with PK and add value to the drug product and patent life.

Powders or granules Formulations

Though limited in therapeutics, the usage of powders or granules is extensive in the preparation of the dosage forms. Powdered drugs may be used to fabricate solid dosage forms as tablets and capsules, or they could be dissolved in solutions or suspended in liquid vehicles or solvents. They could also be incorporated into creams and ointments. Medicated powders are intended to be used internally or externally. Often the solid needs to be processed in order to acquire the desired features for the efficient production of the final dosage form or to ensure optimal therapeutic efficacy. One of the characteristics that can be adjusted and controlled is the powder particle size. The major goal of particle size reduction is to increase the rate of dissolution and increase the surface area.

When taken internally, powders and granules are usually mixed with water and taken orally, or in specialized cases they are inhaled into the lungs for local or systemic effects. These types of simple formulations are advantageous from the point of view of the patient that has difficulty swallowing. When intended for systemic use, powders and granules may have faster rates of dissolution and absorption relative to more complex formulation due to immediate contact with gastric fluids. The major disadvantages are the undesirable taste of the drug (usually bitter) and the inconvenience of handling the powder during compounding and administration. This type of formulation is flexible and easy to formulate but not suitable for hygroscopic materials or drugs that deliquesce. Granules are normally prepared as agglomerates of the smaller particles of powder. They may be prepared by wet or dry methods. The flow properties of granules are important for tablets and capsules as the flow is easier from the feeding container into the tableting presses or capsules fillers. They tend to wet easily if they are constituted into solutions and suspensions. Granules are usually better than powders in terms of flow and because of the smaller surface area (relative to powders); they are typically more stable as well.

Tablets and Capsules Formulations

By far the most frequently administered oral solid dosage forms—tablets and capsules—are the choice for formulators as well as patients. The preparation of the powders to be processed for tablets and capsules are similar, although the actual process involved in filling capsules or compressing tablets is different. Tablets are solid pharmaceutical dosage forms containing drug substance with or without suitable diluents (excipient fillers) and other excipients (including those to help processing, dosing or pharmacokinetics) prepared by either compression or molding methods. Capsules are solid dosage forms in which the drug substance is enclosed in either a hard or soft, soluble container or shell, typically composed of gelatin. Most pharmaceutical companies strive to formulate tablets for the first time in human studies, but capsules are typically easier to prepare. The usual preference for tablets and capsules by physicians and patients is mainly due to convenience in carrying them, the readiness of identification, and ease of administration. Manufacturers usually impart characteristic shapes and colors as well as the manufacturer's name and product code number embossed or imprinted on the surface tablets and capsules for identification purposes. Tablets and capsules typically come in a variety of dosage strengths and are usually tasteless when swallowed. They are normally efficiently and productively manufactured, as well as packaged and shipped at a lower cost. The shelf life is usually longer than for liquids and powders once they are formulated in tablets and capsules. The main advantage for capsules compared to tablets is the flexibility to add ingredients during compounding of capsules in a retail pharmacy setting or compounding pharmacy. Extemporaneous compounding of capsules for pharmacists is easier than for tablets because the solid ingredients just need to be mixed with the drug or active ingredients and filled in empty hard gelatin capsules.

In general, the major disadvantage of tablets and capsules is that formulation is more complex than powders. Thorough understanding of the physicochemical properties of the materials (API and excipients) as well as the mechanical properties is necessary to develop a process that would allow compaction in tablets and filling in capsules. In addition, the formulator should keep in mind that when formulating the final pharmaceutical dosage form there is an intricate relationship between the bioavailability of the product, its chemical and physical stability, and the technical feasibility of producing it. Any changes in one of the properties will likely affect the other two parameters. Many studies are needed to test for compatibility of drug substance with excipients, mechanical properties (such as flowability, compactibility, lubricity), appearance, disintegration, chemical and physical stability and dissolution properties of tablets. For capsules some of the challenges of tablets are removed since there is no need to form a robust compact dosage form. However, capsules are typically more expensive because filling speeds are lower, some materials are unsuitable for filling capsules because of their hygroscopicity and incompatibility with capsule shells, and capsules tend to stick in the esophagus more than tablets.

Solution and Suspension Formulations

In pharmaceutical terms, solutions are liquid preparations that contain one or more chemical substances dissolved in a suitable solvent or mixture of mutually miscible solvents.[32] In general, the major advantage of solutions and suspensions is the ease of formulation. Furthermore, in the administration of a range of doses, solutions and suspensions are flexible and convenient for people with difficulty swallowing (such as the pediatric population). Solutions typically have a faster rate of absorption than suspensions or solid oral dosage forms, since the latter dosage forms have to go through an additional dissolution step before being solubilized. One of the key disadvantages of solutions is that they are limited by the solubility and stability of the drug substance in the solvent and solvent system used to prepare the solutions. However, in cases where the solubility is a limiting factor, suspensions may be able to overcome this issue. Suspensions also have the added advantage over solutions of typically having better chemical stability in addition to some taste masking properties as undissolved particles are administered and hence more difficult to taste. Additional general disadvantages for suspensions relative to solutions forms include the more pronounced effect that chemical and physical interactions have on therapeutic quality and pharmaceutical stability. Physical stability (sedimentation and aggregation) is a typical issue with suspension formulations since complete and uniform redistribution is essential for accuracy of each dose.

Modified-Release Formulations

As discussed earlier, modified-release dosage forms are dosage forms that, by virtue of formulation and product design, have modified drug-release profiles. The most common ones are delayed-release products and extended-release products. A common approach for delayed release is enteric coating. Typically, enteric coating implies that the formulation is coated with a material that permits transit through the stomach to the small intestine before the medication is released. In this process, coated tablets or capsules are designed to pass through the stomach unaltered and release their medication within the intestinal tract. Many of these coatings are polymeric and involve pH-dependent profiles such that the coating is unaffected at low pHs, but dissolves in the higher pH of the small intestine. One of the more common excipients used for this is the EUDRAGIT® family from Degussa, which are based on polymethacrylates. For example, EUDRAGIT® L 30 D-55 targets the drug delivery in the duodenum (pH ~5.5), while EUDRAGIT® FS 30 D (pH ~7) targets the colon. Other products help target the jejunum (pH ~6.0) as well as the ileum (pH ~7).

In extended release formulations, the tablet or capsule is designed to release the medication in a controlled manner, at a predetermined rate, duration, and location to achieve the maximum therapeutic effect. In terms of advantages, modified release dosage forms typically have all the advantages of capsules and tablets listed above, but in addition, there are several characteristics that make this type of formulation a specialized delivery system that, when successful, offers tremendous rewards. Controlled-released dosage forms have reduced drug blood level fluctuations and decreased frequency in dosing. This usually results in enhanced patient convenience as well as better adherence to therapy. Because of the constant therapeutic drug levels, there are typically fewer adverse side effects. In this fashion, despite increased expenditure in development and significant time and effort in formulation development, controlled-release dosage forms can result in overall reduction of health care costs. The key disadvantages of modified-release dosage forms include the loss of flexibility in adjusting the drug dose and/or dosage regimen, the increased risk of sudden and total drug release due to failure of the technology of the dosage unit, and the cost and time burden required for additional development.

Parenteral Formulations

By definition, parenterals are nonoral products, but in pharmaceutical convention the term is used interchangeably with injectables. Drugs can be injected into any organ or area of the body, but usually they are injected into a vein, into a muscle, or into/under the skin. Injections need to be sterile and pyrogen (a substance that produces a rise in body temperature)-free. Parenteral routes of administration are used when a rapid effect is desired such as in emergency situations, when the patient is uncooperative, unconscious, or unable to accept or tolerate the medication given by mouth or when the drug itself is ineffective given by other routes (for example, limited absorption due to poor solubility, permeability, or stability). As mentioned earlier, immediate physiological action is the usual main advantage. In addition, the onset and duration of action can be modified depending on selection of route of administration. Intravenous injection usually has the fastest onset and is rapidly cleared out of the body. Injections into the muscle or into/under the skin tend to have delayed onset and a longer duration of action because they have to reach the systemic circulation before having an effect. The major disadvantage of parenteral formulations is the need for sterilization. Formulation of parenterals is furthermore complicated due to the limited number of solvent systems compatible with physiological fluids and the solubility of the drug substance in the solvent system. Typically, injectables also need to be adjusted to physiologically compatible osmolarity and pH. Finally, with the main exception of insulin injection by diabetic patients, most injectables are administered in hospital settings or by a trained health care professional. They are usually painful and difficult to self-administer, and great care has to be exercised to avoid infecting the site of injection, which makes this type of dosage unpopular with patients.

Inhaled Formulations

Drugs or solutions/suspensions of drugs may be administered via the nasal or oral respiratory route. The drugs may be administered to achieve local action on the bronchial tree or systemic effects through absorption from the lungs. Some gases are administered by inhalation such as ether and oxygen. Typically, the major advantage is the rapid onset of action since absorption is as rapid as the drug can be delivered into the alveoli of the lungs. Many drugs are highly permeable through the alveolar and vascular epithelial membranes, which, in addition to high blood flow and very large surface for absorption, makes inhaled dosage forms very attractive to the formulator. The key major disadvantages are that in many cases it is very difficult to control how much of the drug is being absorbed; it is difficult to maintain constant drug levels and the amount inhaled even with a metered dose varies patient to patient. Another complicating factor is cost and time of development. Deposition of drug in the lungs is very sensitive to the drug particle size (from a dosed suspension). An optimal particle size needs to be achieved to prevent the API from settling in the bronchioles and bronchi or prevent them from being exhaled back out. Finally, it needs to be stressed that most inhalers are complex products, which require a huge development cost and time and also contain a high degree of embedded engineering. In many cases, the delivery device is more critical than the formulation. Development times for this type for formulation are long, and any changes to the device component can add significant delays much more significant than other types of dosage forms.

Topicals

For the most part when applied to the affected site, topical dermatological products exert a localized effect as the drug penetrates into the underlying layers of the skin or mucous membrane. The intended target is the skin in this case. It is also possible to achieve systemic drug absorption in which case transdermal delivery systems are designed for the drug substance to achieve systemic therapeutic levels (in which case the skin is not the targeted organ). Clearly, the major advantage of topical delivery when local effect is sought is that systemic absorption is avoided or occurs at subtherapeutic levels, which may help minimize systemic toxicities or adverse effects. The major disadvantage of this type of delivery is that the drug substance needs to penetrate and be retained in the skin for a period of time. Since the normal unbroken skin acts as a natural barrier, it is not an easy task to achieve the requisite rate and degree of drug penetration and therefore attain the desired therapeutic effect. Drug penetration into the skin depends on a number of factors including the properties of the drug molecule (such as permeability and lipophilicity), the vehicle, and the condition of the skin itself. Most drugs are not suitable for topical delivery. Another major disadvantage commonly seen in patients is the propensity for irritation at the site of application and allergic sensitivities. One other potential disadvantage with topical formulations is that additional photosafety testing is required during development because light-induced reactions, photoallergens, or other toxic effects are more prevalent on the surface of skin compared to other dosing routes.

Summary of Advantages and Disadvantages of Formulation Routes. Table 14.2 summarizes the advantages and disadvantages of the major routes of dosing.

Decision Tree for Formulating for Phase I Dosing in Humans

As discussed earlier in the chapter, there are many questions and issues that may arise when formulating for early clinical studies. In Figure 14.1, a strategy for formulating these studies is presented. This is a generalized and simplified view and can help the chemist understand the formulation rationale.

In general, one should aspire to start with an active pharmaceutical ingredient with perfect physicochemical properties. The first key step is to evaluate the bioavailability in model species from the crystalline solid state. While amorphous material may be used at this stage, it is important to keep the potential liabilities of chemical and physical stability as well as the latent changes in solubility and dissolution rate which could impact the investigation. If the bioavailability is good, then a very simple formulation may be developed. If the bioavailability is poor, then the cause should be explored. Possible causes of poor bioavailability include first-pass metabolism or absorption. If the problem lies in metabolism, it would be recommended that the team look into alternative APIs. However, if the issue is

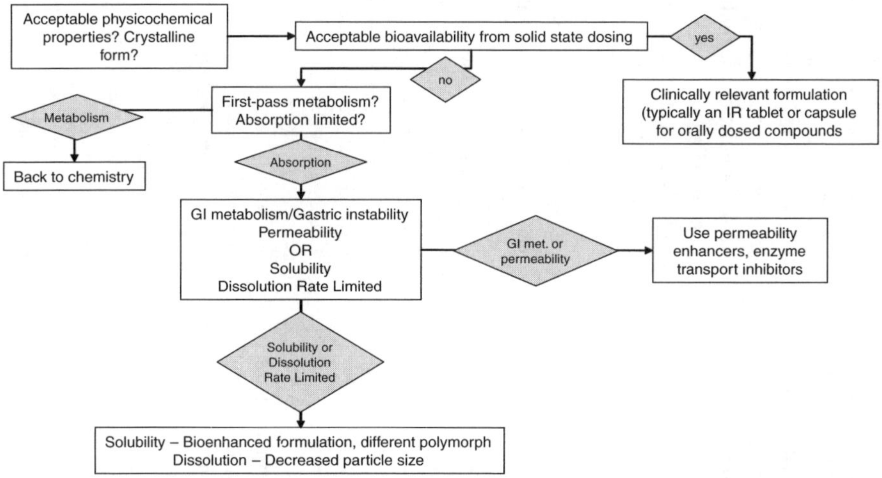

Figure 14.1. Decision tree for formulation for FTIH dosing.

TABLE 14.2. Summary of General Advantages and Disadvantages of Formulation Routes

Formulation Route	Advantages	Disadvantages	PK-PD Response
Powder in a bottle or granules	Most powders are taken orally after only mixing with water or suitable vehicle Ease of formulation, flexibility in compounding Relatively good chemical stability Convenient for people with difficulty swallowing	Compounding is time-consuming Inaccuracy of dose. No taste masking—undesirable taste of the drug Not suitable for hygroscopic or deliquescent drugs	May have faster rates of dissolution and absorption
Tablets	Conveniently carried, readily identified, easily taken by patient with little supervision Tasteless when swallowed Variety of dosage strengths Efficient and productively manufactured Lower cost of packaging and shipping. More stable and longer shelf life than liquids and powders	Formulation is more involved and complex Powder properties need to be suitable for robust compaction Tablets require several excipients to render the material suitable for compression Excipient compatibility and stability testing requirements	Absorption of the drug is dependent on tablet disintegration and drug dissolution
Capsules (hard and soft gelatin)	Conveniently carried, readily identified, and easily taken Tasteless when swallowed Flexibility in ingredients during compounding Efficient and productively manufactured Lower cost of packaging and shipping. More stable and longer shelf life than liquids and powders	Formulation may be more involved and complex More expensive than tablets due to lower filling speeds and special techniques needed for soft gelatin capsules Tends to stick to esophagus more than tablets Not suitable for hygroscopic materials	Improved bioavailability for soft gelatin capsule since drug is already solubilized within the capsule

TABLE 14.2. (*Continued*)

Formulation Route	Advantages	Disadvantages	PK-PD Response
Solutions	Ease of formulation but maximum dose is limited by solubility of the drug substance in the solvent and solvent system Flexibility in the administration of a range of doses. Convenient for people with difficulty swallowing (pediatric administration)	Solubility and stability of the solutes in relation to the solvent or solvent system employed is a major limiting factor Chemical and physical interactions affect the therapeutic quality and pharmaceutical stability Undesirable taste of the drug	Absorption is faster than from suspension or solid dosage forms
Suspensions	Overcomes solubility issues of drug substances Better chemical stability than liquids for some drugs Flexibility in the administration of a range of doses Convenient for people with difficulty swallowing (pediatric administration) Taste is masked somewhat when administering undissolved particles	Physical stability (sedimentation and aggregation) is an issue in formulation and stability of the drug in the solvent or solvent system employed Complete and uniform redistribution (shaking or mixing) is essential for accuracy of dose.	May have faster rates of dissolution and absorption than solid dosage forms due to immediate contact with GI fluids depending on particle size.
Modified release	Maintains all the advantages of capsules and tablets Reduction in drug blood level fluctuations Reduction in dosing frequency Enhanced patient convenience and adherence to therapy Fewer adverse side effects Reduction in overall health care costs	Formulation is time consuming and requires significant effort compared to immediate release dosage forms Loss of flexibility in adjusting the drug dose and or dosage regimen Increased risk of sudden and total drug release due to failure of the technology of the dosage unit	Specific release profile for maximization of PK-PD advantage Increase $t_{1/2}$ Constant therapeutic blood levels

(*continued*)

TABLE 14.2. ***(Continued)***

Formulation Route	Advantages	Disadvantages	PK-PD Response
Parenterals	Immediate physiological action Modification of the onset and duration of action depending on selection of route of administration	Formulation is complicated by the need for sterilization Solubility of the drug substance determines maximum possible dose Necessity to adjust osmolarity and pH Limited number of solvent systems compatible with physiological fluids Painful and difficult to self administer	Accurate dosing possible. Immediate action
Inhalation	Rapid physiological action Local or systemic effects High surface area for the drug to be absorbed	Difficult to achieve consistent drug concentration in systemic circulation Particle size is critical to reach alveoli for absorption into circulatory system Complex products with high degree of formulation and engineering development	Localized effect in the lungs or systemic action Accurate dosing is difficult
Topicals	Mostly used for localized effects at the site of their application Avoids systemic toxicities when local effect is sought	Drug substance needs to penetrate and be retained in the skin for a period of time Most drugs not suitable for topical delivery due to physicochemical properties, pharmaceutical vehicle and condition of skin Irritation at the site of application and allergic sensitivities are common Additional photosafety testing is often required	Localized drug levels to site of action.

being caused by absorption, there are many factors that should be investigated experimentally. The issue might lie in the kinetics of dissolution, the solubility in gastric media, or the gastric metabolic routes or permeability. If the poor bioavailability is caused by poor solubility or dissolution, then a change in particle size or bioenhanced formulation should be attempted. On the other hand, if it is the GI metabolism or permeability that is leading to the reduced bioavailability, permeability enhancers or enzyme transporters should be used as part of the formulation. While the issues are complex, this flowchart may be used to classify the various issues that crop up during formulation.

SUMMARY

Formulation is an enabling capability in the pharmaceutical industry. It has the ability to convert a nondevelopable candidate that is known to possess good biological efficacy and a safe toxicology profile into a drug that can save lives. The pharmaceutical formulator has many exquisite and cunning prospective techniques in which to help enable a candidate. Chief among them are excipients, but the choice of excipients is directed by studies investigating physical and chemical stability, solubility, permeability, and dissolution. Ultimately, it is the interaction between pharmaceutical scientists, chemists, biologists, toxicologists, and pharmacokineticists that enables a drug substance of an active pharmaceutical ingredient candidate to be turned into drug product that can be dosed in the clinic.

REFERENCES

1. Washington, N., Washington, C., and Wilson, C. G. *Physiological Pharmaceutics: Barriers to Drug Absorption*, Taylor and Francis, New York, 2001.
2. Lindahl, A., Ungell, A.-L., Knutson, L., and Lennernäs, S. *Pharm. Res.* **1997**, *14*, 497–502.
3. Charman, W. N., Porter, C. J. H., Mithani, S., and Dressman, J. B. *J. Pharm. Sci.* **1997**, *86*, 269–282.
4. Ungell, A.-L., and Abrahamsson, B. Biopharmaceutical Support in Candidate Drug Selection, in *Pharmaceutical Preformulation and Formulation: A Practical Guide from Candidate Drug Selection to Commercial Dosage Form*, Gibson, M., Ed. IHS Health Group, Englewood, CO, 2001, pp. 97–153.
5. Brittain, H. G. *Polymorphism in Pharmaceutical Solids*, Marcel Dekker, New York, 1999.
6. Hörter, D., and Dressman, J. B. *Adv. Drug Delivery Rev.* **1997**, *25*, 3–14.
7. Remington, J. P. *Remington's Pharmaceutical Sciences*. Mack Publishing Company, New York, 1985.
8. Ansel, H. C., Allen, L. V., and Popovich, N.G. *Pharmaceutical Dosage Forms and Drug Delivery Systems*, Lippincott Williams & Wilkins, New York, 1999.
9. Gould, P. L. *Int. J. Pharm.* **1986**, *33*, 201–217.

10. Broadhead, J. Parenteral Dosage Forms, in *Pharmaceutical Preformulation and Formulation: A Guide from Candidate Drug Selection to Commercial Dosage Form*; Gibson, M., Ed. IHS Health Group: Englewood, CO, 2001, pp. 331–354.
11. Strickley, R. G. *PDA J. Pharm. Sci. Technol.* **1998**, *53*, 324–349.
12. Strickley, R. G. *Pharm. Res.* **2004**, *21*, 201–230.
13. Powell, M. F., Nguyen, T., and Baloian, L. *PDA J. Pharm. Sci. Technol.* **1998**, *52*, 238–311.
14. Sweetana, S., and Akers, M. J. *PDA J. Pharm. Sci. Technol.* **1996**, *50*, 330–342.
15. Stella, V. J., and Rajewski, R. A. *Pharmaceutical Research* **1997**, *14*, 556–567.
16. Thompson, D. O. *Crit. Rev. Ther. Drug Carrier Syst.* **1997**, *14*, 1–104.
17. http://www.cydexinc.com/captisol.html, 2005.
18. www.cargillfoods.com/pdfs/starch/82005.pdf, 2005.
19. Felton, L. A., and McGinity, J. W. *Drug Dev. Ind. Pharm.* **2002**, *28*, 225–243.
20. Sevda, S., and Hýncal, A. A. *J. Controlled Release* **2001**, *72*, 133–144.
21. Van Hoogdalem, E. J., De Boer, A. G., and Breimer, D. D. *Pharm. Therapeut.* **1989**, *44*, 407–443.
22. Artursson, P., Ungell, A.-L., Löfroth, J.-E. *Pharm. Res.* **1993**, *10*, 1123–1129.
23. Walter, E., Kissel, T., Reers, M., Dickneite, G., Hoffman, D., and Stüber, W. *Pharm. Res.* **1995**, *12*, 360–365.
24. Rege, B., Yu, L. X., Hussain, A. S., and Polli, J. E. *J. Pharm. Sci.* **2001**, *90*, 1776–1786.
25. Lee, V. H. L. *Eur. J. Pharm. Sci.* **2000**, *11*, S41–S50.
26. Orlowski, S., and Garrigos, M. *Anti-Cancer Res.* **1999**, *19*, 3109.
27. Dahan1, A., and Altman, H. *Eur. J. Clin. Nutrition* **2004**, *58*, 1–9.
28. Bogman, K., Zysset, Y., Degen, L., Hopfgartner, G., Gutmann, H., and Drewe, *J. Clin. Pharm. Ther.* **2005**, *77*, 24–32.
29. Wang, S.-W., Monagle, J., McNulty, C., Putnam, D., and Chen, H. *J. Pharm. Sci.* **2004**, *93*, 2755–2767.
30. Cornaire, G., Woodley, J., Hermann, P., Cloarec, A., Arellano, C., and Houin, G. *Int. J. Pharm.* **2004**, *278*, 119–131.
31. Wu, C.-Y., and Benet, L. *Pharm. Res.* **2005**, *22*, 11–23.
32. *US Pharmacopeia-28 NF 23*, United States Pharmacopeial, 2005.

15

INTELLECTUAL PROPERTY AND EARLY DEVELOPMENT

MARIA I. SHCHUKA
Johnson & Johnson Law Department, Spring House, PA 19477

INTRODUCTION

During early (preclinical and/or clinical) development a number of research and development activities may lead to the discovery of new and patentable inventions including, but not limited to, new synthetic routes, new crystalline or polymorphic forms, new salt forms, and new delivery formulations. Patenting any of these discoveries can provide extended exclusivity within the market—exclusivity beyond that provided by the patent on the composition of matter (to the extent that the compound is now encompassed by the new form). Thus it is important from a business perspective to understand the nature, scope, and value of patents and to be able to identify and adequately protect these inventions.

To properly understand the potential benefits of patents on new synthetic processes, crystalline or polymorphic forms, salt forms, or formulations, it is important to begin with an understanding of what patents are and what type of protection/exclusivity they provide. This chapter will cover basic definitions and intellectual property concepts, as well as provide fundamental information on the patentability of inventions likely to be discovered during the course of early preclinical and clinical development. Issues related to freedom-to-operate, freedom-to-market, regulatory exclusivity extensions, and so on, will not be covered. (Except in a brief discussion of the 271(e)(1) exemption as it relates to early development activities.) These intellectual property-related issues tend to involve complex and highly

Fundamentals of Early Clinical Drug Development: From Synthesis Design to Formulation, edited by Ahmed F. Abdel-Magid and Stéphane Caron

fact-specific analyses; thus, general guidelines are often inadequate. For this reason it is recommended that any questions related to freedom-to-operate be addressed in consultation with an attorney.

Additionally, the author would like to caution that the application and interpretation of patent law is a constantly changing, complex, and often highly fact-specific area of legal practice. It is strongly recommended that legal counsel (i.e., a patent attorney) always be consulted whenever analysis related to patents, patentability, freedom-to-operate, infringement, and so on, is required.

DEFINITIONS

The terms *patent memo (memorandum), invention record,* and *invention disclosure record* are often used to refer to an internally (within the company, university, or other research organization) generated record of a new invention. The record typically includes a listing of those persons who worked on the invention, a date of conception for the invention, a date of reduction to practice for the invention, details on written records of the invention (e.g., notebook page listings), details of any disclosures of the invention (including disclosures at company meetings, to consultants, at technical meetings, and/or symposia, etc.), a brief description of the invention itself, and optionally a listing of closest art. Although each organization/company may have its own specialized policies for completing and evaluating such invention disclosures, these forms are generally used as a formal documentation of the conceptions and/or reduction to practice of an invention and as a trigger for beginning the process of determining the patentability of the invention, as well as a trigger for beginning the process of drafting and filing a patent application.

Two types of patent-related documents are generally available to the public: *patent publications* and *issued patents*. The United States and most foreign countries publish all filed patent applications 18 months after the initial filing document (with some limited exceptions, for example, for nation security/secrecy reasons). In general, the first document to publish is a publication of the pending patent application, along with the initially submitted patent claims. The published application is therefore either a copy of the US nonprovisional application, a copy of a foreign national application, or a copy of the application filed under the PCT (Patent Cooperation Treaty) as a WO patent application publication.

In the United States, for most pharmaceutical applications, the filing and publication sequence for patent applications is as follows. The first application to be filed is the so-called "provisional application." To successfully use the date of the provisional application as the priority date of the invention, the provisional application should contain all the information required by the patent laws. The provisional application is therefore often written as a complete application, including claims. This application sets an initial filing date for the invention but does not start the clock on patent term. Within 1 year of the filing of the provisional application, a "nonprovisional application" and any foreign applications—including the PCT application and select national applications (for countries not part of the PCT)—are filed, claiming priority to the provisional application. Six months after filing the nonprovisional, PCT, and foreign applications (18 months from the filing of the initial application), the patent application publishes in the United States and as a WO publication.

In the United States, there is no requirement to file a provisional application (which is not examined, but merely sets an initial filing date within the patent office). However, for many companies and businesses, the provisional application provides a number of benefits with respect to exclusivity. First, new matter may be added to the provisional application at the time the nonprovisional application is filed. Alternatively, any number of subsequent provisional applications can be filed, each of which builds on the subject matter of the initial provisional application. A nonprovisional application is then filed at the one-year date, encompassing all the matter added within the subsequent provisional application(s). This provides a period of one year during which time the inventors can continue to perfect and refine the invention through continued experimentation. The ability to add new matter also provides the opportunity to expand or broaden the scope of the claimed invention. Second, the provisional application provides a one-year grace period, since the filing of a provisional application does not start the clock with respect to patent term—the patent term begins with the filing of the nonprovisional application. This effectively adds one year of patent term (or exclusivity) at the end of the patent's life. This is especially attractive for pharmaceutical inventions (drugs) where an additional year of patent term after the drug is on the market can mean significant profits. (Note that for inventions which enter the market quickly and remain on the market for a relatively shorter period of time—for example, medical devices—it may be strategically more important to file a nonprovisional application initially and begin prosecution to issuance sooner.)

It is important in reading and analyzing patent documents to be able to recognize and distinguish between publications of patent applications and publication of issued patents, because the rights associated with each of these types of documents are significantly different. The patent application publication notifies the public that patent rights are being sought for the invention as defined in the claims. The patent application publication may convey some provisional rights to the patentee; however, these rights cannot be enforced until the patent has actually issued. By contrast, upon issuance and publication, the issued patent provides notice to the public that the patentee has been granted patent rights to the invention, as defined by the issued claims.

In the United States, the published patent application will look very similar to an issued patent, but will contain the heading "Patent Application Publication" in the upper left-hand corner and the formal information of "Publication Number" (of the format US+year+7 digit publication number) and "Publication Date" in the upper right-hand corner of the title page. By contrast, the issued patent publishes with the headings "United States Patent" in the upper left-hand corner and the formal information of "Patent Number" and "Date of Patent" in the upper right-hand corner of the title page. For patent applications filed under the PCT or Patent Cooperation Treaty, a "WO" document will publish. This is a publication of the application as filed with the PCT. The front page of this publication will include, in addition to formal filing information, a listing of elected countries. The PCT permits the filing of one application and the election of all or some of the over 100 participating countries for future national filing and prosecution. The PCT was designed to facilitate foreign filings and provide a more uniform patent practice. The PCT does not, however, issue patents; instead, a patent is ultimately issued by each nation into which a national phase application is filed and prosecuted to issuance. Issued patents filed nationally or under the PCT are published by the individual countries or country blocks (e.g., the EP for European countries) and will be designated as such on the title pages. Additionally, citations and title pages of both US and foreign patent publications will include codes such as A, B, A1, B1, and so on. These designate whether the publication is an application, issued patent, correction, search report, and so on. Although not universal, "A" codes typically identify applications, whereas "B" codes typically identify issued patents.

Both the published patent application and the issued patent will include a "title page" and the "specification," which includes the body of the patent and concludes with a listing of claims. The "title page" lists the formal information related to patent or application, including the type of publication, the patent or publication number, the date of filing, the date of issuance, a listing of inventors, a listing of the assignee (if appropriate), a listing of the examiner(s), a listing of the search parameters used in examining the claims (if appropriate), information on related applications, a listings of references cited during prosecution (if appropriate), and an abstract of the invention.

The "patent specification" is the body of the application or issued patent and will typically include the following sections: (1) cross-reference information and information on government support/funding of the research disclosed in the application (if applicable); (2) a background of the invention (often including a description of the problem to be solved, the area or science to which the invention pertains, and/or closest relevant art or publications); (3) a summary of the invention(s) (generally a copy of the broadest independent claim(s)); (4) brief description of any diagrams; (5) a detailed description of the invention; (6) a listing of claims (either those submitted for examination in the application or those allowed in the issued patent); and (7) figures or diagrams (if applicable). The detailed description of the invention contains teachings on how to make and use the invention, and in the United States also contains a disclosure of the best mode for practicing the invention. Typically, the detailed description also includes examples or experimentals. If the examples are written in the past tense, the examples/experiments thus described were completed and are therefore a record of the reduction to practice of the invention. If the examples are written in the present tense, there is no presumption that the experiments were performed. Such experiment write-ups are therefore the so-called prophetic examples and serve to teach how one would run the experiments. In addition to above-described components, the specification may further contain additional information as required by the patent laws of the individual countries.

The specification concludes with a listing of claims. The "claims" define that which the inventor believes is the invention and for which patent protection is sought. The claims can be independent, or dependent on a previous claim. The claims will vary in their scope to provide different levels of protection—defining the invention with different limitations, both in number and in scope. Upon allowance of one or more claims, the issued patent is published with the claims as allowed. These claims define the scope and invention for which patent rights have been awarded. The issued claims may be the same as or different from those filed and published with the application (both in scope and in limitations), depending on the course of prosecution of the application. Additionally, it is important to remember that the claims that are ultimately issued in different countries, from the same, single PCT filing, may also vary depending on the patent laws of each country and the prosecution carried out in each country.

A word about *inventorship*. In the United States, the patent laws specify that the patent application must be filed by the original and first inventor(s). US patent law further defines an inventor as one who "*contributed intellectually, in some degree, to the conception of the claimed invention*." Thus inventorship is directly tied to the claims listed at the end of the application and can change as prosecution of those claims leads to additions, cancellations and changes. Since inventorship is defined and mandated in the patent laws, the determination of who is an inventor is a legal determination, completed by the filing patent attorney. The patent attorney typically interviews each person who worked on the claimed invention and evaluates his/her contribution to the conception of the invention in each claim. During this process, it is important to remember that *inventorship is not authorship*. Contributors may have completed significant amounts of work on reducing the invention to practice, but not be considered inventors because they did not contribute to the conception of the invention. This is often a difficult distinction to establish and may not be representative of the efforts of each individual in the development of the invention. Although the determination of inventorship may be contentious and may not properly acknowledge individual contributions to the reduction to practice of the invention, it is important and often helpful to understand that it is in the company's and attorney's interest to list the true and correct inventors (without ulterior or politically motivated additions or omissions). Intentionally including or excluding an inventor is considered inequitable conduct before the patent office. A finding of inequitable conduct renders the patent unenforceable (exclusivity rights to the invention are permanently and irrevocably lost) and may further result in sanctions or a loss of license to practice for the filing patent attorney.

It is important to note that although inventors are generally listed on foreign patents and applications (for example, on the PCT publication), most countries outside the United States allow the application to be filed by an assignee or company in the name of the inventors. Furthermore, most countries outside the United States grant patent rights using a first-to-file system (i.e., the patent is granted to the first entity to file a patent application on the invention), rather than the first-to-invent system used in the United States (where the patent is granted to the first person to conceive of the invention). Nonetheless, inventions made abroad, when filed in the United States, must be filed by and name the original and true inventors.

BASIC PRINCIPLES OF PATENTS AND INTELLECTUAL PROPERTY

A patent is a grant, from the government, to *exclude others from making, using, selling, or offering to sell the invention that is patented*—that is, the invention as defined by the claims of the issued patent. Patent rights do not, however, provide a patent owner with the right to practice the claimed invention, only to exclude others from doing so. Furthermore, patent rights—the rights to exclude others—are extended only to the patented invention as defined by the claims. The patented invention is defined by the claims in much the same way as a land deed or title defines the property rights of the owner. Thus, a patent owner can exclude others from practicing the patented invention only as it is defined in each claim individually—with all the limitation as defined in that claim. This concept will be discussed further in conjunction with specific invention types later within this chapter.

To obtain a patent on a new invention, the invention must meet basic patentability requirements as outlined in the patent laws of each country. The three basic requirements of patentability in the United States are as follows: (a) novelty—that is, the invention must be new/has not been previously disclosed in any form—written, internet, oral, poster, and so on; (b) nonobviousness—that is, the invention must have been nonobvious to one skilled in the art in light of *all* that has been previously published or disclosed; and (c) utility—that is, the invention must be useful. (There are some classes of inventions which are not patentable—for example, mathematical laws—but these will generally not be relevant to the pharmaceutical industry and will not be further discussed here.) Note that the requirements for patentability in countries outside the United States may be different—some classes of invention may or may not be allowed and the exact requirements for patentability may be defined in different ways—although they will generally fall within these three broad requirement concepts.

Patent rights are often broad and potentially very lucrative from a business/economic standpoint. In exchange for these rights and the competitive advantage, the patentee must further provide the public (by way of the published specification) with a complete disclosure of the invention. This requirement is particularly stringent in the United States, and therefore US disclosure scope will be discussed here. In the United States, the patentee must disclose information on how to make and use the invention (in sufficient detail so that one skilled in the art could practice the invention), as well as the best mode for practicing the invention (i.e., the best mode as perceived by each listed inventor for practicing each claimed invention). This requirement is the *quid pro quo* for the grant of exclusivity.

In most countries, the term or period of time for which patent rights are granted and enforceable is 20 years from the filing date of the patent application. In the United States, the exclusivity period for applications filed after June 8, 1995 is 20 years from the filing of the nonprovisional application. (For patents in force on June 8, 1995 or filed before June 8, 1995, the exclusivity period is the greater of 20 years from filing or 17 years from issue. Patent applications filed prior to June 8, 1995 generally resulted in issued patents that had a patent term of 17 years from issue.)

In terms of the value of patent rights, particularly as they relate to inventions from preclinical and clinical development—such as new synthetic routes, crystalline or polymorph forms, salt forms, and/or formulations—consider the following example:

Company A discovers and patents a new pharmaceutical agent, with a patent term extending from 2000 to 2020. In 2005, Company A discovers and patents a new process for the preparation of the pharmaceutical agent, with a patent term extending from 2005 to 2025. Between 2000 and 2020, Company A can exclude all others from making, using, selling, or offering to sell the pharmaceutical agent. In 2020, the patent rights on the pharmaceutical agent expire and others can then make, use, sell, or offer to sell the drug. However, between 2005 and 2025, and more importantly in the period between 2020 and 2025, Company A can exclude others from using the new process to make the pharmaceutical agent. If the new process is substantially superior to previous processes for making the drug—for example, in terms of impurities, scale, time efficiency, cost, and so on—then between 2020 and 2025 Company A would enjoy a competitive advantage in the manufacture of the pharmaceutical agent, an advantage that could be expected to extend to a market and/or profit advantage.

The scenario, however, is quite different if in 2005 Company B, rather than Company A, discovers and patents the new process for the preparation of the pharmaceutical agent, with a patent term extending from 2005 to 2025. Then, between 2000 and 2020 Company A can exclude others (including Company B) from making, using, selling, or offering to sell the pharmaceutical agent, but at the same time between 2005 and 2025 Company B can exclude others (including Company A) from using the new process to make the pharmaceutical agent.

A similar analysis can be extended to patents for new crystalline forms, new polymorphic forms, new salt forms, and/or new delivery formulations. With crystalline forms, polymorph forms, salt forms, and formulations, the competitive advantage can be potentially greater, particularly if the crystalline form, polymorphic form, or salt form is the form that is ultimately approved by the FDA or another regulatory body. In effect, the patent protection on the new crystalline, polymorphic, or salt form can serve to extend exclusivity rights of the approved drug.

Finally, any discussion of intellectual property would not be complete without mention of trade secrets. Trade secrets are just what the name suggests—information related to inventions and/or marketed products which the company chooses to keep secret to maintain a competitive advantage in the marketplace. An example of a famous trade secret is the recipe for Coca Cola™. In the pharmaceutical industry, however, regulations require the submission and public disclosure of much information related to the pharmaceutical agent. As such, for the pharmaceutical industry, trade secrets are generally limited to discoveries such as optimization conditions in manufacture, and so on. For this reason, trade secrets will not be further discussed herein.

METHODS OF MANUFACTURE (METHODS OF SYNTHESIS)

In US patent law, processes or synthesis routes are patentable subject matter. The patented process may be directed to making a specific chemical or group of chemicals or may be directed to making a particular type of compound or chemical structure—for example, making a specific crystalline form or a new way of making amide bonds. In general, process patents tend to be directed to new processes for the preparation of a specific pharmaceutical agent and close derivatives thereof. To meet the requirements of patentability, the process must be useful, novel (never disclosed), and nonobvious. (The utility of the pharmaceutical agent creates the utility for the process.) In general the question of whether a new process is nonobviousness is the most difficult to answer.

If a process utilizes "known" chemical steps through "known" intermediates, then the process may be viewed as a new combination of old components. To be patentable, the combination (i.e., the series of steps in the process) must be nonobvious (often a difficult hurdle to overcome) or the process must possess some unexpected property—for example, unexpectedly higher yields, unexpectedly higher purity, unexpected enhanced regioselectivity, unexpected decreased impurities, and so on. Typically, process patents of this type are directed to improvements of old processes and may be difficult to get allowed.

Claims directed to processes that use "new" reaction steps, move through "new" intermediates, and/or employ "new" reagents are also patentable and, because they do not constitute a new combination of old steps, are generally easier to get allowed. As with process improvements, new processes of this type may also possess unexpected properties and/or have significant advantages over previously known processes. Additionally, if the "new" process proceeds through previously unknown intermediates, these intermediates may also be separately patented as new compositions of matter.

Processes to synthesize new chemical entities (e.g., pharmaceutical agents which have not been previously disclosed in any manner) are, by definition, novel and nonobvious and therefore patentable. If the pharmaceutical agent is new and nonobvious, then by definition the process to make the agent must be novel and nonobvious.

The scope of patent protection (exclusivity) accorded to process patents is the same as that accorded to patents claiming compositions of matter. Thus, the patent holder of a process patent is entitled to exclude others from using, selling, or offering to sell the process as defined by the claims, for the term of the patent. From a business perspective, the most useful right is the right of the patentee to exclude others from practicing the patented process. If the patent also includes independent, issued claims to novel intermediates, the patentee can further exclude others from making, using, selling, or offering to sell those intermediate compounds (regardless of how they are made). This means that the patentee can prevent others from using processes different from the patented process, if the alternate process goes through the patented intermediate(s).

In the United States, additional protection for processes was enacted in 1988 and codified as 35 USC §271(g), which reads in part as follows:

> Whoever without authority imports into the United States or offers to sell, sells, or uses within the United States a product which is made by a process patented in the United States shall be liable as an infringer.... A product which is made by a patented process will, for purposes of this title, not be considered to be so made after (1) it is materially changed by subsequent processes or (2) it becomes a trivial and nonessential component of another product.

Under this rule, a US process patent also provides the patentee with the ability to prevent others from using the patented process in a country where patent protection on the process in not available (either because processes are not patentable by law in that country or because a patent on the process has not (yet) issued in that country) and then importing or selling into the United States the product made using that process. In the pharmaceutical arena, this protection is particularly effective during periods of time when the patent on the pharmaceutical agent has expired, but during which the process patents are still in force. More specifically, the rights accorded under the Process Patent Act may be used to prevent a generic or competitor company from making a pharmaceutical agent using a US patented process in a country outside the United States and then importing and selling the product back into the United States.

As with any patent, the issued claims of the process patent define the scope of protection. Process claims are generally written as a series of steps which describe the reaction steps, usually in sequence. Thus in order to infringe a process patent, the steps as defined in the claims (with all the limitations including in a particular claim) must be followed. Process claims may be broad or narrow (i.e., with multiple limitations) and the breath of the claim will, in essence, define the variety of process conditions which will fall within the metes and bounds of the claim and thus within the scope of exclusivity. For example, consider the following two claims to a process for the formation of an amide bond. Claim A is broad and will provide broad exclusivity rights, whereas Claim B is narrower, containing a number of limitations and will provide somewhat narrower exclusivity rights.

Claim A: A process for the preparation of an amide bond comprising reacting a carboxylic acid with an amine.

Claim B: A process for the preparation of an amide bond comprising reacting a carboxylic acid with an amine, wherein the amine is present in an amount equal to or greater than about 2 equivalents, in the presence of an inorganic base and in an organic solvent.

The patent that includes issued Claim A will confer onto the patentee the right to exclude others from reacting a carboxylic acid with an amine—under any conditions. To infringe Claim A, the only limitations are the presence of a carboxylic acid and the presence of an amine, in any amounts, at any temperature, and in the presence of any number and type of additional reagents, solvents, and so on. Therefore this claim provides very broad exclusivity rights. By contrast, the patent that includes issued Claim B will confer onto the patentee narrower exclusivity rights. More specifically, the right to exclude other from reacting a carboxylic acid with an amine, but only under reaction conditions where the amine is present in an amount equal to or greater than about 2 equivalents and where the reaction is run in the presence of an inorganic base and in an organic solvent. In this example, therefore, the exclusivity rights to Claim A are broader and potentially more valuable from a competitive standpoint that those associated with Claim B. For this reason, process claims are often written to include the barest of limitations necessary to make the process succeed—that is, the minimum number of limitations and broadest ranges for these limitations. This is because, to be a valid claim, the broadest claims as written must result in the product as claimed, but does not need to make the product in the most optimal way. Note that as with any patent, process patents typically contain multiple claims with varying numbers and types of limitations to provide various scopes of protection. Additionally, at least one claim in any process patent will be directed to the process to make the pharmaceutical agent of interest, under the optimal conditions.

As discussed above, the potential business or competitive advantage from a process patent is the ability to prevent others from making a patented or off-patent compound using a process that is, in some way, more efficient or otherwise more competitive. For example, a lower cost of manufacture advantage can translate directly into greater profits. Alternatively, a lower cost of manufacture advantage may be used to lower the product sale price, which in turn may result in increased sales and/or increased market share. Finally, if the cost advantages of the patented process are significant in relation to known processes (other processes not patent protected), the patent rights associated with the patented process may discourage competitors from entering the marketplace or at the very least delay their entry into the marketplace until such time as the process patent term expires. If the patent further claims key intermediates, then competitors will not be able to use any process that goes through the patented intermediates. This may further limit the synthetic routes available to competitors and thus provide additional business advantages.

Any discussion of new processes would not be complete without a mention of product-by-process claims. In the United States it is possible to obtain claims directed to the product prepared using a specific process—for example, "Claim 2. A product prepared according to the process of Claim 1." Such a claim is essentially a new composition of matter claim. The usefulness or competitive advantage to be gained from claims directed to product-by-process materials may be significant, if the process produces a product with enhanced purity or a specific, identifiable, and preferred impurity profile. If the patent term or exclusivity on the product extends beyond the term for the pharmaceutical agent (as would be expected for later process patent filings) and if the FDA or other regulatory agency approval can be tied to the preferred impurity profile, then the patent to the product-by-process could hinder or even prevent entry into the marketplace of competitive or generic products for a significant period of time after expiration of the original composition of matter patent to the pharmaceutical agent. This approach, however, is often difficult to achieve and often even more difficult to police.

SALT FORMS, CRYSTALLINE FORMS, POLYMORPHIC FORMS, STEREOISOMERS, PRODRUGS, AND METABOLITES

Salt forms, crystalline forms, polymorphic forms, and metabolites, if not previously disclosed in the original composition of matter application to the pharmaceutical agent, may be viewed as new compositions of matter and thus patentable. By contrast, stereoisomers may have been literally or inherently disclosed in the original composition of matter application to the pharmaceutical agent and, as such, may or may not be separately patentable. Regardless, patent protection on any and all of these new forms can provide significant business and/or competitive advantage. Although each new composition of matter or type of form—crystalline, polymorphic, salt, stereoisomer, or metabolite—must meet the same basic requirements of patentability, each presents different scenarios and challenges in meeting these requirements of patentability and, as such, will be discussed separately below.

Salt Forms

Most patents (and applications) directed to new pharmaceutical agents include disclosures of pharmaceutically acceptable salts and claims that include a phrase such as "or a pharmaceutically acceptable salt thereof." The specification (or body) of the patent may further include a long list of pharmaceutically acceptable salts, a list of acids and bases that can be used to make the pharmaceutically acceptable salts, and/or generic or broad disclosures on how to make pharmaceutically acceptable salts of the claimed compounds. From the standpoint of patentability, this disclosure is equivalent to the disclosure of a genus of salts. Therefore, by definition, each species within the genus would be obvious to one skilled in the art. The question then becomes, Can one get a patent with claims directed to a preferred salt form discovered during early/preclinical development? As is often the case in patent law, the answer is maybe. There is judicially created doctrine which permits an applicant to obtain a patent on a species or small subgenus (small number of species) when the broader genus has been disclosed, provided that the species or small subgenus was not *specifically* exemplified and provided that the species or small subgenus has unexpected properties. This doctrine, referred to as a "selection patent," is supported by court decisions—for example, the decision *In re* Lenin, 332 F.2d 839, 141 U.S.P.Q. 814, 815, 816 (C.C.P.A. 1964), which states:

> There is nothing non-obvious in choosing "some" among "many" indiscriminately. However, obviousness is overcome by showing that the choice is based on a discovery that some compounds, falling within a prior-art genus, have a special significance, particularly where there is nothing in the prior art to establish criticality of the "some".

Applying the analysis suggested in the court decision above to the question of salt forms results in the following conclusions. The genus of disclosed salts is the complete list of pharmaceutically acceptable salts as defined in the specification of the published application or patent. To be patentable, the specific salt form (or small number of salt forms) must not have been specifically exemplified in the published patent or application and must possess some unexpected property(ies) that would not be suggested by the original disclosure (i.e., some property that one skilled in the art would not expect or predict). What then can such unexpected properties constitute? Although it is not possible to state with complete confidence what unexpected properties are required to make a species or small subgenus of salts patentable over a genus (long list) of disclosed salts, some possible unexpected properties are as follows: (a) The species is the only salt out of many prepared which is, for example, crystalline, nonhydroscopic, or a nonsolvate; (b) the species exhibits unexpectedly better biological properties—for example, enhanced bioavailability or permeability—relative to what would be expected and/or relative to other salts; (c) the species exhibits unexpectedly preferred physical properties—for example, melting point or solubility—relative to what would be expected and/or relative to other salts; and so on. Note that this unexpected property should always be disclosed in any patent application to the salt species or subgenus, because it supports the nonobviousness of the species or subgenus and may constitute at least part of the best mode of the invention.

It is important to provide an additional caution on salt forms here. If the original specification disclosing the pharmaceutical agent and pharmaceutically acceptable salts thereof includes examples that disclose and/or describe detailed preparation of the salt form, and specifically the salt form which is in early/preclinical development, then a selection patent is no longer available, because the novelty and nonobviousness of the salt is lost. This is true, regardless of whether or not the unexpected properties of the salt are known at the time of filing of the original patent application.

Although rare, it is possible that the originally filed specification claiming the new pharmaceutical agent does not disclose any pharmaceutically acceptable salts within either the specification, claims, or examples. In this case, by definition, *any* subsequently discovered salt of the pharmaceutical agent is a new composition of matter and therefore patentable.

As has been previously discussed, the claims define the scope or boundaries of the patent rights (exclusivity). It is important to keep this in mind when drafting and prosecuting claims to new salt forms. First, if the claim specifically includes the word "crystalline" as in "Claim 1. A crystalline sodium salt of Compound X," then the patent rights will extend only to crystalline sodium salts of compound X—others will be able to make, use, sell, or offer to sell different "noncrystalline" (amourphous) forms of the sodium salt. Therefore it is important to include claims of all other known forms. For each claimed form, it is important to fully characterize the crystallinity of the salt, to include this information in the patent filing, and to claim the salt forms in multiple ways (e.g., different percents crystallinity) and with the crystallinity limitations fully defined. Similarly, claims that include terms such as "nonhydrate," "trihydrate," "nonhygroscopic," and so on, will be limited to salt forms that bear these limitations. Second, if the claim includes XRD data as a means of identifying the salt form, it is important to remember that each listed peak is a limitation in the claim and must be present in the competitor's product in order to infringe. Thus if the claim reads "Claim 1. A salt characterized by the following XRD pattern . . ." and lists a table of 2θ position, spacing, and relative intensity, then to infringe the competitor's salt, when measured for its XRD, must produce a pattern which includes *all* of the peaks, at the defined spacing and with the defined relative intensities that are listed in the claim. For this reason it is important to understand and identify compound peaks versus noise or interference peaks and to claim the salt by XRD pattern in multiple ways—for example, with a table listing all peaks and a separate claim with an XRD table listing only peaks that exhibit a relative intensity of say greater than 10%.

If a patent can be obtained on a salt form, what then is the business or competitive advantage to be gained? Consider the following timeline scenario. The original patent covering the new pharmaceutical agent has a patent term running from 2005 through 2025. In 2008, a patent application is filed on the preferred salt form. If issued, this patent will have a term extending from 2008 through 2028—a full three years longer than the term on the pharmaceutical agent itself. If the salt form is the ultimate compound approved by the FDA or other regulatory agency, then the advantage is an additional 3 years of patent exclusivity. If the salt form also possesses a critical unexpected property—for example, it is the only salt that is crystalline, or it is the only salt form that is sufficiently bioavailable, or it is the only salt form that can be formulated in solid or liquid dosages—then competitors would additionally be unable to "design around" the salt patent by providing the FDA with an alternative, bioequivalent salt. In this way, the competitor companies would be prevented from entering the market after expiry of the original composition of matter patent and before the salt patent expired.

Crystalline and Polymorphic Forms

An analysis of patents claiming new crystalline and polymorphic forms will follow similar discussion lines as those laid out above for salt forms, with a few differences in terms of meeting the requirements of patentability and competitive or business advantage.

As in the previous analysis of salt forms, the first question is whether any of the newly discovered crystalline or polymorphic forms were previously disclosed in the filings for the new pharmaceutical agent. Note that the forms need not have been characterized (e.g., by DCS or XRD) to be disclosed. If specific examples within the original patent application disclosure inherently made the crystalline or polymorphic form, then by definition the crystalline or polymorphic form was disclosed and therefore not separately patentable. (Note that in such a situation, claims directed to the inherently disclosed crystalline or polymorphic form may be added during prosecution, although their patent term will be the same as those of the compounds claimed in the original composition of matter application and, as such, will not provide additional patent term exclusivity.)

Crystalline and polymorphic forms that have not been previously disclosed, either specifically or inherently, are by definition novel and are generally viewed as nonobvious. In fact, in *In re Cofer* [354 F.2d 664, 148 U.S.P.Q. 268, 271 (C.C.P.A. 1966)], the court held that a new crystalline form of a compound would not have been obvious in the absence of evidence that "the prior art suggests the particular structure or form of the compound of composition as well as suitable methods of obtaining that structure or form." Therefore, if during polymorph screening, for example, multiple crystalline and/or polymorphic forms are discovered for a pharmaceutical agent under development, these forms should be patentable over the previously published patent application to the composition of matter.

The claims for new polymorphic and/or crystalline forms are generally similar in form and wording to those for new salt forms. In addition to claims that define the crystalline or polymorphic forms in terms of their respective XRD patterns, the new forms may be claimed in terms of their DSC or other characteristics. Here again, it is important to remember that the claims define the protected invention and that any limitations included within the claim would necessarily have to be present for a competitor's form to infringe. It is important to carefully draft the claims and accurately present any discussion of the data related to the new solid forms within the patent specification.

As with new salt forms, patents granted on new crystalline or polymorphic forms provide the patent holder with the right to exclude others from making, using, selling, or offering to sell that which is covered by the claims. If, as in the salt example discussed above, the new crystalline or polymorphic form is the one ultimately approved for sale by the FDA or other regulatory agency, and the crystalline/polymorphic form patent is filed several years after the filing of the original patent to the pharmaceutical agent, then the crystalline/polymorphic form patent can provide additional years of patent exclusivity, with all the attendant business/competitive advantages as previously discussed. Additionally, it is often useful to obtain patent protection on all known crystalline and polymorphic forms of a pharmaceutical agent to prevent competitors from selecting a crystalline or polymorphic form different from the marketed one and obtaining approval on the basis of bio-equivalency. By patenting all known crystalline and/or polymorphic forms, additional protection against competitive incursion into the marketplace may be achieved.

Active Stereoisomers

In some cases, the active compound being developed is a mixture of enantiomers or a mixture of diastereomers. If, during early development, the corresponding stereoisomers are separated or individually synthesized and it is determined that only one of the stereoisomers is active, it may be possible to obtain a patent on the active stereoisomer. The courts have concluded—as, for example, in *In re May and Eddy,* 197 U.S.P.Q. 601 (C.P.P.A. 1978)—that the active stereoisomer (in purified form) may be patentable. In fact, such a patent may be viewed as an extreme form of a selection patent—selecting the active stereoisomer from the group of stereoisomers. The difficulty in obtaining a patent to an active stereoisomer is that it is generally accepted in the medical and pharmaceutical arts that one isomer is active or more active than the other(s) and thus any claims directed to the active stereoisomer may be obvious. As with other selection patents, nonobviousness may be imparted by some superior or unexpected property of the active stereoisomer or by the difficulty of synthesizing or purifying the active stereoisomer relative to the racemic compound.

Claims directed to the active stereoisomer will read as composition of matter claims, usually with added limitation(s) related to isolation of the active stereoisomer or a given %ee (% enantiomeric excess) range, and will provide the patent owner with the right to exclude others from making, using, selling, or offering to sell the active stereoisomer for the length of the patent term. If the patent on the active stereoisomer is filed several years after the filing of the patent claiming the racemic compound, and the active isomer is marketed, the result is extended exclusivity in the market along with the associated competitive and business advantage this provides. Note, however, that claims directed to the isolated active stereoisomer will not prevent competitors or generic companies from making, using, selling, or offering to sell the racemic mixture. This is because the limitations in the issued claims—the %ee range and/or any "isolated" language—will limit the scope to active stereoisomer compositions that exactly meet the limitation dimensions. For example, if the issued claim requires the active stereoisomer to be present in an amount of greater than about 90%ee, then competitors or generic companies who make, use, sell, or offer to sell mixtures where the active stereoisomer is present in an amount less than about 90%ee will fall outside the claim and therefore not infringe. Nonetheless, there may be medical and/or competitive advantages to isolating, claiming, and ultimately marketing the active stereoisomer.

Note that the FDA currently requires the submission of efficacy and toxicity data for the individual stereoisomers of an NCE (new chemical entity). Because of this requirement, detailed information on the synthesis and properties of the individual stereoisomers will usually be available when the original composition of matter patent is filed. Thus, the individual stereo isomers will likely be disclosed and covered by the original patent filing.

Prodrugs

During early development, research efforts may focus on the synthesis and/or identification of prodrugs. From a patent perspective, prodrugs are chemically different from known compounds, including the pharmaceutical agent being developed. If the prodrugs are not disclosed and claimed as part of the genus in the original composition of matter patent, then as new compositions of matter, they are separately patentable. A new patent application can therefore be filed claiming the prodrugs and, when the patent issues, will have a patent term extending 20 years from filing. If synthesis and identification of the prodrug occurs well after the original composition of matter patent is filed and the prodrug is the chemical entity that is ultimately developed, approved, and marketed, then the later filing date of the patent claiming the prodrug will result in an extended patent term. For example, if the original composition of matter application has a patent term extending from 2005 through 2025, while a later discovered and patented prodrug has a patent term extending from, for example, 2008 through 2028, then the marketed drug will have an additional three years of patent exclusivity, with all of the associated business and competitive advantages.

Active Metabolites

Another key activity during early development is the isolation, synthesis, and identification of all major metabolites. In some cases, the active metabolite may have better physical and/or biological properties than the pharmaceutical agent under development, and it may be desirable to switch development to the active metabolite.

If the chemical structure and method of synthesis of the active metabolite have not been previously disclosed (for example, in the patent filing claiming the pharmaceutical agent), then the metabolite will be novel, nonobvious, and useful. The active metabolite will therefore be separately patentable as a new composition of matter. If the patent on the active metabolite is filed several years after filing of the patent claiming the pharmaceutical agent and the metabolite is the chemical entity ultimately approve by the FDA or other regulatory agency, the result may be extended exclusivity in the market along with the associated competitive and business advantage this provides.

Note, however, that the ability to patent an active metabolite may be lost if the metabolite is inherently disclosed or used by the public. Recently, in *Schering Corp. v. Geneva Pharm., Inc.* 67 U.S.P.Q.2d 1664 (Fed. Cir. 2003) the courts held that a patent to the active metabolite of loratadine was invalid because of inherency. More specifically, the court stated that

> . . . claims 1 and 3 broadly encompass compounds defined only by structure. Such bare compound claims include within their scope the recited compounds as chemical species in any surrounding, including within the human body as metabolites of a drug. [T]hese broad compound claims are inherently anticipated by a prior art disclosure of a drug that metabolizes into the claimed compound.

In this case, the "parent" compound (loratadine) had been patented and sold in the United States and other markets prior to filing of the metabolite patent. As such, the courts ruled that the metabolite was inherently "known" because in administering the parent compound (loratadine) to patients, the metabolite was necessarily and inevitably formed within the body of the patient and that one skilled in the art (if they looked) would have recognized the metabolite and its activity. It is important, however, to recognize that in this case, Schering was attempting to extend their patent protection on loratadine beyond the patent term on the composition of matter by asserting the metabolite patent against generic companies wishing to market the "parent" pharmaceutical agent. The courts ruled that such a *de facto* extension of patent exclusivity on the "parent" pharmaceutical agent is not available. The courts have not, however, commented on nor ruled out the patentability of an active metabolite prior to marketing or public use of the "parent" pharmaceutical agent. Nor have the courts completely precluded claims to isolated metabolites, metabolites formed outside the human body, methods of treatment using metabolites, pharmaceutical compositions containing metabolites or other claim forms. As such, it is advisable to file patent applications with claims directed to the active metabolite, provided that the drug has not been marketed. Because defining patent rights around active metabolites is an active area of patent law and one which can be expected to evolve over the coming years, due consideration and consultation with a patent attorney is strongly recommended if considering substitution or development of a metabolite at any stage during development. If the "parent" pharmaceutical agent has been previously disclosed and/or marketed, the analysis becomes more complicated and the form and limitations in any claims submitted must be carefully defined and considered.

FORMULATIONS AND DOSING REGIMENS

A key activity during early preclinical and clinical research is the development of suitable delivery systems (formulations) for the new pharmaceutical agents. This may include new solid formulations (tablets, etc.) new liquid formulations (for IV and/or pediatric administration), and/or new formulations that overcome or take advantage of specific physical and/or biological properties of the pharmaceutical agent.

As with other inventions, to be patentable, the new formulation must meet the requirements of novelty, nonobviousness, and utility. In general, the requirement of nonobviousness is often the more difficult one to overcome. As in the case of the salts, there is usually general language in the original patent specification claiming the pharmaceutical agent, boilerplate language that describes standard formulations and how to make them. Additionally, it can be broadly stated that the formulation of a pharmaceutical agent using known excipients and carriers, in known amounts and in known ways, to generate "standard-type" formulations would be obvious to one skilled in the art (in this case a formulation chemist). However, a formulation may meet the requirements of nonobviousness in a number of ways. For example, (a) the formulation may use nonobvious excipients or carriers (although this is not likely because these excipients and/or carriers would have to be separately, approved by the FDA), or (b) the formulation may impart some unexpected desirable properties (e.g., enhanced stability, solubility, permeability, etc.), and then the formulation should be patentable.

Additionally, even if broad generic disclosure of formulations is included in the original patent claiming the pharmaceutical agent, specific formulation dosages, amounts and/or regimens may, nonetheless, be nonobvious and therefore patentable. Although this is often a difficult argument to sustain, the courts have clearly indicated that formulations are patentable inventions. For example, in *In re Wiggins*, the court stated the following:

> The patentability of a composition comprising a known compound and a carrier is not precluded, particularly when claims define the amount or concentration of the compound, are limited to dosage units or specified dosage forms, or otherwise distinguish over prior art teachings.

As with any new invention, it is important to include claims of varying scope to cover the formulation in different ways. Typically, the broadest claim should define the formulation with the minimum number of necessary components and with the broadest ranges of each component amount or concentration. The claims should, however, always include the components and/or component amounts which impart any unexpected or enhanced property to the formulation. Additionally, claims should include more narrow formulations—with more components and/or more narrow amount and/or concentration ranges. Finally, claims to the specific formulations and dosages to be marketed should always be included. As has been stated before, it is important to remember that for a competitor's formulation to infringe a given claim, all the limitations of the claim (i.e., all components listed, within the range of amounts listed) must be present.

The potential business or competitive advantage from an issued patent claiming a formulation depends largely on how critical the formulation is to the suitable delivery of the pharmaceutical agent. For example, if the pharmaceutical agent is difficult to formulate, the developed and patented formulation is somehow unique, and the FDA or other regulatory agency ultimately approves the formulation for marketing, then the extended patent term associated with the formulation patent may provide the company with extended exclusivity in the marketplace. Similarly, if the formulation that is covered by a patent imparts some biological or delivery advantages, the extended patent term may also provide competitive advantage and hinder competitive formulations from entry into the market. If, however, the pharmaceutical agent is not difficult to formulate and the formulation does not impart any specific advantage unachievable with alternate formulations, then the patent rights associated with the formulation patent will provide only limited advantage. If a competitor can easily formulate the pharmaceutical agent with different components that lie outside the claimed formulation and prove bioequivalence, they will be able to enter the marketplace undeterred and unlimited by the formulation patent.

SECOND METHOD OF USE OR METHOD OF TREATMENT

During clinical trials, it is not unusual to discover unexpected secondary effects of the pharmaceutical agent. The most celebrated example of this is the use of PDEV inhibitor compounds for erectile dysfunction and the resulting blockbuster drug Viagra®. Pfizer had initially studied the drug in clinical trials for cardiac indications. The clinical trial results were not as positive as expected and Pfizer attempted to close the trial and retrieve remaining dosages from study patients. The fanciful story goes that patients participating in the study refused to return the study samples, even upon request. As a result, clinicians began asking questions of why, and the effect on erectile dysfunction was discovered. Regardless of how the discovery was made, the effect on erectile dysfunction was not originally studied. Upon recognition of this unexpected second medical use and the underlying mechanism of PDEV inhibition, Pfizer filed patent applications on the underlying mechanism of action (the use of a PDEV inhibitor for the treatment of erectile dysfunction) and on the second use of the pharmaceutical agent. These patents significantly extended the patent term/exclusivity for Viagra®. Additionally, the broad claims to the PDEV inhibition mechanism, if upheld by the courts in pending litigation, could provide Pfizer with significant royalties from competitive, marketed compounds such as Levitra® and Cialis®.

To be patentable, second-method-of-use inventions must meet the same requirements of novelty, nonobviousness, and utility as other patentable inventions. For second-methods-of-treatment inventions, the key to novelty is that the pharmaceutical agent is not yet marketed. Once marketed, the second medical use may be considered inherent: The activity or biological effect would be viewed as occurring in the general population being treated with the drug and therefore inherent and known (or would be recognized by one skilled in the art). The courts have sustained this interpretation, stating that

> Inherent properties of known compositions are not patentable. *General Electric Co. v. Heochst Celanese Corp.,* 683 F. Supp. 305, 16 U.S.P.Q.2d 1977, 1985 (Del. 1990)

and further that

> Anticipation by inherency requires that 1) the missing descriptive matter be "necessarily present" in the prior art reference and that 2) it would be so recognized by persons of ordinary skill in the art. *Continental Can Co. v. Monsanto Co.,* 948 F.2d 1264, 13268, 20 U.S.P.Q.2d 1746, 1749 (Fed. Cir. 1991)

Clinical trials, however, are generally considered by the patent office to be confidential experimentation carried out to reduce to practice the invention (here the pharmaceutical agent) and thus not novelty destroying prior art (in this case prior use). Therefore it is important that timely patent applications are filed on any side effects and/or unexpected biological effects which are recognized during clinical trials (that is, prior to marketing of the pharmaceutical agent).

Claims directed to second methods of use are simple in their reading and typically include few limitations. A second method of use claim for treating, for example, obesity may read as follows: "Claim 1. The use of Compound X for the treatment of obesity." These claims, when issued, will then provide the patentee with the right to exclude others from making, using, selling, or offering to sell (including promoting in sales literature or by other means) Compound X for the treatment of obesity.

The business or competitive advantages from second-method-of-use patents vary depending on whether or not the second method of use is ultimately approved by the FDA or other regulatory agency. At the very least, the second method of use patent will provide the patent holder with the right to prevent others from marketing or promoting the pharmaceutical agent for the second use, regardless of whether or not the patent on the pharmaceutical agent is in force or expired. A somewhat larger advantage could be obtained if the pharmaceutical agent is developed and approved for the claimed second use, as in the following scenario example:

A patent claiming a new pharmaceutical agent for the treatment of diabetes is filed in 2005, with a patent term extending from 2005 through 2025. During clinical trials in 2012, it is discovered that the pharmaceutical agent is also useful for treating high lipid levels. A patent application is filed and issues with claims directed to the second method of use, with a patent term extending from 2012 through 2032. If the pharmaceutical agent is approved for both the treatment of diabetes and for lowering lipids, then during the period between 2025 and 2032 competitor or generic companies would not be able to market or promote their drugs for the lowering of lipids. This would not, however, prevent physicians from prescribing the competitor or generic drug for lowering lipids and may thus decrease the competitive advantage to some extent.

There is an important difference between US and foreign patent laws relating to method of use claims which needs to be noted here. Claims directed to the use of a compound for the treatment of specific medical conditions are generally not permitted outside the United States. Rather, what are known as "Swiss-type" claims are drafted and may ultimately be issued. These typically read as follows:

> Claim 1. The use of Compound X in the preparation of a medicament for the treatment of diabetes or a related disorder, in a subject in need thereof.

These types of claims generally provide the same level of protection and competitive advantage outside the United States as second-method-of-use claims do within the United States.

THE ORANGE BOOK AND IP RIGHTS

In the United States, FDA regulations provide that an applicant who submits an NDA (New Drug Application) or amended or supplemental NDA must submit information on each issued patent, which includes claims to the pharmaceutical agent or method of using the pharmaceutical agent that is the subject of the NDA or amended or supplemental NDA. The applicant must then submit the following information related to each patent: (1) patent number and expiration date (including any extensions or exclusivities (e.g., Hatch–Waxman extension, pediatric extension, etc.); (2) type of patent (compound, formulation, method of use, etc.); (3) name of patent owner or agent residing in or having a business address in the United States; and (4) a signed declaration (subject to charges of perjury) that the listed patent(s) cover the pharmaceutical agent or method of use for which the NDA or amended on supplemental NDA is filed. The patents are then listed, with the relevant information, in the so-called Orange Book, and serve as public notice of patent rights and any other patent term extensions and/or regulatory exclusivity periods. Note that in addition to requiring that the patents be listed in the Orange Book, the regulations also require that changes or corrections to the listings be made in a timely manner.

Although companies are required by law to submit patent information for listing in the Orange Book, it is important to understand the specific competitive/business advantages in doing so. The listing of patents that cover the approved pharmaceutical agent provide competitors or generic companies wishing to make and sell the pharmaceutical agent with notice as to the patents and exclusivity rights associated with the pharmaceutical agent. The governing regulations further require that as part of any competitor or generic company's ANDA (Abbreviated New Drug

Application) filing, there must be included either a declaration that the patents covering the pharmaceutical agent have expired, a declaration that the product will not launch until after the patents have expired, or a declaration that the patents in question are invalid or unenforceable for some reason that's detailed in what's known as a Paragraph IV Certification. If a Paragraph IV Certification is made (with notice to the patent holder), and the patent is listed in the Orange Book, then the patent holder has 45 days in which to file suit against the party making the certification, at which point approval of the generic or competitor company's ANDA is delayed for 30 months or until the court decision, whichever is shorter. This effectively prevents the competitor from launching the competing product during the 30-month time frame (even if FDA approval is complete). For this reason it is advantageous to list any and all patents that cover an approved drug, formulation, regimen, second method of use, and so on, in the Orange Book.

The following types of patents may be listed in the Orange Book: (1) patents covering the approved chemical entity (pharmaceutical agent), (2) patents covering the approved salt, crystalline, or polymorphic form, (3) patents covering any approved indication, and (4) patents covering an approved formulation or composition. Note that although patents containing product-by-process claims may be listed in the Orange Book, there are additional declarations that must be made with such listings. The following types of patents may not be listed in the Orange book: (1) patents covering processes to make the pharmaceutical agent and/or formulation, (2) patents covering nonapproved salts, crystalline, or polymorphic forms, and (3) patents covering nonapproved indications.

THE 271(e)(1) EXEMPTION

The so-called 271(e)(1) exemption (sometimes also referred to as the "research exemption"), which allows for certain activities on a patented pharmaceutical agent (including manufacture and testing) during the patent term covering the pharmaceutical agent, was established as part of the Hatch–Waxman Act of 1984. Prior to the act, a generic drug maker could not manufacture and/or test a patented pharmaceutical agent for any reason while the patent remained in force. This prohibition extended even to those activities required by the FDA and other regulatory agencies for approval to sell a competitive or generic version of the pharmaceutical agent. The result of this limitation was a *de facto* patent term extension for drug maker, an extension that could run months and perhaps years. The US Congress decided that this limitation was tantamount to an unfair advantage for the patent holder and an inherent disadvantage to the public (which could not get competitive generic drugs in a timely manner) and sought to remedy the situation by enacting the Hatch–Waxman Act of 1984.

Under the Hatch–Waxman Act, certain activities related to gaining regulatory approval for a generic or competitor drug (for example, activities required for filing an ANDA) are excluded from patent infringement, even if they are carried out during the time when the patent on the pharmaceutical agent is in force. In exchange, the patentee holding a patent on a pharmaceutical agent is entitled to a patent term extension, not to exceed 5 years, on a day-by-day basis, to compensate for any delay or loss of patent term as a result of the time elapsed while obtaining regulatory approval to market the drug.

The protection from patent infringement was codified in 35 U.S.C. 271(e)(1), and it reads in part as follows (with emphasis added by the author):

> It shall not be an act of infringement to make, use, offer to sell, or sell, within the US or import into the US a patented invention . . . *solely* for *uses reasonably related* to the development and submission of *information under a Federal Law* which regulates the manufacture, use, or sale of drugs . . . .

The exemption can be generally summarized as permitting the manufacture and use (including testing) of patented pharmaceutical agents, if and only if the reason for the manufacture and/or use is solely related to the development of information that would be required for submission to the FDA or other regulatory agency. The difficulty in applying this exemption to any particular situation or activity stems from the definitions of the terms "solely" and "reasonably related." The term "solely" is the easier of the two to understand. Activities such as manufacture, use, sale, or offer to sell of a patented compound which are commercial in nature—not related to preclinical or clinical activities in support of FDA filings—are clearly outside the exemption. Similarly, the manufacture of a patented compound for demonstration of manufacture control or for clinical trial supplies may be covered under this exemption, but the later sale of any of this material would not be allowed.

The more difficult limitation spelled out in the language of the 271(e)(1) exemption is the requirement that the activity be "reasonably related" to the development and submission of information required by the FDA or other regulatory agency. Many lawsuits have been fought over whether or not a particular activity falls under this umbrella. Most recently, in Merck KgaA v. Integra Lifesciences I, LTD., 125 Supreme Court 2372, 74 U.S.P.Q.2d 1801 (June 13, 2005), the question of whether or not specific activities fall under the 271(e)(1) exemption was considered by the US Supreme Court. The opinion of the court, handed down on June 13, 2005, ruled that some preclinical work with a patented compound may fall under the 271(e)(1) exemption, but gave no hard-and-fast guidelines for determining which activities fall under the exemption and which do not. (The case was further remanded to the lower courts for re-trial based on the US Supreme Court's interpretation of the 271(e)(1) exemption language and at press time was still undecided.) Furthermore, the US Supreme Court did not comment on the use of so-called research tools—biological assays and the like—and provided no guidance on whether or not the use of research tools in preclinical and/or clinical activities would be protected under the 271(e)(1) exemption.

Thus it is still a highly fact-specific analysis that is required in evaluating any proposed activity for its possible protection under the 271(e)(1) exemption. Given this high level of uncertainty in the interpretation of the limitations and scope of the 271(e)(1) exemption, the author strongly advises obtaining legal evaluation and counsel whenever there is a question as to whether or not specific activities would be barred by an issued patent and/or whether or not the activities, and to what extent, would fall under the 271(e)(1) exemption.

REFERENCES AND ADDITIONAL SOURCES OF INFORMATION

The discussion herein has focused on basic information around patentability and relatively straightforward exclusivity scenarios. Patent law and the environment of preclinical and clinical development in the highly competitive pharmaceutical arena often provide more complex situations, with highly fact-specific scenarios and required analyses. In any situation where a patent opinion is required, the author suggests consultation with a patent attorney.

Additional information for the independent inventor (not associated with company or firm) may be obtained from multiple texts and Web sites. For example, information related to patentability and the requirements of patentability may be found at the USPTO Web site www.uspto.gov or at the Web sites for foreign patent offices such as the WIPO (PCT filing organization) at www.wipo.int/index.html.en, the European Patent Office at www.european-patent-office.org/index.en.php, or its public access site at my.epoline.org/portal/public, and so on. Downloads of patent publications, including applications and issued patents, may be obtained through MicroPatent™ at http://www.micropatent.com/static/index.htm or individual patent office websites.

Orange Book Listings for approved pharmaceutical agents and explanation of codes and regulations may be found on the Web at www.fda.gov/cder/ob/default.htm. US Federal Circuit court decisions may be downloaded from the Federal Circuit Web site at http://www.fedcir.gov/ or the US Supreme Court Web site at http://www.supremecourtus.gov/.

INDEX

Fundamentals of Early Clinical Drug Development: From Synthesis Design to Formulation, edited by Ahmed F. Abdel-Magid and Stéphane Caron